Operationalization of Remote Sensing Solutions for Sustainable Forest Management

Operationalization of Remote Sensing Solutions for Sustainable Forest Management

Editors

Gintautas Mozgeris
Ivan Balenović

MDPI • Basel • Beijing • Wuhan • Barcelona • Belgrade • Manchester • Tokyo • Cluj • Tianjin

Editors
Gintautas Mozgeris
Institute of Forest Management
and Wood Science,
Agriculture Academy,
Vytautas Magnus University
Lithuania

Ivan Balenović
Croatian Forest Research Institute,
Division for Forest Management
and Forestry Economics
Croatia

Editorial Office
MDPI
St. Alban-Anlage 66
4052 Basel, Switzerland

This is a reprint of articles from the Special Issue published online in the open access journal *Remote Sensing* (ISSN 2072-4292) (available at: https://www.mdpi.com/journal/remotesensing/special_issues/sustainable_forest_management).

For citation purposes, cite each article independently as indicated on the article page online and as indicated below:

LastName, A.A.; LastName, B.B.; LastName, C.C. Article Title. *Journal Name* **Year**, *Volume Number*, Page Range.

ISBN 978-3-0365-0982-2 (Hbk)
ISBN 978-3-0365-0983-9 (PDF)

Contents

About the Editors . . . . **vii**

Gintautas Mozgeris and Ivan Balenović
Operationalization of Remote Sensing Solutions for Sustainable Forest Management
Reprinted from: *Remote Sens.* **2021**, *13*, 572, doi:10.3390/rs13040572 . . . . **1**

Shingo Obata, Chris J. Cieszewski, Roger C. Lowe III and Pete Bettinger
Random Forest Regression Model for Estimation of the Growing Stock Volumes in Georgia,
USA, Using Dense Landsat Time Series and FIA Dataset
Reprinted from: *Remote Sens.* **2021**, *13*, 218, doi:10.3390/rs13020218 . . . . **7**

Markus Löw and Tatjana Koukal
Phenology Modelling and Forest Disturbance Mapping with Sentinel-2 Time Series in Austria
Reprinted from: *Remote Sens.* **2020**, *12*, 4191, doi:10.3390/rs12244191 . . . . **25**

Piotr Janiec and Sébastien Gadal
A Comparison of Two Machine Learning Classification Methods for Remote Sensing Predictive
Modeling of the Forest Fire in the North-Eastern Siberia
Reprinted from: *Remote Sens.* **2020**, *12*, 4157, doi:10.3390/rs12244157 . . . . **51**

Martina Deur, Mateo Gašparović and Ivan Balenović
Tree Species Classification in Mixed Deciduous Forests Using Very High Spatial Resolution
Satellite Imagery and Machine Learning Methods
Reprinted from: *Remote Sens.* **2020**, *12*, 3926, doi:10.3390/rs12233926 . . . . **69**

Ivan Pilaš, Mateo Gašparović, Alan Novkinić and Damir Klobučar
Mapping of the Canopy Openings in Mixed Beech–Fir Forest at Sentinel-2 Subpixel Level Using
UAV and Machine Learning Approach
Reprinted from: *Remote Sens.* **2020**, *12*, 3925, doi:10.3390/rs12233925 . . . . **87**

**Luri Nurlaila Syahid, Anjar Dimara Sakti, Riantini Virtriana, Ketut Wikantika, Wiwin
Windupranata, Satoshi Tsuyuki, Rezzy Eko Caraka and Rudhi Pribadi**
Determining Optimal Location for Mangrove Planting Using Remote Sensing and Climate
Model Projection in Southeast Asia
Reprinted from: *Remote Sens.* **2020**, *12*, 3734, doi:10.3390/rs12223734 . . . . **117**

**Angel Fernandez-Carrillo, Zdeněk Patočka, Lumír Dobrovolný, Antonio Franco-Nieto and
Beatriz Revilla-Romero**
Monitoring Bark Beetle Forest Damage in Central Europe. A Remote Sensing Approach
Validated with Field Data
Reprinted from: *Remote Sens.* **2020**, *12*, 3634, doi:10.3390/rs12213634 . . . . **147**

**Paweł Hawryło, Saverio Francini, Gherardo Chirici, Francesca Giannetti, Karolina Parkitna,
Grzegorz Krok, Krzysztof Mitelsztedt, Marek Lisańczuk, Krzysztof Stereńczak, Mariusz
Ciesielski, Piotr Wężyk and Jarosław Socha**
The Use of Remotely Sensed Data and Polish NFI Plots for Prediction of Growing Stock Volume
Using Different Predictive Methods
Reprinted from: *Remote Sens.* **2020**, *12*, 3331, doi:10.3390/rs12203331 . . . . **167**

João Rocha, André Duarte, Margarida Silva, Sérgio Fabres, José Vasques, Beatriz Revilla-Romero and Ana Quintela
The Importance of High Resolution Digital Elevation Models for Improved Hydrological Simulations of a Mediterranean Forested Catchment
Reprinted from: *Remote Sens.* **2020**, *12*, 3287, doi:10.3390/rs12203287 187

Angel Fernandez-Carrillo, Antonio Franco-Nieto, Erika Pinto-Bañuls, Miguel Basarte-Mena and Beatriz Revilla-Romero
Designing a Validation Protocol for Remote Sensing Based Operational Forest Masks Applications. Comparison of Products Across Europe
Reprinted from: *Remote Sens.* **2020**, *12*, 3159, doi:10.3390/rs12193159 205

André Duarte, Luis Acevedo-Muñoz, Catarina I. Gonçalves, Luís Mota, Alexandre Sarmento, Margarida Silva, Sérgio Fabres, Nuno Borralho and Carlos Valente
Detection of Longhorned Borer Attack and Assessment in Eucalyptus Plantations Using UAV Imagery
Reprinted from: *Remote Sens.* **2020**, *12*, 3153, doi:10.3390/rs12193153 223

Anjar Dimara Sakti, Adam Irwansyah Fauzi, Felia Niwan Wilwatikta, Yoki Sepwanto Rajagukguk, Sonny Adhitya Sudhana, Lissa Fajri Yayusman, Luri Nurlaila Syahid, Tanakorn Sritarapipat, Jeark A. Principe, Nguyen Thi Quynh Trang, Endah Sulistyawati, Inggita Utami, Candra Wirawan Arief and Ketut Wikantika
Multi-Source Remote Sensing Data Product Analysis: Investigating Anthropogenic and Naturogenic Impacts on Mangroves in Southeast Asia
Reprinted from: *Remote Sens.* **2020**, *12*, 2720, doi:10.3390/rs12172720 243

Hyeongkeun Kweon, Jung Il Seo and Joon-Woo Lee
Assessing the Applicability of Mobile Laser Scanning for Mapping Forest Roads in the Republic of Korea
Reprinted from: *Remote Sens.* **2020**, *12*, 1502, doi:10.3390/rs12091502 273

About the Editors

Gintautas Mozgeris is a Professor at the Institute of Forest Management and Wood Science, Faculty of Forests and Ecology of Agriculture Academy, Vytautas Magnus University. His research interests and competencies cover GIS, spatial modelling, aerial photography using all types of aircraft platforms, hyperspectral imaging, image and LiDAR data processing, orthophoto production, natural resource inventories and management planning, land use change and habitat modelling, and forestry decision support systems. He took an active part in the development and implementation of geographic information systems and remote sensing in Lithuanian forestry. He has more than 25 years of experience in university teaching, supervising Ph.D. students, and research.

Ivan Balenović is a researcher, Head of the Division for Forest Management and Forestry Economics, and Assistant Director at the Croatian Forest Research Institute. His line of research is focused on the application of remote sensing methods in forest inventory and forest management with a special focus on LiDAR data (airborne, unmanned airborne, terrestrial, personnel laser scanning) and digital photogrammetry data (satellite, airborne, unmanned airborne). He is Vice-President of the Scientific Council for Remote Sensing in the Croatian Academy of Sciences and Arts. He has participated in a number of national and international projects with the aim of implementing the research findings in the practice of forestry.

Editorial

Operationalization of Remote Sensing Solutions for Sustainable Forest Management

Gintautas Mozgeris [1,*] and Ivan Balenović [2]

[1]   Vytautas Magnus University, K. Donelaičio str. 58, LT-44248 Kaunas, Lithuania
[2]   Croatian Forest Research Institute, Division for Forest Management and Forestry Economics, Trnjanska cesta 35, HR-10000 Zagreb, Croatia; ivanb@sumins.hr
*   Correspondence: gintautas.mozgeris@vdu.lt; Tel.: +370-37-752-291

Citation: Mozgeris, G.; Balenović, I. Operationalization of Remote Sensing Solutions for Sustainable Forest Management. *Remote Sens.* **2021**, *13*, 572. https://doi.org/10.3390/rs13040572

Received: 26 January 2021
Accepted: 3 February 2021
Published: 5 February 2021

Publisher's Note: MDPI stays neutral with regard to jurisdictional claims in published maps and institutional affiliations.

The pre-requisite for sustainable management of natural resources is the availability of timely, cost-effective, and comprehensive information on the status and development trends of the management object. An essential source of such information has always been the application of remote sensing. Data and algorithms to use remote sensing in forestry are discussed in numerous texts. Nevertheless, the level of operationalization of research results needs to be either improved or further tested. On the other hand, remote sensing researchers are often aimed at purely academic objectives, thus lacking support and guidance from practical forestry, which does influence the quality of scientific exercises. Thus, science-driven solutions for knowledge transfer between researchers and stakeholders/policy makers/end-users are becoming increasingly important.

This Special Issue aimed to compile research papers dealing both with methodologies of remote sensing and implementation of research results to facilitate sustainable forest management. The focus was on the development of algorithms for the characterization of forest and forestry; however, with the emphasis also on the operationalization of remote sensing for natural resource management through the integration of scientific research and its practical utilization. Even though all authors identified further research needs, some of their developments had already been implemented or were ready for operational use. Altogether, there were 13 papers published in this Special Issue. The studies were implemented in three continents, using multiple remote sensing platforms, ranging from mobile laser scanning (MLS) devices or unmanned aerial vehicle (UAV) and airborne laser scanning (ALS) systems up to various satellite systems. Among the methods to process the remotely sensed data, the increasing focus on machine learning algorithms should be mentioned. Table 1 summarizes the key message of all published papers. More detailed information on the individual articles published in this Special Issue is given below in alphabetical order according to the name of the first author.

Deur at al. [1] introduced a study aiming to identify three deciduous tree species in a relatively very complex deciduous forest utilizing information available from conventional multispectral satellite imagery (WorldView-3). Two machine learning algorithms were tested—random forest and support vector machine, with the first method resulting in relatively better performance. High overall classification accuracy (85%) was reported using spectral satellite image characteristics only as the input. The authors managed to improve the accuracy of classification by introducing textural features from the satellite imagery and using them in combination with the spectral ones.

Table 1. Summarizing facts on the topics covered in the Remote Sensing Special Issue on "operationalization of remote sensing solutions for sustainable forest management".

Reference	Region of Study	Remote Sensing Data/Equipment	Focus of Study	Implementation Status
Deur at al. [1]	Croatia	WorldView-3	Classification of deciduous tree species using machine learning algorithms	Further research needed
Duarte et al. [2]	Portugal	eBee SenseFly drone with a Parrot SEQUOIA camera	Detection of pest-introduced damages using multispectral images acquired from unmanned aerial vehicle (UAV) and a variety of image processing approaches	Partly ready for operational use, some aspects need further research
Fernandez-Carrillo et al. [3]	Europe	Multiple satellite data	Development of a protocol to validate remote sensing derived forest/non-forest classification maps across Europe	Approach ready for operational use, some aspects need further research
Fernandez-Carrillo et al. [4]	Czech Republic	Sentinel-2	Development of methodology for automated bark beetle damage mapping using satellite images	Further research needed
Hawryło et al. [5]	Poland	Landsat 7 and airborne laser scanning (ALS) data	Predicting growing stock volume using remotely sensed data and ground reference data from National Forest Inventories (NFIs) with uncertain georeferencing accuracies	Approach ready for operational use under some conditions, some aspects need further research
Janiec et al. [6]	Republic of Sakha, Russian Federation	Multiple satellite, geographic information systems (GIS) and bioclimatic data	Modelling the forest fire risk using multiple predictors as an option for automated fire management system	Approach ready for operational use, some aspects need further research
Kweon et al. [7]	Republic of Korea	Mobile laser scanning (MLS) device	Evaluating the use of an MLS device mounted on a vehicle for forest road inventories	Approach ready for operational use, some aspects need further research
Löw et al. [8]	Austria	Sentinel-2	Time series analysis (TSA) framework for phenology modelling and forest disturbance mapping	Approach ready for operational use
Obata et al. [9]	Georgia, United States of America	Landsat	Predicting growing stock volume using Landsat time series and publicly available ancillary information	Approach ready for operational use, some aspects need further research
Pilaš et al. [10]	Croatia	Sentinel-2 and DJI INSPIRE 2 drone with a ZENMUSE X5S camera	Bi-sensor approach to map gaps in forest cannopy	Approach ready for operational use, some aspects need further research
Rocha et al. [11]	Portugal	Synthetic aperture radar (SAR), global positioning systems (GPS), and ALS	What should be the resolution of digital elevation models for eco-hydrological simulations	Approach ready for operational use, some aspects need further research
Sakti et al. [12]	Southeast Asia	Multiple satellite and derived data	The role of remote sensing data products to investigate the drivers behind degradation of mangroves in Southeast Asia	Approach ready for operational use, some aspects need further research
Syahid at al. [13]	Southeast Asia	Multiple satellite and derived data	Mapping the land suitability for mangrove restoration using remote sensing under different climate scenarios	Approach ready for operational use, some aspects need further research

Detection and quantification of damages by Eucalyptus Longhorned Borers in eucalyptus stands using widely available UAV-based equipment was investigated in a study by Duarte et al. [2]. They used five spectral indices and nonparametric Otsu thresholding analysis to identify the damage. Individual treetops were detected using local maxima filtering; the crowns were extracted using large-scale mean-shift segmentation and then they were classified using random forest algorithm, resulting in a very high overall classification accuracy (98.5%). Finally, classified crowns were used to elaborate on the forest density maps. The findings were discussed as having high importance for forest practitioners, as they could support full area surveys and timely identification of the most critical hotspots of damaged trees.

Fernandez-Carrillo et al. [3] presented a validation methodology to evaluate remote sensing-based products under conditions, when homogeneous field reference data sets were available. Their approach was based on using stratified random and the development of a reference data set based on independent visual interpretation of satellite images. The approach was checked testing the accuracy of forest masks derived for 16 test sites in six European countries (Spain, Portugal, France, Croatia, Czech Republic, and Lithuania) using Sentinel-2 images. The overall forest/not-forest classification accuracies ranged from 76 to 96.3%. The checks were then discussed with external local forestry stakeholders. To facilitate the application of remote sensing in operational forestry, the authors suggested to consider a creation of unified reference data sets at continental or global scales.

Sentinel-2 images were tested to map bark beetle damages by Fernandez-Carrillo et al. [4]. Classification accuracies over 80% were yielded using multi-temporal regression models to detect and map the severity of pest outbreaks in spruce forests. Suggesting their approach for operational use and underlining the cost-effectiveness and ability to derive forest vitality maps on a regular base, the authors, however, accepted further research needs aimed at early detection and forecasting of the outbreaks.

The issue raised by Hawryło et al. [5] could be used to explain the limited use of remotely sensed data in operational National Forest Inventories (NFIs) in many countries. That is, the NFIs often lack accurately georeferenced field plots. The authors tested the performance of prediction algorithms (multiple linear regression, k-nearest neighbours, random forest, and deep learning fully connected neural network) to estimate the growing stock volume using Landsat 7 and ALS data with the Polish NFI sample plots as the ground reference. The positional errors of the NFI plots could be in the order of several to about 15 m. Nevertheless, the possibility to achieve desirable outputs was proven in coniferous forests with a high number of NFI plots. Landsat-derived predictors did not increase much the accuracies of growing stock volume predictions if used together with ALS data. Moreover, the deep learning algorithm did not outperform other more traditional approaches under the conditions of the conducted experiments.

Janiec et al. [6] evaluated the impacts of various factors on fire occurrence in North-Eastern Siberia. Then, they tested two machine learning algorithms (maximum entropy and random forest) to estimate the forest fire risks using satellite images (Landsat TM, Modis TERRA, GMTED2010, VIIRS) as well as geographic and bioclimatic data. They also suggested the key role of remote sensing in operational forest fire monitoring and management system.

Kweon et al. [7] evaluated the accuracy and efficiency aspects of using Trimble MX2 MLS device mounted on a vehicle to produce 3D maps within the frames of forest road inventories. They also compared the performance of MLS against the solutions based on surveys using the global navigation satellite system and total station. The key finding was that forest roads could be mapped using MLS device at precision levels suitable to produce high resolution 3D maps.

A novel time series analysis approach based on Sentinel-2 data and a dynamic Savitzky–Golay phenology modelling algorithm was presented by Löw et al. [8]. They proposed a ready for operational use solution for forest monitoring with functionality to locate and date forest disturbances. The key points of their approach were the modelling phenology courses using Sentinel-2 time series; calculating of several vegetation indices and detecting deviations from the modelled phenology courses to be associated with forest disturbances; and, finally, producing the forest phenology and disturbance maps.

Obata et al. [9] introduced forest growing stock estimation solution using long Landsat time series and two ancillary data bases on land covers and disturbances, available for USA. They also tackled the well-known problem in application of satellite remote sensing in forest inventories—bias in large and small volume classes. The data of Forest Inventory and Analysis (FIA) unit field inventory measurements were used as the ground reference. Random forest algorithm was applied for the predictions. The authors managed to improve the prediction accuracies by employing the Landsat time series; however, incorporation of ancillary spatial data did not improve the estimations. The authors suggested to use their approach for volume estimations at sub-county level.

Pilaš et al. [10] demonstrated a novel approach to map the openings in forest canopy using information from two sensors and machine learning algorithms. They considered the UAV images as a potential source of ground truth and took the advantages of Sentinel-2 as a source of information for mapping over large areas. The authors explored various approaches to improve the predictive performance, like testing the use of single versus multi date satellite images, several vegetation and biophysical indices and textural features as predictors, and eight processing algorithms. Their main finding was that, by combining data from two sensors, the limitations of each sensor, namely, the coarser Sentinel-2 spatial resolution and limited flying range of the UAV, were minimized.

Rocha et al. [11] questioned the spatial resolution of digital elevation models (DEMs) developed using synthetic aperture radar (SAR) interferometric data, global positioning systems (GPS)-based surveys, and ALS data. The key finding was that the finer resolution DEMs improved the performance of the soil and water assessment tool.

Two studies dealt with facilitating the management of mangroves using remote sensing and modelling tools. Sakti et al. [12] used multi-source remote sensing data products together with derived land cover, geophysical, climatic, and vegetation data to evaluate the factors leading to degradation of mangroves in Southeast Asia. They concluded that the predominant factors of mangrove degradation were agriculture and fisheries. Remote sensing was suggested as a powerful tool in evidence-based policy making.

Syahid at al. [13] introduced a land suitability map for mangrove plantations in Southeast Asia assuming different climate scenarios. They used an analytical hierarchy process with remote sensing, geomorphological, hydrodynamic, climatic, and socio-economic data to assess the lands suitable for mangrove restoration. They suggested the availability of near 400,000 ha of land in Southeast Asia suitable for mangrove planting with Indonesia accounting for the largest share.

All the above-mentioned studies confirmed the great potential of various remote sensing technologies for operational use in sustainable forest management. We hope that the obtained results and findings will encourage further research and convince forestry practitioners of the importance and benefits of better integration of remote sensing in operational forestry.

Author Contributions: Conceptualization, G.M. and I.B.; Writing—original draft, G.M.; Writing—review & editing, I.B. All authors have read and agreed to the published version of the manuscript.

Funding: This Special Issue is partly linked with the MySustainableForest project, which has received funding from the European Union's Horizon 2020 research and innovation program under grant agreement No 776045.

Institutional Review Board Statement: Not applicable.

Informed Consent Statement: Not applicable.

Data Availability Statement: Data available on request.

Acknowledgments: As the Guest Editors we would like to thank all the authors who accepted the challenge to share their research results and ideas in this Special Issue. Special thanks to the reviewers, who were anonymous to the authors, but not to us—it was amazing how much you helped the authors to improve their manuscripts. Thanks to the editorial staff of Remote Sensing for igniting the idea and support to have it "operationally implemented". Thanks to MySustainableForest people for helping to spread the message and pressing to publish.

Conflicts of Interest: The authors declare no conflict of interest.

References

1. Deur, M.; Gašparović, M.; Balenović, I. Tree Species Classification in Mixed Deciduous Forests Using Very High Spatial Resolution Satellite Imagery and Machine Learning Methods. *Remote. Sens.* **2020**, *12*, 3926. [CrossRef]
2. Duarte, A.; Acevedo-Muñoz, L.; Gonçalves, C.; Mota, L.; Sarmento, A.; Silva, M.; Fabres, S.; Borralho, N.M.; Valente, C. Detection of Longhorned Borer Attack and Assessment in Eucalyptus Plantations Using UAV Imagery. *Remote. Sens.* **2020**, *12*, 3153. [CrossRef]
3. Fernandez-Carrillo, A.; Franco-Nieto, A.; Pinto-Bañuls, E.; Basarte-Mena, M.; Revilla-Romero, B. Designing a Validation Protocol for Remote Sensing Based Operational Forest Masks Applications: Comparison of Products Across Europe. *Remote. Sens.* **2020**, *12*, 3159. [CrossRef]
4. Fernandez-Carrillo, A.; Patočka, Z.; Dobrovolný, L.; Franco-Nieto, A.; Revilla-Romero, B. Monitoring Bark Beetle Forest Damage in Central Europe A Remote Sensing Approach Validated with Field Data. *Remote. Sens.* **2020**, *12*, 3634. [CrossRef]
5. Hawryło, P.; Francini, S.; Chirici, G.; Giannetti, F.; Parkitna, K.; Krok, G.; Mitelsztedt, K.; Lisańczuk, M.; Stereńczak, K.; Ciesielski, M.; et al. The Use of Remotely Sensed Data and Polish NFI Plots for Prediction of Growing Stock Volume Using Different Predictive Methods. *Remote. Sens.* **2020**, *12*, 3331. [CrossRef]
6. Janiec, P.; Gadal, S. A Comparison of Two Machine Learning Classification Methods for Remote Sensing Predictive Modeling of the Forest Fire in the North-Eastern Siberia. *Remote. Sens.* **2020**, *12*, 4157. [CrossRef]
7. Kweon, H.; Seo, J.I.; Lee, J.-W. Assessing the Applicability of Mobile Laser Scanning for Mapping Forest Roads in the Republic of Korea. *Remote. Sens.* **2020**, *12*, 1502. [CrossRef]
8. Löw, M.; Koukal, T. Phenology Modelling and Forest Disturbance Mapping with Sentinel-2 Time Series in Austria. *Remote. Sens.* **2020**, *12*, 4191. [CrossRef]
9. Obata, S.; Cieszewski, C.J.; Iii, R.; Bettinger, P. Random Forest Regression Model for Estimation of the Growing Stock Volumes in Georgia, USA, Using Dense Landsat Time Series and FIA Dataset. *Remote. Sens.* **2021**, *13*, 218. [CrossRef]
10. Pilaš, I.; Gašparović, M.; Novkinić, A.; Klobučar, D. Mapping of the Canopy Openings in Mixed Beech–Fir Forest at Sentinel-2 Subpixel Level Using UAV and Machine Learning Approach. *Remote. Sens.* **2020**, *12*, 3925. [CrossRef]
11. Rocha, J.; Duarte, A.; Silva, M.; Fabres, S.; Vasques, J.; Revilla-Romero, B.; Quintela, A. The Importance of High Resolution Digital Elevation Models for Improved Hydrological Simulations of a Mediterranean Forested Catchment. *Remote. Sens.* **2020**, *12*, 3287. [CrossRef]
12. Sakti, A.D.; Fauzi, A.; Wilwatikta, F.N.; Rajagukguk, Y.S.; Sudhana, S.A.; Yayusman, L.F.; Syahid, L.N.; Sritarapipat, T.; Principe, J.; Trang, N.T.Q.; et al. Multi-Source Remote Sensing Data Product Analysis: Investigating Anthropogenic and Naturogenic Impacts on Mangroves in Southeast Asia. *Remote. Sens.* **2020**, *12*, 2720. [CrossRef]
13. Syahid, L.N.; Sakti, A.D.; Virtriana, R.; Wikantika, K.; Windupranata, W.; Tsuyuki, S.; Caraka, R.E.; Pribadi, R. Determining Optimal Location for Mangrove Planting Using Remote Sensing and Climate Model Projection in Southeast Asia. *Remote. Sens.* **2020**, *12*, 3734. [CrossRef]

remote sensing

Article

Random Forest Regression Model for Estimation of the Growing Stock Volumes in Georgia, USA, Using Dense Landsat Time Series and FIA Dataset

Shingo Obata [1,*], Chris J. Cieszewski [2], Roger C. Lowe III [2] and Pete Bettinger [2]

[1] National Institute for Mathematical and Biological Synthesis, 1122 Volunteer Blvd., University of Tennessee, Knoxville, TN 37996, USA

[2] Warnell School of Forestry and Natural Resources, University of Georgia, 180 E Green St., Athens, GA 30602, USA; biomat@uga.edu (C.J.C.); lowe@uga.edu (R.C.L.III); pbettinger@warnell.uga.edu (P.B.)

[*] Correspondence: sobata@utk.edu; Tel.: +1-865-974-4943

Abstract: The forest volumes are essential as they are directly related to the economic and environmental values of the forests. Satellite-based forest volume estimation was first developed in the 1990s, and the accuracy of the estimation has been improved over time. One of the satellite-based forest volume estimation issues is that it tends to overestimate the large volume class and underestimate the small volume class. Free availability of the major satellite imagery and the development of cloud-based computational platforms facilitate an immense amount of satellite imagery in the estimation. In this paper, we set three objectives: (1) to examine whether the long Landsat time series contributes to the improvement of the estimation accuracy, (2) to explore the effectiveness of forest disturbance record and land cover data as ancillary spatial data on the accuracy of the estimation, and (3) to apply the bias correction method to reduce the bias of the estimation. We computed three Tasseled-cap components from the Landsat data for preparation of short (2014–2016) and long (1984–2016) time series. Each data entity was analyzed with harmonic regressions resulting in the coefficients and the fitted values recorded as pixel values in a multilayer raster database. Data included Forest Inventory and Analysis (FIA) unit field inventory measurements provided by the United States Department of Agriculture Forest Service and the National Land Cover Database and disturbance history data added as ancillary information. The totality of the available data was organized into seven distinct Random Forest (RF) models with different variables compared against each other to identify the ones with the most satisfactory performance. A bias correction method was then applied to all the RF models to examine the effectiveness of the method. Among the seven models, the worst one used the coefficients and fitted values of the short Landsat time series only, and the best one used coefficients and fitted values of both short and long Landsat time series. Using the Out-of-bag (OOB) score, the best model was found to be 34.4% better than the worst one. The model that used only the long time series data had almost the same OOB score as the best model. The results indicate that the use of the long Landsat time series improves model performance. Contrary to the previous research employing forest disturbance data as a feature variable had almost no effect on OOB. The bias correction method reduced the relative size of the bias in the estimates of the best model from 3.79% to −1.47%, the bottom 10% bias by 12.5 points, and the top 10% bias by 9.9 points. Depending on the types of forest, important feature variables were differed, reflecting the relationship between the time series remote sensing data we computed for this research and the forests' phenological characteristics. The availability of Light Detection And Ranging (LiDAR) data and accessibility of the precise locations of the FIA data are likely to improve the model estimates further.

Keywords: remote sensing; landsat time series; growing stock volume; forest inventory; harmonic regression; random forest

Citation: Obata, S.; Cieszewski, C.J.; Lowe, R.C., III; Bettinger, P. Random Forest Regression Model for Estimation of the Growing Stock Volumes in Georgia, USA, Using Dense Landsat Time Series and FIA Dataset. *Remote Sens.* **2021**, *13*, 218. https://doi.org/10.3390/rs13020218

Received: 26 November 2020
Accepted: 31 December 2020
Published: 10 January 2021

1. Introduction

Forest densities and volumes are the most important forest attributes used by the forest product industry in forest management and planning. The forest volumes are directly related to the economic benefits of the forest operations, while the forest densities, which directly determine the piece-size of logged timber and associated with it investment returns, are also important elements of various ecosystem functions and wildlife habitats, such as the maximum basal area suitable for the habitat of the red-cockaded woodpecker (*Picoides borealis*) is suggested to be 18.4 m^2/ha [1]. Furthermore, the timber volume and density are directly related to carbon sequestration [2] and sustainability analysis [3].

In the United States, both private and public organizations manage forest inventories and their measurements. The United States Department of Agriculture (USDA) Forest Service's Forest Inventory and Analysis (FIA) unit provides access to data from their large-scale, continuous forest inventory. The main objectives of the USDA Forest Service FIA unit are to determine the extent, condition, volume, and growth of forests and the estimation of the changes in their landbase [4]. The inventory splits the conterminous United States into 28,000 constituent hexagons, with their centers approximately 27.4 km apart. The centers of the constituent hexagons serve as the field survey points, with each established FIA plot representing about 2428 hectares. In addition to the central point, three additional satellite sample points are located around the central point of each hexagonal (Figure 1).

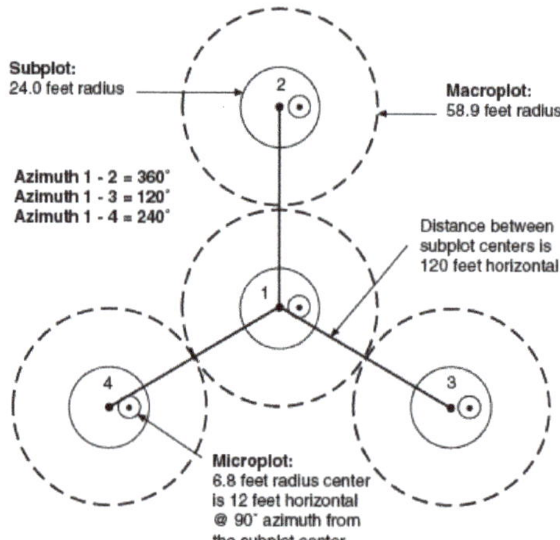

Figure 1. Sampling plot design of Forest Inventory and Analysis (FIA) [5].

All the trees within a plot whose diameter at breast height are greater than 12.7 cm are measured [5]. Although these data provide a good estimation of volume at the state-level, they are not suitable for sub-county-level estimations of the stand volumes. Many public and private landowners conduct separate inventory assessments using ground measurements, forest information systems, and various associated field data. Their inventories may provide higher-resolution volume estimations, but these systems are spatially limited to their individual property boundaries, and as privately owned information, they are generally treated as company assets and are not available publicly.

Satellite-based forest volume estimation was first developed in the 1990s for the purpose of building national-scale forest inventories. This approach combines field inventory data with satellite or other airborne sensor measurements represented by the imagery. Statistical models are applied to estimate the volume for each pixel in the raster database associated with each image. Landsat imagery is the most frequently used type of imagery with k-nearest neighbors (kNN) methods modeling estimated volumes onto the pixels spatially corresponding to the ground measurement locations and propagating the same information onto other spectrally similar pixels. The combination of FIA ground measurements and estimates of volume mapped on the Landsat TM imagery enables us to develop a distribution of forest resources at a pixel-level spatial resolution. The first operational application based on this type of approach was developed in 1990 in Finland [6]. The product derived through this process was a 30 m resolution raster database with the pixel based growing stock volume estimates. Following the Finnish example, a similar inventory was created in Sweden [7,8]. After these successful implementations of the kNN approach to forest inventory spatial estimates in Finland and Sweden, similar approaches were applied in many regions and national forest inventories in the USA, Norway, Ireland, and Japan [9–13]. Recently, Random Forest (RF) algorithm has obtained popularity as another statistical estimation method.

The developments of various methods based on the use of satellite imagery data and the advancements in the satellite sensor technology have led to various new developments of improved kNN-based approaches. In the earlier research involving kNN methods, satellite imagery was used singularly at an individual date posing problems with cloud coverage at given times, which was subsequently addressed by creating composite images spanning over a year [14,15]. Since the release of Landsat imagery to the public domain in October 2008, there has been a marked increase in the quantity of satellite imagery used in research [16,17]. A notable improvement was the use of all available Landsat imagery for the estimations of the land use changes and disturbance tracking [18–20]. In this type of research, multiple images acquired within established spatial and temporal boundaries are employed collectively to construct Landsat Time Series (LTS) data, allowing for the tracking of stand changes over time. The time series datasets are usually decomposed trends, seasonal changes, and noise components, prior to the analysis of the land use changes. The raw pixel values of satellite imagery are regarded to be a quasi-systematic reflection of the land surface [20]. The derivatives of the raw pixel values are then used as inputs into the land use change analysis. Although the considerable computational power necessary to analyze all the available satellite imagery has made it more difficult to perform these types of analyses, the rise of cloud-based computational platforms has provided the ability for such large-scale geospatial analyses. Google Earth Engine (GEE), for instance, serves as one of the most prominent platforms for the implementation of large-scale geospatial analyses [21].

Nguyen et al. [22] discuss two major advantages of using LTS: The first advantage is that it extracts the records of the spectral information regarding disturbances and re-generations [23–26] Second, it fills spatial and temporal data gaps in the estimation. The incorporation of the forest dynamics is proven to significantly improve the accuracy of the model estimations. Most of the research based on LTS imagery composite suggests choosing the best available pixel from the various annual images in each year [27,28]. This process creates an annual composite of LTS over the time period. Some researchers also suggest creating a composite LTS using more imagery per year than annual or near-annual LTS. The utility of seasonal LTS is explored and found to be able to improve the volume estimation accuracy [29–31]. Wilson et al. [32] performed a harmonic regression on all available Landsat imagery and found that the estimations showed a two- to three-fold increase in the explained variance. Wilson et al. [32] thoroughly examined the advantage of using relatively short LTS (2013–2016) and did not inspect the advantage of using the longer LTS.

In the study described in this paper, we set the following three objectives on the forest growing stock volume estimation: (1) to examine whether the long Landsat time series contributes to the improvement of the estimation accuracy, (2) to explore the effectiveness of forest disturbance record and land cover data as ancillary spatial data on the accuracy of the estimation, and (3) to apply the bias correction method to reduce the bias of the estimation. We developed models that estimate the growing stock volume of forests in the state of Georgia, United States, using an RF regression. The models' accuracy were evaluated by the Out-of-bag (OOB) score and the relative RMSE (rRMSE). The bias of the models was evaluated by the relative bias (rB).

2. Materials and Methods

2.1. Research Overview

The workflow of this study is illustrated in Figure 2. The following analyses were completed for each objective. For the first objective, we prepared two types of LTS with imagery originating from a distinct time range in each data. Subsequently, we transformed the time series data into Fourier series via harmonic regression. From each series, we retrieved the key values that were used as the feature variables of the RF regression and created a multilayer raster, in which pixel values represent the key values. We combined the publicly available field inventory data with the multilayer raster data to create the tabular data used in the RF regression with various combinations of the feature variables to determine the best combination of them and to test the importance of individual features. For the sake of the second objective, we added two types of ancillary raster data to the combination of the feature variables derived from LTS. The first type of data was raster data, in which the pixel values represent the last disturbance year of the forest stands since 1987. Another ancillary data was land cover data. Next, we examined the contribution of these two ancillary databases to the estimation. For the third objective, we examined the impact of the bias correction model on the predictions when it was applied to the best RF model of all the models we built. Finally, based on the results of the work, we considered the differences between all the various situations and the factors contributing to the improvements of the estimations.

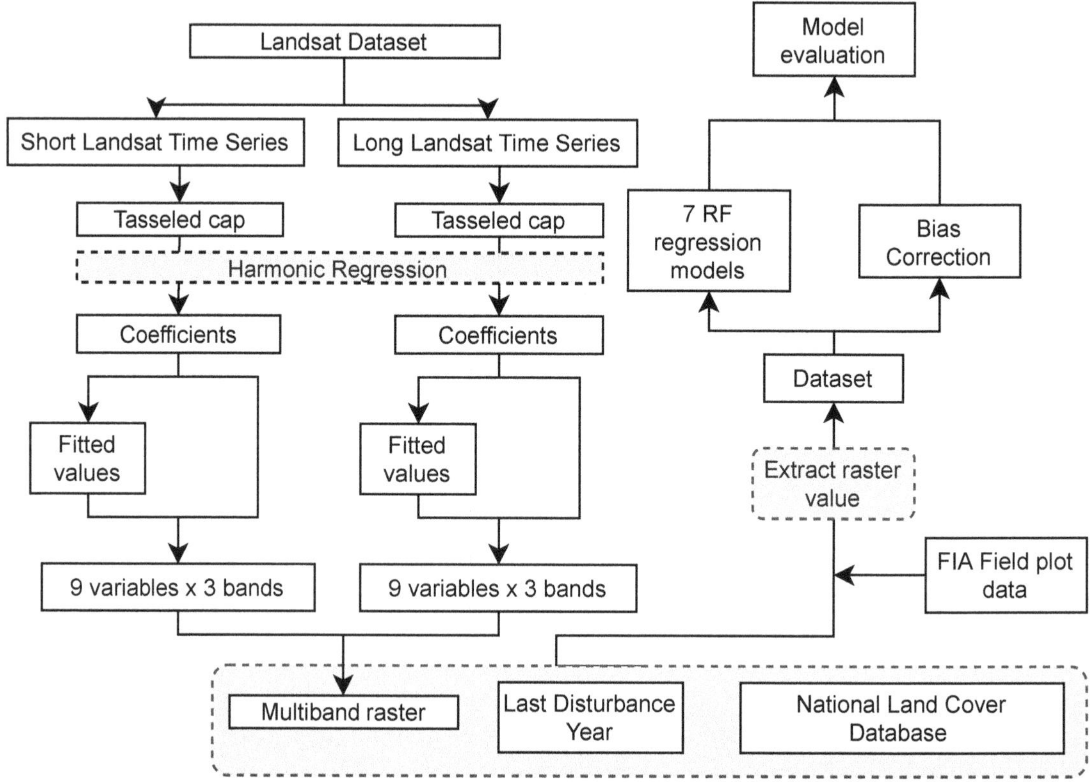

Figure 2. Flow chart of the research.

2.2. Study Area

Our study area is the state of Georgia, United States. Georgia is located in the southeastern region of the United States and contains approximately 15 million ha of land area (Figure 3). In the Southern Coastal Plain plantation forests are often intensively managed. The main plantation species is loblolly pine (*Pinus taeda*), which has a rotation age of 20–25 years under intensive management [33,34]. The Southeastern Plains ecoregion is covered by a mosaic of cropland, pasture, woodland, and forest. The Piedmont ecoregion is located between the Appalachian Mountains and the Southeastern Plains, and it includes the Atlanta metropolitan area, where more than 50% of the Georgia population resides. In the Appalachian mountain area, most of the forests are hardwood or mixed forest that are less frequently disturbed [35].

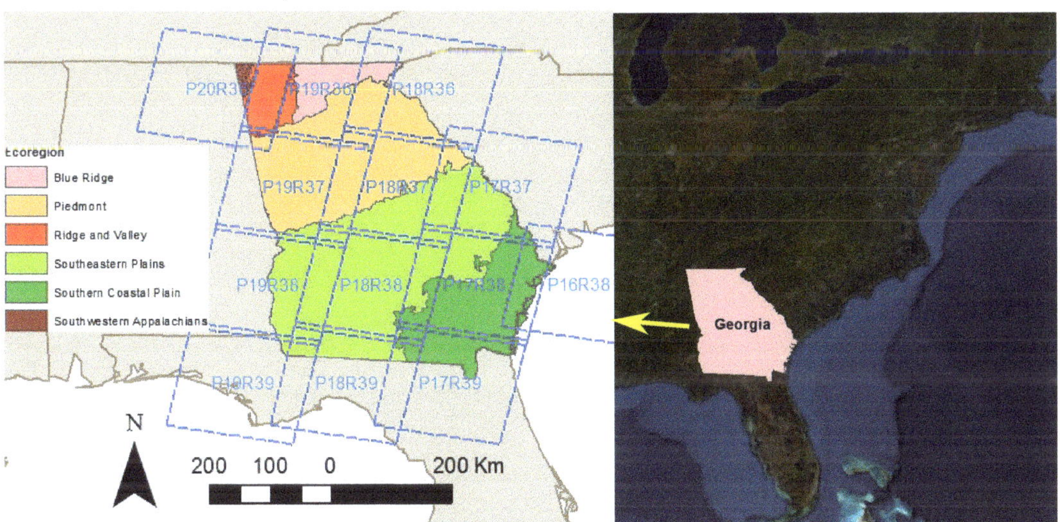

Figure 3. Study area and its ecoregions.

2.3. Satellite Data

All of the satellite data in this study were queried and processed using the GEE platform. All of the Landsat data were selected from the Level-1 Precision Terrain corrected product (L1TP) for 13 path/row combinations, as shown in Figure 3. The L1TP satisfies both radiometric and geometric criteria set by the United States Geological Survey (USGS) [36]. From the L1TP collection, we selected Landsat 5 TM and Landsat 7 ETM+ Surface Reflectance data, which were generated using the Landsat Ecosystem Disturbance Adaptive Processing System (LEDAPS) algorithm [37].

We compiled two different time ranges to create distinctive LTS. The short range was limited to 3 years and ranged from the beginning of 2014 to the end of 2016. The long range was set to 33 years, spanning from 1984 to 2016. All available images were queried for each time range. For the two sets of images, clouds, cloud shadows, water, and snow interference were masked out using the C Function of Mask (CFMask) algorithm [38–40]. For convenience, we refer to the long Landsat time series as the "Long Landsat Time Series" (LLTS) data. Similarly, we call the short Landsat time series as the "Short Landsat Time Series" (SLTS) data. Additionally, we computed the Tasseled Cap Brightness (TCB), Tasseled Cap Greenness (TCG), and Tasseled Cap Wetness (TCW) using surface reflectance data. The coefficients calculated in [41] were applied to compute the TCB, TCG, and TCW. Subsequently, these values were input into a multilayer time series raster. Additionally, an ordinary least squares, harmonic regression was performed to fit the Fourier series to each Tasseled Cap band for both SLTS and LLTS (Figure 4). The form of the Fourier curve is derived from [42] and is as follows,

$$\hat{Y}_t = \beta_0 + \beta_1 t + \beta_2 \cos(2\pi\omega t) + \beta_3 \sin(2\pi\omega t) \tag{1}$$

where $\hat{Y}_t$: Fitted values for the imagery taken at t, β_0: Intercept, β_1: Slope, β_2: Cosine term coefficient, and β_3: Sine term coefficient.

We fixed $\omega = 1$ so that the Fourier curve has a single cycle in a year, although there is previous research that assigns ω a greater value than 1 to obtain multiple cycles in a year [43]. All four coefficients of the Equation (1) are stored as the raster values. The amplitude is computed as follows.

$$amplitude = \sqrt{\beta_2^2 + \beta_3^2} \tag{2}$$

The impact of the *amplitude* is on the height of the wave. Fitted values were computed for all the dates for which LTS was acquired. Next, we calculated the maximum, minimum, mean, and RMSE from both the fitted and the observed values (Figure 5). Consequently, 9-band imagery was created for each band by stacking all derived metrics. Then, each of the bands generated from SLTS and LLTS was compiled to create the single raster layer used for the subsequent analysis.

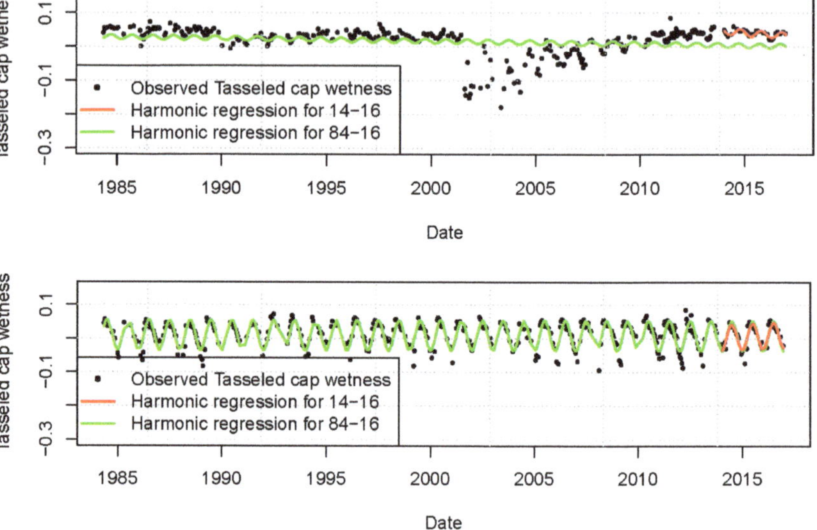

Figure 4. Harmonic regression on Tasseled Cap Wetness (TCW) of Landsat Time Series (LTS). Top: evergreen forest (Lon. −81.790, Lat. 31.063). Bottom: decidous forest (Lon. −82.052, Lat. 31.8389).

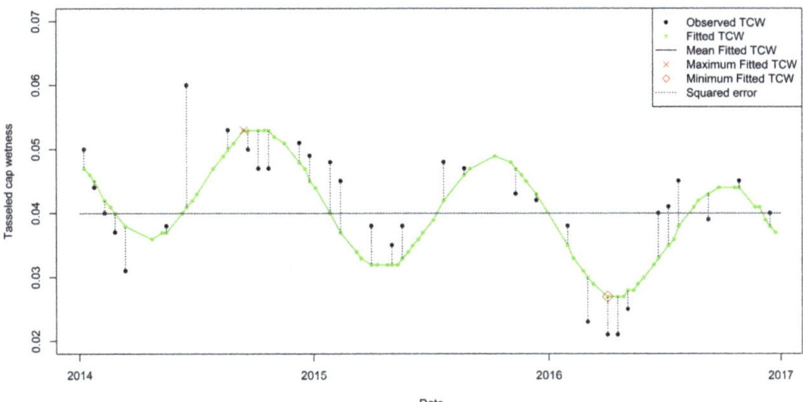

Figure 5. Harmonic regression on TCW of LTS of an evergreen forest (Longitude: −81.789827, 31.062906).

2.4. Ancillary Databases

In addition to the remote sensing data, we used two ancillary databases, which had the potential to help to improve the RF predictions. The first of these was the 2016 National Land Cover Database (NLCD) Land Cover products for the conterminous United States [44]. The Multi-Resolution Land Characteristics consortium created the 2016 NLCD to provide consistent multi-temporal land cover, and land cover change maps for the conterminous United States at 30 m spatial resolution. The 2016 NLCD classifies the land into 16 classes. Out of these 16 classes, the land where shrubs or trees cover more than 20% of the area is classified either as deciduous forest, evergreen forest, mixed forest, or woody wetlands. We note that more than 20% of the FIA field inventory data are positioned on locations where the land cover class is not forested Table 1. We included all field inventory data that is classified as non-forest in a later analysis, as the NLCD misclassifies some of the forest pixels as a non-forest class. The second ancillary database used in this research was the last disturbance year map of Georgia. This map depicts the most recent disturbance that occurred between 1984 and 2016 for every land area in the entire state of Georgia at 30 m spatial resolution [45]. Regardless of the current land use, a pixel without any disturbance record between 1984 and 2016 is classified as undisturbed.

Table 1. National Land Cover Database (NLCD) 2016 Land Cover Class on the FIA field plots.

Land Cover Class	Class [1]	# of Plots	Mean Volume (m³/ha)	# of Plots Disturbed [2]
Water	0	4	220.61	2
Developed	0	43	330.14	6
Barren land	0	2	146.99	1
Deciduous forest [3,4]	2	191	433.5	21
Evergreen forest [3,4]	1	274	431.61	80
Mixed forest [3,4]	2	75	416.8	10
Shrubland	0	32	138.43	16
Herbaceous	0	28	148.46	19
Planted/Cultivated	0	51	132.47	2
Woody wetlands [3]	2	185	462.28	32

[1] 0: Non-forest. 1: Evergreen Forest. 2: Non-Evergreen. [2] Disturbance record for each plot was retrieved from [45]. [3] Areas where forest or shrubland vegetation accounts for greater than 20% of vegetative cover. [4] Areas dominated by trees generally greater than 5 m tall.

2.5. Growing Stock Estimation

We used the FIA dataset as our ground inventory measurements. Satellite data and ancillary data were stacked into a multilayer raster that contained 56 bands (54 bands from satellite data and two bands from ancillary data). To integrate the raster and the FIA inventory ground measurements data, we requested the USDA Forest Service to extract the pixel values of the raster data onto the field plot points; they extracted our raster data onto the point data and provided us with tabular data containing plot ID numbers and the pixel values of our raster data. We note that all information potentially allowing the data user to detect the exact coordinate of the plot data was removed by USDA Forest Service in compliance with the Privacy Act in 1974 [46]. Thus, we do not know the exact locations of the plots.

The tabular data were aggregated with the FIA's original database available from the FIA DataMart (https://apps.fs.usda.gov/fia/datamart/datamart.html). The individual tree measurement data were available for each plot ID. Although individual tree measurements are available from the four subplots shown in Figure 1, only the data from the central subplot of a plot were used for the volume calculation, in order to avert the problem of spatial correlation among subplot observations [10]. Based on the code found in [47], individual tree data were aggregated into the plot-level growing stock volume per acre as follows,

$$V_i = \left(\sum_{j=1}^{m_i} v_{ij} \right) \times k \tag{3}$$

where V_i: Per hectare growing stock volume of plot i, v_{ij}: Net m^3 volume of jth tree in plot i equivalent to the net volume of wood in the central stem, m_i: The number of trees in plot i, and k: Expansion factor to convert the total growing stock volume of the plot to per hectare growing stock volume. The distribution of the volume for each plot is illustrated in Figure 6.

RF is an algorithm that handles large volumes of data within a relatively short computation time [48]. RF regression is widely used for making data-based predictions, including forest attribute estimation [32,49,50]. One of the primary advantages of using an RF model is that it can determine the importance of a variable, which indicates the contribution of each feature variable to the model prediction. The mean reduction in prediction accuracy evaluates the importance. One of the known issues of RF involves a potential bias in the model predictions. Breiman [51] argues that bagging could diminish the extent of the variance of regression predictors, yet it does not reduce the magnitude of the bias. On the other hand, because extreme observations are estimated using the average of the estimation of each tree, large observations close to the maximum value within the data are underestimated and small values of the regression function are overestimated [52]. When data are imbalanced, estimations using the RF algorithm are more susceptible to the risk of bias [53]. Zhang and Lu [52] propose a method to correct the bias in RF.

For the RF regression, the data were split into the dependent variable, which is the growing stock volume per hectare, and the independent variables that are all derived from Landsat imagery and ancillary data. We calculated the $rRMSE$, the relative bias (rB), and the OOB score. $RMSE$ has been used as the primary determinant of the model performance [54]. As the absolute value of the RMSE is incomparable between research conducted in different study areas, the $rRMSE$, calculated by the following formula, is used in favor of the $RMSE$.

$$rRMSE = \frac{RMSE}{\bar{y}} \times 100 \tag{4}$$

Knowledge of the bias is required to know the direction of the error. Subsequently, rB is calculated in the same way as rRMSE; they are formulated as follows.

$$Bias = \frac{\sum_{n_i=1} (y_i - \hat{y}_i)}{n} \tag{5}$$

$$rB = \frac{Bias}{\hat{y}} \times 100 \tag{6}$$

In addition to the evaluation of the entire data, we focused on the smallest and largest volume group as it is known that the error of the nonparametric estimation of the volume is usually heteroskedastic [55]. We grouped the field inventory data into deciles based on the observed growing stock volume. The bottom 10% ranged between 0.1 m^3/ha to 12.9 m^3/ha, while the top 10% group ranged between 249.1 m^3/ha to 682.1 m^3/ha. We calculated the bias for the bottom 10%, middle 80%, and top 10%, separately. The RF regressor was trained using the training data. Scikit-learn, a Python module that provides machine learning algorithms for medium-scale supervised and unsupervised problems, was used to perform the RF regression [56]. The number of decision trees created in the RF algorithm was set to 500. The mean squared error was selected as the function to measure the quality of a split in the individual decision trees. Individual decision trees were trained by the data bootstrapped from the original training data.

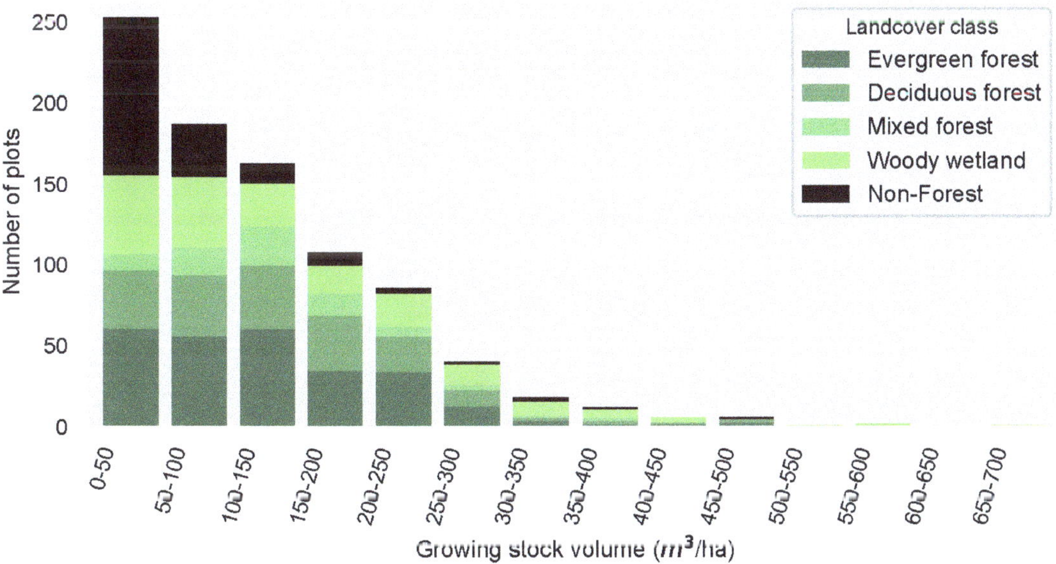

Figure 6. Growing stock volume of FIA plots in Georgia.

First, we built a base model that used the coefficients of the harmonic regression on SLTS, LLTS and the last disturbance year record as the feature variables (C_{SL} in Table 2). Next, the fitted values of SLTS, LLTS, and the last disturbance year data were selected as the feature variables of the second model (F_{SL} in Table 2). In the third model, we selected both the fitted values and the coefficients of SLTS and the last disturbance year data (CF_S in Table 2). To make a comparison with the third model, the fourth model included both the fitted values and the coefficients of LLTS, along with the last disturbance year data (CF_L in Table 2). CF_{SL}, meanwhile, used all of the feature variables from the previous models (CF_{SL} in Table 2). After determining the best combination of the variables, as derived from the remote sensing data, we divided the data by the forest type, as defined in the 2016 NLCD. The first group contained only the evergreen forest and was denoted as the Evergreen data. The second group contained the remaining forest groups listed in Table 1 and was denoted as the Non-Evergreen data. Then, the RF model was trained and evaluated separately (E_{SL} and NE_{SL} in Table 2). Following this, predictions for each data were aggregated to compute the OOB score and rRMSE as the eighth model ($E_{SL} + NE_{SL}$ in Table 3). To compare OOB between models, we calculated the rate of change as follows,

$$ROC(\%) = \frac{B - A}{A} \times 100 \tag{7}$$

where ROC: rate of change; A: rRMSE, OOB, or rB in model A; and B: rRMSE, OOB, or rB in model B. Percentage point (% point) was used to compare rRMSE and rB.

Table 2. Summary of the feature variables.

	Vegetation Index /Data Source	Time Range	# of Variables	Values	C_{SL}	F_{SL}	CF_S	CF_L	CF_{SL}	E_{SL}	NE_{SL}
Features	Landsat TCB	1984–2016	4	Regression co-efficients		✓			✓	✓	✓
		1984–2016	5	Fitted values				✓	✓	✓	✓
		2014–2016	4	Regression coefficients	✓	✓			✓	✓	✓
		2014–2016	5	Fitted values			✓	✓	✓	✓	✓
	Landsat TCG	1984–2016	4	Regression co-efficients		✓			✓	✓	✓
		1984–2016	5	Fitted values				✓	✓	✓	✓
		2014–2016	4	Regression coefficients	✓	✓			✓	✓	✓
		2014–2016	5	Fitted values			✓	✓	✓	✓	✓
	Landsat TCW	1984–2016	4	Regression co-efficients		✓			✓	✓	✓
		1984–2016	5	Fitted values				✓	✓	✓	✓
		2014–2016	4	Regression coefficients	✓	✓			✓	✓	✓
		2014–2016	5	Fitted values			✓	✓	✓	✓	✓
	NLCD	2016	1	Land use class						✓	✓
	Last disturbance	1984–2016	1	Disturbance year	✓	✓	✓	✓	✓	✓	✓
Response	FIA dataset	2016	1	Growing stock volume	✓	✓	✓	✓	✓	✓	✓
			57 variables		14	26	17	32	56	57	57

The bias correction method proposed in [52] was applied to each model to reduce the bias observed in the top and bottom 10% of the volume classes. In the model, we conjectured that bias would be attributed to the response variables. To inspect the effect of the bias correction, the data was split into training and test data, respectively. The ratio of the training to test data was then set to a 2:1 ratio. We created the RF model for the training data and computed the residual of the RF regression (e) as follows,

$$e = Y - \hat{f}(\mathbf{X}) - B(Y) + \epsilon \tag{8}$$

where Y: The growing stock volume of the observations in the training data, $f(\mathbf{X})$: Predicted values of the RF regression using feature variables of the training data, $B(Y)$: Regression bias, and ϵ: The error term. $\epsilon \sim N(0, \sigma^2)$.

$$\hat{B}(Y) = \alpha + \beta_1 Y^2 + \beta_2 Y \tag{9}$$

The bias-corrected prediction ($\hat{f}_{bc}$) was calculated by subtracting the estimated bias.

$$\hat{f}_{bc} = \hat{f} - \hat{B}(Y) \tag{10}$$

The effect of the correction was evaluated for the test data.

3. Results

We have trained and evaluated the eight models described in the previous section (Table 3). Among the models using all species for the field inventory data, in terms of the OOB score, the best model was CF_{SL}. This result, when compared to the OOB score of CF_S, was found to be 35.2% better. The OOB score of F_{SL} was found to be better than C_{SL} by 2.2%. rRMSE of F_{SL} was 1.6% points better than C_{SL}. Between the models that used both coefficients and fitted values, CF_L showed a better result than CF_S. The rRMSE of CF_L was improved by 6.4% points, while the OOB score was improved by 34.4%. Figure 7 shows the feature importance of the top 10 variables in CF_{SL}. The maximum value of the TCW generated from SLTS had the highest feature importance. E_{SL} and NE_{SL} were trained for the smaller sample sizes, as the data were split based on the forest type. The evergreen forest had a better OOB score than CF_{SL} by 9%, whereas NE_{SL}, which takes field inventory data from non-evergreen samples, returned a lower OOB score than CF_{SL}. The OOB predictions of E_{SL} and NE_{SL} were aggregated to compute the rRMSE and OOB score for the entire data ($E_{SL} + NE_{SL}$ in Table 3). The rRMSE for the aggregated prediction was similar to that of CF_{SL}, while the OOB score for the aggregated prediction was worse than that of CF_{SL}. The feature importance of E_{SL} and NE_{SL} was presented in Figure 7. For evergreen forests, the six most important features were either the TCW or the TCB of LLTS. For the rest of the species, the maximum fitted values of the harmonic regression derived from the TCW of SLTS. The second important feature was the maximum fitted values of the harmonic regression on the TCW of LLTS. In comparison with E_{SL} and NE_{SL}, the maximum fitted values are given greater importance in E_{SL} than in NE_{SL}.

Table 3. Summary of the RF models.

	C_{SL}	F_{SL}	CF_S	CF_L	CF_{SL}	E_{SL}	NE_{SL}	$E_{SL} + NE_{SL}$
Observation Mean	121.21	121.21	121.21	121.21	121.21	113.39	128.16	-
rRMSE	68.93	67.38	71.48	65.09	64.42	59.66	70.50	65.67
rB	4.19	3.48	3.72	2.66	3.79	3.09	4.82	-
OOB_score	34.8	35.87	23.39	35.63	36.11	16.52	34	39.15
Species	all	all	all	all	all	Evergreen	non-Evergreen	all

The inclusion of LLTS into the set of feature variables was effective. This result coincides with the result shown in previous research [57]. As is shown in the comparison between CF_S and CF_L, the inclusion of LLTS into the set of feature variables contributed to improving the OOB score. Table 4 shows how many times a variable was selected as being one of the 10 most important variables in terms of feature importance for CF_{SL}, E_{SL}, and NE_{SL}. The number of features created from LLTS is more than SLTS in CF_{SL}, E_{SL}, and NE_{SL}. In E_{SL}, features derived from LLTS were more important than in the NE_{SL} model. On the other hand, SLTS maintains a degree of importance for the NE_{SL} model. The different effects of LLTS and SLTS on the two models resulting from the different ratios of the field inventory data with disturbance (Table 1). While 29% of the field inventory data of evergreen forest has a disturbance record, only 14% of the field inventory data of the non-Evergreen forest has a disturbance record. As LLTS convolutes the time series trajectory of Landsat spectral values over long periods of time, features from LLTS gained importance in E_{SL}, of which field inventory data was taken from the relatively dynamic and young forest. As SLTS captures recent trends more precisely than LLTS, SLTS gained a degree of importance for NE_{SL}, of which field plot data relate to the relatively stable and mature forest.

The bias correction method was applied to each model. The relationship between the observed growing stock volume and the estimated bias for CF_{SL} is illustrated in Figure 8. For each RF model, we subtracted the estimated bias from predicted volumes to acquire the bias-corrected prediction. Bias-corrected prediction reduced rB, bottom 10% bias, and top 10% bias from the original prediction in all models (Table 5).

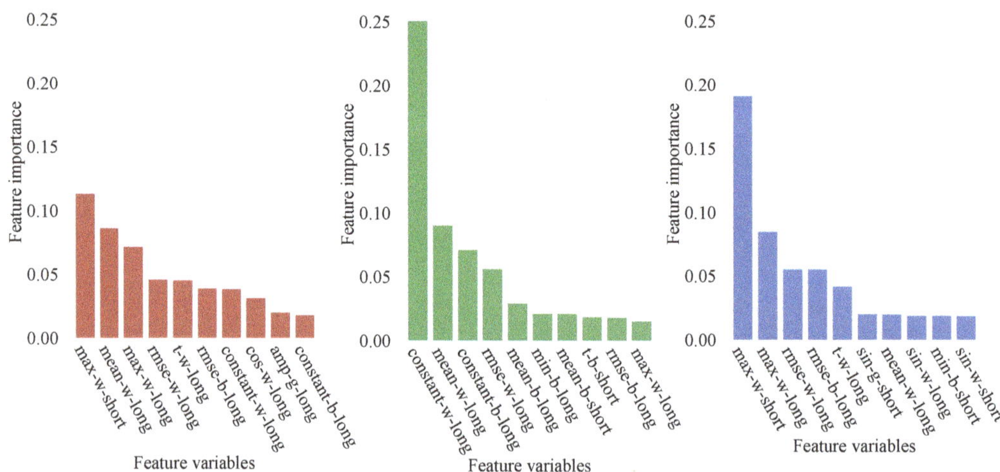

Figure 7. Feature importances of three RF models. Left: CF_{SL}, Center: E_{SL}, and Right: NE_{SL}. The abbreviated name for the feature variable represents <Type of variable>-<Type of tasseled cap component used>-<Length of the Landsat data used> (i.e., max-w-short represents the maximum value of the fitted values of the harmonic regression on LLTS of TCW).

Table 4. Number of top 10 feature variables for CF_{SL}, E_{SL}, and NE_{SL}.

Length	Model	Max	Mean	Min	RMSE	Sin	Slope	Intercept	Total
SLTS	CF_{SL}	1	-	-	2	-	-	-	3
	E_{SL}	-	1	-	-	-	1	-	2
	NE_{SL}	1	-	1	-	2	-	-	4
LLTS	CF_{SL}	1	1	-	2	1	1	1	7
	E_{SL}	1	2	1	2	-	-	2	8
	NE_{SL}	1	1	-	2	1	1	-	6

Table 5. Summary of the relative bias for each volume group. rB: relative bias, rB_corr: relative bias with bias correction, middle80: relative bias of the middle 80% volume class, middle80: relative bias of the middle 80% volume class with bias correction, bottom10: relative bias of the bottom 10% volume class, bottom10_corr: relative bias of the bottom 10% volume class after bias correction, top10: relative bias of the top 10% volume class, and top10_corr: relative bias of the top 10% volume class with bias correction.

	C_{SL}	F_{SL}	CF_S	CF_L	CF_{SL}	E_{SL}	NE_{SL}
rB	4.19	3.48	3.72	2.66	3.79	3.09	4.82
rB_corr	−2.73	−1.98	−1.48	−0.69	−1.47	−1.26	−2.86
middle80	−16.63	−15.42	−16.19	−11.94	−14.1	−15.17	−15.857
middle80_corr	−14.41	−13.08	−13.6	−9.01	−10.91	−12.61	−13.23
bottom10	−69.16	−65.42	−71.37	−71.51	−69.89	−53.39	−71.03
bottom10_corr	−56.43	−52.74	−58.14	−60.13	−57.43	−42.03	−56.31
top10	151.93	146.08	159.81	138.88	140.93	141.87	142.24
top10_corr	139.55	132.90	152.97	127.93	131.01	128.26	128.88

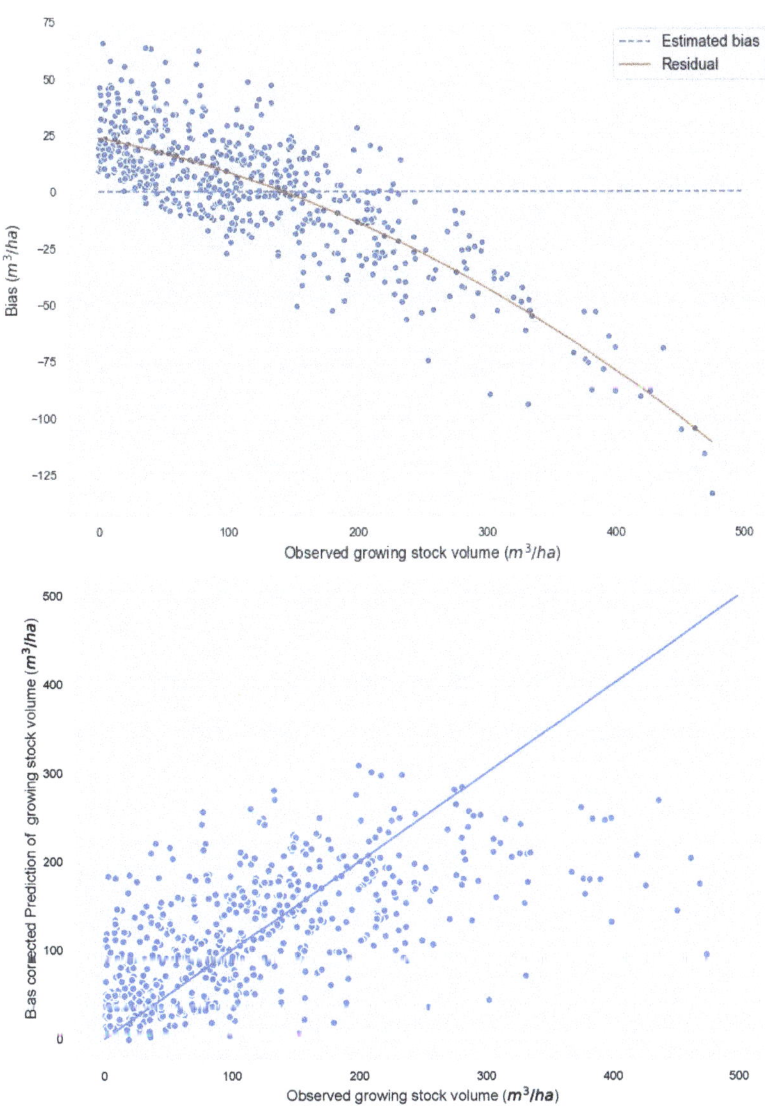

Figure 8. (**Top**) Bias correction for CF_{SL}. (**Bottom**) Observed volume vs. bias corrected prediction in CF_{SL}.

4. Discussion

We constructed multiple models and evaluated them in the previous section for the three objectives. Regarding the first objective, comparison between CF_S and CF_L contrasted the effect of the length of LTS as the difference between these models is only the length of LTS employed as the feature variables. CF_L showed 34.4% better OOB than CF_S that it is reasonable to conclude that using LLTS as feature variables contributes to the improvement of the estimation accuracy. In addition, CF_L reduced the bias of the top 10% volume class by 21% points from CF_S. This difference might be caused by the characteristics of feature variables derived from LLTS that are less likely to spectrally saturate. The saturation of the spectral reflectance value of satellite imagery refers to the situation whereby spectral reflectance values mimic the values normally seen in forest vegetation with dense canopy cover. This phenomenon is the decisive factor in the low estimation accuracy of the forest aboveground biomass and volume estimation, especially when the volume or aboveground biomass is high [58,59]. The second objective was examined by focusing on disturbance year and NLCD data. In any model that used disturbance year and NLCD data as feature variables, these variables did not have importance more than 0.02. This fact indicates that the contribution of two ancillary spatial data was less important than LTS. The difference between the OOB score of the model without the disturbance year record and the model with the disturbance year record was less than 0.01, unlike the previous research that showed the importance of the disturbance metrics on the model performance [23–25]. The plausible reason for the relative unimportance of the last disturbance year's data is that 80% of our field inventory data does not have any disturbance records. To make the information about the dynamics of the forest stands more relevant to the changes, combining more metrics acquired from the change detection algorithms (i.e., magnitude of disturbance and start of regeneration) is necessary. Effectiveness of NLCD was examined by the comparison between CF_{SL} and aggregated model of E_{SL} and NE_{SL} ($E_{SL} + NE_{SL}$ in Table 3) as $E_{SL} + NE_{SL}$ is constructed by adding only NLCD land cover class to the feature variables (Table 2). While the OOB socre of $E_{SL} + NE_{SL}$ was better than CF_{SL}, rRMSE of $E_{SL} + NE_{SL}$ was slightly worse than CF_{SL}. Concerning the third objective, the bias correction method reduced the absolute value of the relative bias for all the models. rB changed from 3.79% to -1.47% in CF_{SL}, which was the best model among all the models. The size of the overestimation in the bottom 10% data was reduced by 12.5% point in CF_{SL}. In addition, the underestimation in the top 10% data was reduced by 9.9% point. These results coincide with the reported findings in [52].

Our models were compared with the previous research dedicated for the similar purpose as ours. Although the direct comparison of the accuracy of the model is difficult as the metrics used to evaluate the model performance depend on the study area, remote sensing data, and field plot data [54,60], it is possible to make a comparison of the metrics with the similar research using LTS and the FIA dataset. The accuracy of the best model of this research (rRMSE = 65%) was better than the estimation shown in [32] (rRMSE = 170% for total aboveground biomass (kg/ha)), which used FIA dataset and all available Landsat imagery. Deo et al. [61] built and evaluated aboveground biomass estimation models for various regions in the U.S. Among the models, the rRMSE for the generic model, which pools all the data from the regions and use only LTS data as satellite imagery, was 60.8%. The rRMSE for the site-specific model that used the data only from South Carolina, and which used only LTS data like satellite imagery, was 73.1%. We note that majority of the recent research, which employs LTS also used LiDAR data as feature variables [25,57,61–63]. rRMSEs for those research ranged between 15% and 50% if models employed LiDAR data. The difference between our research, and others, in the rRMSE, is attributed to the fact that the LiDAR-derived variables have a higher correlation with aboveground biomass and growing stock volume [64].

The variation of the feature importance among the models shown in Figure 7 reflects the relationship between the time series remote sensing data we computed for this research and the forests' phenological characteristics. More specifically, the importance of the features was different between species. In NE_{SL}, which used only non-evergreen forest data, the mean fitted values were given lower feature importances than in E_{SL}, which used evergreen forest data only. On the other hand, the maximum fitted values were given lower feature importance in E_{SL} than in NE_{SL}. The fitted values for the Tasseled cap indices of the LTS generally correlate to the vegetation density. As the vegetation density correlates to the growing stock volume, the fitted values can be important feature variables in our models. The leaf-off season's fitted values do not have a clear difference between the large-volume class and the small volume class in non-evergreen forests. Therefore, the mean fitted values which combine the fitted value of the leaf-off season and the leaf-on season cannot be an important variable for NE_{SL}. On the contrary, the maximum fitted value captures the highest value at the middle of the leaf-on season that it was given higher feature importance for NE_{SL}. In E_{SL}, the mean fitted values was important since the evergreen forest has smaller seasonality than the non-evergreen forest.

5. Conclusions

Forest densities and volumes are the most principal variables used by forest management and planning. The developed growing stock volume estimation models using RF regression for the forest in the state of Georgia, United States, were examined to explore the variables and the method potentially improve the estimation accuracy. The results of this research showed that using the long Landsat time series (LLTS) for the predictor variables of the estimation model improves the OOB of the estimation by 34.4%. Furthermore, using the bias correction method that attempts to reduce the size of the bias contributes by decreasing the bias in the small volume class and the large volume class. However, incorporation of the ancillary spatial data did not improve the accuracy of the model. Therefore, it is inferred that the ecophisiological variations in each forest are explained better by the variables derive from LTS. As the RF model presented in this research can estimate the growing stock volume of the forest stand with 30 m spatial resolution, it is expected that the data can be used for sub-county areas volume estimations, which is an important functionality for the forest product industry and land owners in the state of Georgia.

Finally, to further improve our model in our area of interest, two issues should be addressed. The first issue is the lack of readily available public LiDAR data. As freely available LiDAR data cover only a partial area of Georgia [65], we could not incorporate these data for Georgia. If the availability and coverage of LiDAR were to be improved in the future, it is expected that a better estimation can be made available. The second issue is the inaccessibility of the FIA plot location information. Due to this, we could not inspect the location of the forest, allowing the possibility that some of the sampling plots were located at the edge or outside of the forest stand boundaries.

Author Contributions: Conceptualization, C.J.C.; methodology, C.J.C. and S.O.; software, S.O.; validation, S.O.; formal analysis, S.O.; investigation, S.O.; resources, S.O.; data curation, S.O.; writing—original draft preparation, S.O.; writing—review and editing, C.J.C., R.C.L.III and P.B.; visualization, S.O.; supervision, C.J.C. and P.B.; project administration, C.J.C.; funding acquisition, P.B. All authors have read and agreed to the published version of the manuscript.

Funding: This research was funded by the United States Department of Agriculture, National Institute of Food and Agriculture, McIntire-Stennis project administered by the Warnell School of Forestry, and Natural Resources at the University of Georgia, grant number GEOZ-0195-MS.

Institutional Review Board Statement: Not applicable.

Informed Consent Statement: Not applicable.

Data Availability Statement: Restrictions apply to the availability of FIA dataset. Data were obtained from USDA Forest Service and are available from https://apps.fs.usda.gov/fia/datamart/datamart.html with the permission of USDA Forest Service. The rest of the data presented in this study are available on request from the corresponding author. The data are not publicly available due to the data size.

Acknowledgments: We would like to mention our appreciation for the support and assistance of the Google Earth Engine Development team.

Conflicts of Interest: The authors declare no conflict of interest.

References

1. U.S. Department of the Interior, Fish and Wildlife Service. *Recover Plan for the Red-Cockaded Woodpecker (Picoides borealis)*; Technical Report; U.S. Department of the Interior, Fish and Wildlife Service: Washington, DC, USA, 2003.
2. Miksys, V.; Varnagiryte-Kabasinskiene, I.; Stupak, I.; Armolaitis, K.; Kukkola, M.; Wojcik, J. Above-ground biomass functions for Scots pine in Lithuania. *Biomass Bioenergy* **2007**, *31*, 685–692. [CrossRef]

3. Cieszewski, C.J.; Zasada, M.; Borders, B.E.; Lowe, R.C.; Zawadzki, J.; Clutter, M.L.; Daniels, R.F. Spatially explicit sustainability analysis of long-term fiber supply in Georgia, USA. *For. Ecol. Manag.* **2004**, *187*, 349–359. [CrossRef]
4. Brandeis, T.J.; Hartsell, A.J.; Bentley, J.W.; Brandeis, C. *Economic Dynamics of Forests and Forest Industries in the Southern United States*; Technical Report SRS-152; U.S. Department of Agriculture, Forest Service, Southern Research Station: Asheville, NC, USA, 2012.
5. Bechtold, W.A.; Patterson, P.L. *The Enhanced Forest Inventory and Analysis Program: National Sampling Design and Estimation Procedures*; Technical Report SRS-GTR-80; U.S. Department of Agriculture, Forest Service, Southern Research Station: Asheville, NC, USA, 2015. [CrossRef]
6. Tomppo, E. *Designing a Satellite Image-Aided National Forest Survey in Finland*; Swedish University of Agricultural Sciences: Umea, Sweden, 1990; pp. 43–47.
7. Reese, H.; Nilsson, M.; Sandström, P.; Olsson, H. Applications using estimates of forest parameters derived from satellite and forest inventory data. *Comput. Electron. Agric.* **2002**, *37*, 37–55. [CrossRef]
8. Reese, H.; Nilsson, M.; Pahén, T.G.; Hagner, O.; Joyce, S.; Tingelöf, U.; Egberth, M.; Olsson, H. Countrywide estimates of forest variables using satellite data and field data from the National Forest Inventory. *J. Hum. Environ.* **2003**, *32*, 542–548. [CrossRef] [PubMed]
9. Franco-Lopez, H.; Ek, A.R.; Bauer, M.E. Estimation and mapping of forest stand density, volume, and cover type using the k-nearest neighbors method. *Remote Sens. Environ.* **2001**, *77*, 251–274. [CrossRef]
10. McRoberts, R.E.; Nelson, M.D.; Wendt, D.G. Stratified estimation of forest area using satellite imagery, inventory data, and the k-Nearest Neighbors technique. *Remote Sens. Environ.* **2002**, *82*, 457–468. [CrossRef]
11. Maselli, F.; Chirici, G.; Bottai, L.; Corona, P.; Marchetti, M. Estimation of Mediterranean forest attributes by the application of k-NN procedures to multitemporal Landsat ETM+ images. *Int. J. Remote Sens.* **2005**, *26*, 3781–3796. [CrossRef]
12. Tanaka, S.; Takahashi, T.; Nishizono, T.; Kitahara, F.; Saito, H.; Iehara, T.; Kodani, E.; Awaya, Y. Stand volume estimation using the k-NN technique combined with forest inventory data, satellite Image data and additional feature variables. *Remote Sens.* **2014**, *7*, 378–394. [CrossRef]
13. Barrett, F.; McRoberts, R.E.; Tomppo, E.; Cienciala, E.; Waser, L.T. A questionnaire-based review of the operational use of remotely sensed data by national forest inventories. *Remote Sens. Environ.* **2016**, *174*, 279–289. [CrossRef]
14. Moody, A.; Johnson, D.M. Land-surface phenologies from AVHRR using the discrete fourier transform. *Remote Sens. Environ.* **2001**, *75*, 305–323. [CrossRef]
15. Hird, J.N.; McDermid, G.J. Noise reduction of NDVI time series: An empirical comparison of selected techniques. *Remote Sens. Environ.* **2009**, *113*, 248–258. [CrossRef]
16. Woodcock, C.E.; Allen, R.; Anderson, M.; Belward, A.; Bindschadler, R.; Cohen, W.; Gao, F.; Goward, S.N.; Helder, D.; Helmer, E.; et al. Free access to Landsat imagery. *Science* **2008**, *320*, 1011. [CrossRef] [PubMed]
17. Roy, D.P.; Wulder, M.A.; Loveland, T.R.; Woodcock, C.E.; Allen, R.G.; Anderson, M.C.; Helder, D.; Irons, J.R.; Johnson, D.M.; Kennedy, R.; et al. Landsat-8: Science and product vision for terrestrial global change research. *Remote Sens. Environ.* **2014**, *145*, 154–172. [CrossRef]
18. Kennedy, R.E.; Yang, Z.; Cohen, W.B.; Pfaff, E.; Braaten, J.; Nelson, P. Spatial and temporal patterns of forest disturbance and regrowth within the area of the Northwest Forest Plan. *Remote Sens. Environ.* **2012**, *122*, 117–133. [CrossRef]
19. Zhu, Z.; Woodcock, C.E. Continuous change detection and classification of land cover using all available Landsat data. *Remote Sens. Environ.* **2014**, *144*, 152–171. [CrossRef]
20. Brooks, E.B.; Wynne, R.H.; Thomas, V.A.; Blinn, C.E.; Coulston, J.W. On-the-fly massively multitemporal change detection using statistical quality control charts and Landsat data. *IEEE Trans. Geosci. Remote Sens.* **2014**, *52*, 3316–3332. [CrossRef]
21. Gorelick, N.; Hancher, M.; Dixon, M.; Ilyushchenko, S.; Thau, D.; Moore, R. Google Earth Engine: Planetary-scale geospatial analysis for everyone. *Remote Sens. Environ.* **2017**, *202*, 18–27. [CrossRef]
22. Nguyen, T.H.; Jones, S.; Soto-Berelov, M.; Haywood, A.; Hislop, S. Landsat time-series for estimating forest aboveground biomass and Its dynamics across space and time: A review. *Remote Sens.* **2020**, *12*, 98. [CrossRef]
23. Pflugmacher, D.; Cohen, W.B.; Kennedy, R.E. Using Landsat-derived disturbance history (1972–2010) to predict current forest structure. *Remote Sens. Environ.* **2012**, *122*, 146–165. [CrossRef]
24. Pflugmacher, D.; Cohen, W.B.; Kennedy, R.E.; Yang, Z. Using Landsat-derived disturbance and recovery history and lidar to map forest biomass dynamics. *Remote Sens. Environ.* **2014**, *151*, 124–137. [CrossRef]
25. Kennedy, R.E.; Ohmann, J.; Gregory, M.; Roberts, H.; Yang, Z.; Bell, D.M.; Kane, V.; Hughes, M.J.; Cohen, W.B.; Powell, S.; et al. An empirical, integrated forest biomass monitoring system. *Environ. Res. Lett.* **2018**, *13*, 025004. [CrossRef]
26. Liu, L.; Peng, D.; Wang, Z.; Hu, Y. Improving artificial forest biomass estimates using afforestation age information from time series Landsat stacks. *Environ. Monit. Assess.* **2014**, *186*, 7293–7306. [CrossRef] [PubMed]
27. Hermosilla, T.; Wulder, M.A.; White, J.C.; Coops, N.C.; Hobart, G.W. An integrated Landsat time series protocol for change detection and generation of annual gap-free surface reflectance composites. *Remote Sens. Environ.* **2015**, *158*, 220–234. [CrossRef]
28. Matasci, G.; Hermosilla, T.; Wulder, M.A.; White, J.C.; Coops, N.C.; Hobart, G.W.; Zald, H.S.J. Large-area mapping of Canadian boreal forest cover, height, biomass and other structural attributes using Landsat composites and lidar plots. *Remote Sens. Environ.* **2018**, *209*, 90–106. [CrossRef]
29. Nguyen, H.C.; Jung, J.; Lee, J.; Choi, S.U.; Hong, S.Y.; Heo, J. Optimal atmospheric correction for above-ground forest biomass estimation with the ETM+ remote sensor. *Sensors* **2015**, *15*, 18865–18886. [CrossRef]

30. Nguyen, T.H.; Jones, S.; Soto-Berelov, M.; Haywood, A.; Hislop, S. A Comparison of imputation approaches for estimating forest biomass using Landsat time-series and inventory data. *Remote Sens.* **2018**, *10*, 1825. [CrossRef]

31. Zhu, X.; Liu, D. Improving forest aboveground biomass estimation using seasonal Landsat NDVI time-series. *ISPRS J. Photogramm. Remote Sens.* **2015**, *102*, 222–231. [CrossRef]

32. Wilson, B.T.; Knight, J.F.; McRoberts, R.E. Harmonic regression of Landsat time series for modeling attributes from national forest inventory data. *ISPRS J. Photogramm. Remote Sens.* **2018**, *137*, 29–46. [CrossRef]

33. Fox, T.R.; Jokela, E.J.; Allen, H.L. The development of pine plantation silviculture in the Southern United States. *J. For.* **2007**, *105*, 337–347. [CrossRef]

34. D'Amato, A.W.; Jokela, E.J.; O'Hara, K.L.; Long, J.N. Silviculture in the United States: An amazing period of change over the past 30 years. *J. For.* **2017**, *116*, 55–67. [CrossRef]

35. Obata, S.; Cieszewski, C.J.; Bettinger, P.; Lowe, R.C., III; Bernardes, S. Preliminary analysis of forest stand disturbances in Coastal Georgia (USA) using Landsat time series stacked imagery. *Formath* **2019**, *18*, 1–11. [CrossRef]

36. U.S. Geological Survey. Landsat Levels of Processing. 2020. Available online: https://www.usgs.gov/land-resources/nli/landsat/landsat-levels-processing (accessed on 9 February 2020).

37. Masek, J.; Vermonte, E.; Saleous, N.; Wolfe, R.; Hall, F.; Huemmrich, F.; Gao, F.; Kulter, J.; Lim, T. A Landsat surface reflectance data set for North America, 1990–2000. *Geosci. Remote Sens. Lett.* **2006**, *3*, 68–72. [CrossRef]

38. Zhu, Z.; Woodcock, C.E. Object-based cloud and cloud shadow detection in Landsat imagery. *Remote Sens. Environ.* **2012**, *118*, 83–94. [CrossRef]

39. Zhu, Z.; Wang, S.; Woodcock, C.E. Improvement and expansion of the Fmask algorithm: Cloud, cloud shadow, and snow detection for Landsats 4–7, 8, and Sentinel 2 images. *Remote Sens. Environ.* **2015**, *159*, 269–277. [CrossRef]

40. Foga, S.; Scaramuzza, P.L.; Guo, S.; Zhu, Z.; Dilley, R.D.; Beckmann, T.; Schmidt, G.L.; Dwyer, J.L.; Hughes, M.J.; Laue, B. Cloud detection algorithm comparison and validation for operational Landsat data products. *Remote Sens. Environ.* **2017**, *194*, 379–390. [CrossRef]

41. Crist, E.P. A TM Tasseled Cap equivalent transformation for reflectance factor data. *Remote Sens. Environ.* **1985**, *17*, 301–306. [CrossRef]

42. Shumway, R.H.; Stoffer, D.S. Spectral analysis and filtering. In *Time Series Analysis and Its Applications: With R Examples*, 4th ed.; Springer Texts in Statistics; Springer Science + Business Media: New York, NY, USA, 2017; pp. 165–172.

43. Zhu, Z.; Fu, Y.; Woodcock, C.E.; Olofsson, P.; Vogelmann, J.E.; Holden, C.; Wang, M.; Dai, S.; Yu, Y. Including land cover change in analysis of greenness trends using all available Landsat 5, 7, and 8 images: A case study from Guangzhou, China (2000–2014). *Remote Sens. Environ.* **2016**, *185*, 243–257. [CrossRef]

44. Yang, L.; Jin, S.; Danielson, P.; Homer, C.; Gass, L.; Bender, S.M.; Case, A.; Costello, C.; Dewitz, J.; Fry, J.; et al. A new generation of the United States national land cover database: Requirements, research priorities, design, and implementation strategies. *ISPRS J. Photogramm. Remote Sens.* **2018**, *146*, 108–123. [CrossRef]

45. Obata, S.; Bettinger, P.; Cieszewski, C.J.; Lowe, R.C., III. Mapping forest disturbances between 1987–2016 using all available time series Landsat TM/ETM+ Iimagery: Developing a reliable methodology for Georgia, United States. *Forests* **2020**, *11*, 335. [CrossRef]

46. Smith, W. Forest inventory and analysis: A national inventory and monitoring program. *Environ. Pollut.* **2002**, *116*, 233–242. [CrossRef]

47. Burrill, E.A.; Wilson, A.M.; Turner, J.A.; Pugh, S.A.; Menlove, J.; Christensen, G.; Conkling, B.L.; David, W. *The Forest Inventory and Analysis Database: Database Description and User Guide for Phase 2 (Version 7.2)*; Technical Report; U.S. Forest Service: Washington, DC, USA, 2018.

48. Breiman, L. Random Forests. *Mach. Learn.* **2001**, *45*, 5–32. [CrossRef]

49. Gigović, L.; Pourghasemi, H.R.; Drobnjak, S.; Bai, S. Testing a new ensemble model based on SVM and random forest in forest fire susceptibility assessment and its mapping in Serbia's Tara National Park. *Forests* **2019**, *10*, 408. [CrossRef]

50. Tompalski, P.; White, J.C.; Coops, N.C.; Wulder, M.A. Demonstrating the transferability of forest inventory attribute models derived using airborne laser scanning data. *Remote Sens. Environ.* **2019**, *227*, 110–124. [CrossRef]

51. Breiman, L. Bagging predictors. *Mach. Learn.* **1996**, *24*, 123–140. [CrossRef]

52. Zhang, G.; Lu, Y. Bias-corrected random forests in regression. *J. Appl. Stat.* **2012**, *39*, 151–160. [CrossRef]

53. Chen, C.; Liaw, A.; Breiman, L. *Using Random Forest to Learn Imbalanced Data*, Technical Report 666, Department of Statistics, University of California Berkeley: Berkley, CA, USA, 2004.

54. Chirici, G.; Mura, M.; McInerney, D.; Py, N.; Tomppo, E.O.; Waser, L.T.; Travaglini, D.; McRoberts, R.E. A meta-analysis and review of the literature on the k-Nearest Neighbors technique for forestry applications that use remotely sensed data. *Remote Sens. Environ.* **2016**, *176*, 282–294. [CrossRef]

55. McRoberts, R.E. Diagnostic tools for nearest neighbors techniques when used with satellite imagery. *Remote Sens. Environ.* **2009**, *113*, 489–499. [CrossRef]

56. Pedregosa, F.; Varoquaux, G.; Gramfort, A.; Michel, V.; Thirion, B.; Grisel, O.; Blondel, M.; Prettenhofer, P.; Weiss, R.; Dubourg, V.; et al. Scikit-learn: Machine Learning in Python. *J. Mach. Learn. Res.* **2011**, *12*, 2825–2830.

57. Bolton, D.K.; White, J.C.; Wulder, M.A.; Coops, N.C.; Hermosilla, T.; Yuan, X. Updating stand-level forest inventories using airborne laser scanning and Landsat time series data. *Int. J. Appl. Earth Obs. Geoinf.* **2018**, *66*, 174–183. [CrossRef]

58. Foody, G.M.; Boyd, D.S.; Cutler, M.E.J. Predictive relations of tropical forest biomass from Landsat TM data and their transferability between regions. *Remote Sens. Environ.* **2003**, *85*, 463–474. [CrossRef]
59. Lu, D.; Chen, Q.; Wang, G.; Liu, L.; Li, G.; Moran, E. A survey of remote sensing-based aboveground biomass estimation methods in forest ecosystems. *Int. J. Digit. Earth* **2016**, *9*, 63–105. [CrossRef]
60. Tinkham, W.T.; Mahoney, P.R.; Hudak, A.T.; Domke, G.M.; Falkowski, M.J.; Woodall, C.W.; Smith, A.M. Applications of the United States Forest Inventory and Analysis dataset: A review and future directions. *Can. J. For. Res.* **2018**, *48*, 1251–1268. [CrossRef]
61. Deo, R.K.; Russell, M.B.; Domke, G.M.; Woodall, C.W.; Falkowski, M.J.; Cohen, W.B. Using Landsat time-series and LiDAR to inform aboveground forest biomass baselines in Northern Minnesota, USA. *Can. J. Remote Sens.* **2017**, *43*, 28–47. [CrossRef]
62. Matasci, G.; Hermosilla, T.; Wulder, M.A.; White, J.C.; Coops, N.C.; Hobart, G.W.; Bolton, D.K.; Tompalski, P.; Bater, C.W. Three decades of forest structural dynamics over Canada's forested ecosystems using Landsat time-series and lidar plots. *Remote Sens. Environ.* **2018**, *216*, 697–714. [CrossRef]
63. Nguyen, T.H.; Jones, S.D.; Soto-Berelov, M.; Haywood, A.; Hislop, S. Monitoring aboveground forest biomass dynamics over three decades using Landsat time-series and single-date inventory data. *Int. J. Appl. Earth Obs. Geoinf.* **2020**, *84*, 101952. [CrossRef]
64. Deo, R.K.; Russell, M.B.; Domke, G.M.; Andersen, H.E.; Cohen, W.B.; Woodall, C.W. Evaluating site-specific and generic spatial models of aboveground forest biomass based on Landsat time-series and LiDAR strip samples in the Eastern USA. *Remote Sens.* **2017**, *9*, 598. [CrossRef]
65. U.S. Geological Survey. 3D Elevation Program. 2020. Available online: https://www.usgs.gov/core-science-systems/ngp/3dep (accessed on 19 February 2020).

Article

Phenology Modelling and Forest Disturbance Mapping with Sentinel-2 Time Series in Austria

Markus Löw * and Tatjana Koukal

Department of Forest Inventory, Austrian Research Centre for Forests (BFW), 1131 Vienna, Austria; tatjana.koukal@bfw.gv.at
* Correspondence: markus.loew@bfw.gv.at; Tel.: +43-1-87838-1325

Received: 28 October 2020; Accepted: 17 December 2020; Published: 21 December 2020

Abstract: Worldwide, forests provide natural resources and ecosystem services. However, forest ecosystems are threatened by increasing forest disturbance dynamics, caused by direct human activities or by altering environmental conditions. It is decisive to reconstruct and trace the intra- to transannual dynamics of forest ecosystems. National to local forest authorities and other stakeholders request detailed area-wide maps that delineate forest disturbance dynamics at various spatial scales. We developed a time series analysis (TSA) framework that comprises data download, data management, image preprocessing and an advanced but flexible TSA. We use dense Sentinel-2 time series and a dynamic Savitzky–Golay-filtering approach to model robust but sensitive phenology courses. Deviations from the phenology models are used to derive detailed spatiotemporal information on forest disturbances. In a first case study, we apply the TSA to map forest disturbances directly or indirectly linked to recurring bark beetle infestation in Northern Austria. In addition to spatially detailed maps, zonal statistics on different spatial scales provide aggregated information on the extent of forest disturbances between 2018 and 2019. The outcomes are (a) area-wide consistent data of individual phenology models and deduced phenology metrics for Austrian forests and (b) operational forest disturbance maps, useful to investigate and monitor forest disturbances to facilitate sustainable forest management.

Keywords: phenology modelling; forest disturbance; forest monitoring; bark beetle infestation; forest management; time series analysis; remote sensing; satellite imagery; Sentinel-2

---

1. Introduction

Worldwide forests are increasingly affected by changes and dynamics of various origin and at different scales [1–3]. Shifting patterns in timber demands [4,5] or in silvicultural perceptions [6] can constitute "sustainable" forest management, but can also trigger a change in timber harvest practises, including illegal logging and vast deforestation processes. Further, climate change effects on forest ecosystems accelerate forest mortality worldwide [7]. Forest biomes are the main terrestrial carbon stock [8]. Without doubt there is an urgent need to safeguard forested areas worldwide and trace dynamics of altering site conditions caused by climate change [7,9]. At the same time, forest product supply must be ensured, despite an increased multiuse demand concerning forest ecosystem functions [10]. Therefore, the monitoring of unobtrusive and small scale land cover changes such as those caused by natural events (e.g., pest infestation, higher mortality due to altering site conditions) or forest management practices (e.g., thinning or selective timber extraction) becomes more and more crucial [8,11]. Recent studies underlined the importance of a vital forest at stand or even single-tree level [12]. Forest disturbances can decrease the capability of forest ecosystems to protect against natural hazards, which is a major regulating function, especially of mountainous forests [13]. From a global perspective, it will not be sufficient to avoid deforestation to meet global climate change mitigation goals. Small-scale forest management has to be guided by the principles of sustainability too, because forest management has an unexpectedly large impact on standing biomass and related carbon sequestration [14]. On the one hand, the sustainable extraction of various forest products guarantees a young age structure, which can increase carbon sequestration rates up to 25% [15]. On the other hand, in mountainous terrain, unmanaged forests show a higher capacity of climate and erosion self-regulation compared to managed forests. Therefore, natural forests are more resilient to altering environmental conditions and will provide valuable regulating ecosystem services in the future [16]. The monitoring of such small-scale forest management practices will be crucial to guarantee sustainable forestry, not only in Austria.

Earth observation (EO) data has proved to be a comprehensive source to continuously assess the state of forests and to detect disturbances globally. Today image processing and analysis tools can map these changes and are increasingly capable of tracing slight phenology anomalies on different temporal scales, informing about intra-, inter- and transannual dynamics [7,9].

During the last few decades, EO programmes, such as Landsat [17] or MODIS [18] deliver data, which have enabled the implementation of large scale monitoring systems (e.g., Global Forest Watch [2]), as data provision is continuous and data quality consistent.

However, a phenological time series analysis (TSA) of global satellite imagery must cope with a highly varying topography and seasonal vegetation effects, compared to studies explicitly focusing on selected world regions that are less challenging (e.g., tropical forest or other biomes closer to the equator).

Austria, as an example of a country with diverse landscapes, shows distinct seasonality with low to high sun levels. The illumination conditions strongly vary, especially in alpine regions, due to topographic shadow areas, which affect the remotely sensed signal reflected by the Earth's surface. During winter, snow cover and diverse weather patterns, such as invasive fog that is omnipresent in alpine valleys, significantly reduce the number of useful observations.

Previous research shows different methodological approaches to cope with these challenges. Most of the mainly Landsat-based TSAs rely on an image composite analysis [13,19] or on some variant of a harmonic modelling approach [13,19–24]. Harmonic modelling approaches are robust but show some limitations regarding the quality of temporal information [25] and allow only little detail in reconstructing seasonal vegetation courses. This limits the ability to depict the occurrence and magnitude of changes, which is needed to scrutinise dynamics on a forest management level. The COPERNICUS programme, the European Union's earth observation programme [26], with its satellite twin consisting of Sentinel-2A and Sentinel-2B (S2) [27], provides new opportunities for monitoring forest ecosystems. The multispectral S2 sensor shows a spatial, spectral, and temporal resolution at a level which so far is unique in non-commercial EO. The ground resolution is up to 10 m and the revisit time is less than five days. The use of S2 data that is free of charge is quite established in agricultural monitoring [28,29] and non-forest phenology modelling [30], whereas advanced S2-based TSAs focusing on forest ecosystems are so far rare [31].

The Austrian Research Centre for Forests (in German: Bundesforschungszentrum für Wald—BFW) is, as a central federal research institution, focusing on forest state and forest future (BFW 2020). The BFW's Department of Forest Inventory, responsible for the national forest inventory (NFI) in Austria, gathers, prepares and analyses nationwide information about forest state and dynamics [32]. New earth observation data, such as the S2 imagery, are a good supplement for existing terrestrial inventory data. Nationwide auxiliary data from remote sensing can compensate for weaknesses of sample-based inventory assessments [33,34]. The compilation of facts and figures for stakeholders and decision makers is a main task of the Austrian NFI. National to local forest authorities and other stakeholders increasingly request information concerning hot topics as storm damages, changing forest site conditions (e.g., tree species specific drought stress) or spatial patterns of pest invasions (e.g., the spread of bark beetle infestations) [8,11]. For these reasons, the BFW established and maintains a local archive for nationwide S2 data and develops operational data processing schemes to optimally exploit this data pool [11].

This article describes a novel TSA approach based on Sentinel-2 data, using a dynamic Savitzky–Golay phenology modelling algorithm [35]. The overall aim of the presented study is to develop an operational forest-monitoring approach that locates forest disturbances and accurately determines the date of their occurrence. To achieve this goal, our objectives are (1) to develop a straightforward workflow for modelling phenology courses from dense Sentinel-2 data time series, (2) to examine the suitability of several vegetation indices to deduce forest phenology features (phenology metrics) and to detect deviations from the modelled phenology courses (forest disturbances), and (3) to produce forest phenology and disturbance maps that comprehensibly visualize the information contained in the Sentinel-2 imagery.

2. Material and Methods

2.1. Material

2.1.1. Sentinel-2 Data

The proposed TSA approach relies on multispectral Sentinel-2A and -2B data. The approach uses the four spectral bands with a spatial resolution of 10 m of the top-of-atmosphere product (TOA, Level-1C), i.e., the bands B02 (blue), B03 (green), B04 (red), and B08 (near infrared). In theory, bottom-of-atmosphere (BOA) data are expected to be more suitable for time series analysis than TOA data. However, in previous tests it was found that BOA data produced by Sen2Cor (version 2.5.5) [36] are too error prone for a fully operational approach without any visual image checking and selection. Spectrally distorted pixel observations caused by atmospheric effects are, therefore, removed by outlier detection and filtering techniques, as explained in Section 2.2.4. We only use Sen2Cor's quality grid outputs to derive a granule-wide mask to exclude pixels not useable for the TSA (Section 2.2.2).

We identified all S2 granules that intersect the area of Austria. In total, twenty granules were selected. All L1C datasets available for this area are stored in a local image archive that is updated on a regular basis by searching for and downloading new data with an *oData*-query via the ESA API hub [37]. The image archive contains data from the year 2017 and is updated regularly. For this study, images from January 2017 to December 2019 were available. The approach processes at least 75 images per granule and year.

In the TSA process, we distinguish two periods: (a) the model period (MP), and (b) the detection period (DP). The model period comprises one or more complete years. It is used to compute the reference phenology course. The detection period is the period that is examined for deviations from the reference phenology course. In the study, the model period is set to the period from 1st January 2017 to 31st December 2018 and the detection period ranges from 1st January 2018 to 31st December 2019.

2.1.2. Forest Map

To confine the TSA to areas covered by forest, a national forest map, produced at the BFW according to the Austrian NFI forest definition [38] is used. The vector map was resampled to the 10 m pixel grid of Sentinel-2.

2.1.3. Reference Datasets

For validating the class "Disturbance" of the forest disturbance map, we use field observations (in situ dataset) provided by the forest section of the Federal Government of Lower Austria. The data were collected during on-site inspections according to forest protection regulations between January 2018 and December 2019. The points are located in the north-western part of Lower Austria (Figure 1). The point attributes are the date of creation (in situ date), the number of affected trees and the disturbance type. The most frequent disturbance type is bark beetle infestation, followed by wind throw, wind breakage, snow breakage, and fungal attack. In the validation procedure, described in Section 2.5, all sites with at least three affected trees were considered, resulting in 1500 observations that could be used in the study.

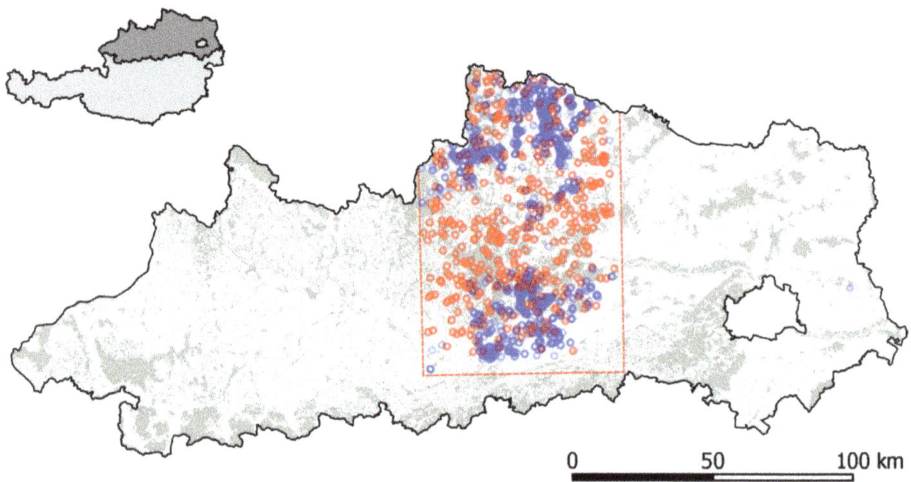

Figure 1. In situ dataset for class "Disturbance" (blue) and random sample dataset for class "No Disturbance" (red) for evaluating the forest disturbance map in the northern region of Austria, i.e., the region also referred to in Section 2.5.

For assessing the accuracy of the map class "No Disturbance", we created a random sample dataset with 271 points for this stratum (Figure 1, red points) within the same area, where in situ data are also available. The number of sample points was chosen according to the recommendations provided by Olofsson et al. 2014 [39] with a target standard error for overall accuracy of 0.01. Each point was checked based on visual image interpretation, as described in Section 2.5.

2.2. Preprocessing

Images with a cloud cover of less than 80%, according to the L1C metadata file, are preprocessed. The preprocessing is divided into two parts: first, steps that are applied image by image (Sections 2.2.1–2.2.3), and second, steps that are applied to sets of images combined to multitemporal image stacks (Sections 2.2.4 and 2.2.5).

2.2.1. Spectral Indices Computation

Because band ratios are less affected by atmospheric and topographic effects than single band values [40] we chose three spectral indices, suggested for vegetation analysis in literature, i.e., the Normalised Difference Vegetation Index (NDVI), the Green Normalised Difference Vegetation Index (GNDVI), and the Red-Green Vegetation Index (RGVI). In addition, we use the near infrared band, scaled by a factor of 5500 to get values between 0 and 1. Table 1 gives an overview of the spectral indices used in this study.

Table 1. Spectral indices used in the study.

Index	Name	Equation	Reference
NDVI	Normalised Difference Vegetation Index	$\frac{B08-B04}{B08+B04}$	[11]
GNDVI	Green Normalised Difference Vegetation Index	$\frac{B08-B03}{B08+B03}$	[42]
RGVI	Red-Green-Vegetation Index	$\frac{B03-B04}{B03+B04}+0.5$	[43], edited
BNIR	Band: Near Infrared	$\frac{B08}{5500}$	own equation

2.2.2. Cloud, Shadow, and Non-Forest Masking

We create masks to exclude pixels affected by clouds and shadows. The masks rely on several quality assessment outputs of ESA's stand-alone atmospheric correction algorithm Sen2Cor (version 2.5.5) [36] that converts Level-1C (TOA) to Level-2A (BOA) data. We use a combination of a near-infrared threshold ($B08_{L2A} > 900$), a cloud probability threshold ($CLD_{L2A} = 0$) and a value selection of the land cover classification product ($4 \leq SCL_{L2A} \leq 5$). Alternatively, cloud and shadow masks from other sources could be easily integrated into our workflow. Contaminated pixels that are not identified in this step are eliminated later by the outlier detection and filtering procedure (Sections 2.2.4 and 2.2.5).

To exclude areas not covered by forest, we use the prepared NFI forest mask (Section 2.1.2).

2.2.3. Multitemporal Layer Stacking

After the granule-wide preprocessing steps, all images are combined to multitemporal layer stacks resulting in one layer stack per spectral index with a layer for each acquisition date. Missing values, e.g. due to clouds or shadows, are supplemented by no-data values. The next preprocessing steps are applied per pixel on time series vectors.

2.2.4. Outlier Filtering

The raw time series vectors are checked for unnatural discontinuities, i.e., an abrupt decrease followed immediately by a significant increase in the spectral signal. In forests, regreening processes (successive recovery) after disturbances occur rather slowly as compared to agricultural land, for example. Thus, such patterns are classified as outliers. They are excluded from the time series vector and are replaced by no-data values. The applied outlier criteria were empirically determined and are illustrated in Figure 2. A data point is classified as an outlier if the first two general criteria (Figure 2, criteria 1 and 2) and one of the remaining criteria (Figure 2, criteria 3a or 3b) are fulfilled. For outlier filtering, one previous and one subsequent data point are needed. Therefore, it cannot be applied to the first and the last data points of the time series.

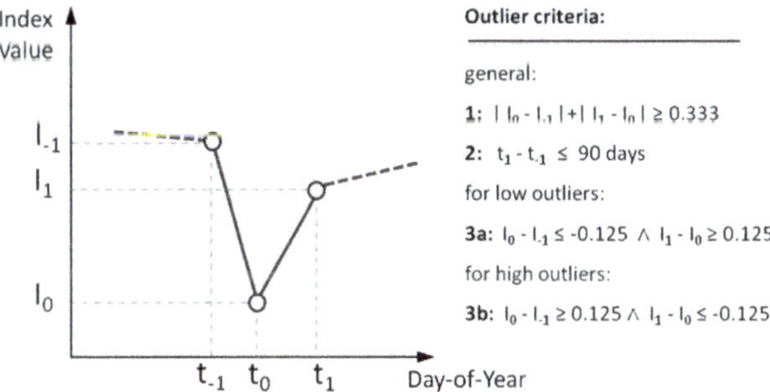

Figure 2. Schematic graph to illustrate criteria for outlier filtering (of I_0).

2.2.5. Interpolation and Smoothing

After outlier filtering, data gaps are populated by linear interpolation to receive a continuous time series vector containing a value for each day of the year. Then a Savitzky–Golay Filter (SGF) [35] is applied to smooth the time series vector. The SGF is a moving window filtering method calculating polynomial functions that do not smooth the data excessively. We use a dynamic SGF-window width, because a fixed window width can lead to insufficient or nonmeaningful smoothing if the data are noisy [44]. The SGF is applied on the model period data and on the detection period data using different window settings, which were empirically determined as specified in Table 2.

Table 2. Savitzky-Golay-Filtering (SGF) settings used in the study.

Parameter	Model Period	Detection Period
Degree of polynomial function	3	2
Window type	dynamic	fixed
Window size (days)	$\left[2 - \left(\frac{P85-P15}{\bar{I}} / 0.55\right)^{\frac{1}{2}}\right]^{\frac{5}{3}}$	31
Minimum window size (days)	31	-
Maximum window size (days)	122	-

For the model period, the window size is chosen in dependence on the index value range between the 85th percentile (P85) and the 15th percentile (P15), relative to the mean index value ($\bar{I}$) of all data points in the model period, as specified in Table 2. In this way, the window size is automatically adapted to the vegetation type's specific phenological variation. The dynamic window size results in a stronger smoothing effect for coniferous pixels (low phenological variation) than for deciduous pixels (high phenological variation).

As a result, we get for each pixel two smoothed time series curves per index with 365 values per year, i.e., one for the model period (MPTS) and one for the detection period (DPTS).

2.3. Phenology Modelling and Phenology Metrics

The smoothed multiyear course covering the whole model period is split into single-year snippets, resulting in a yearly vector for 2017, 2018, and 2019, respectively. Then a set of statistical parameters is computed over the three vectors for each day of the year. So, the information from all years within the model period is aggregated. Optionally, each year can be weighted individually, for example, to reduce the impact of previous years or of years with extreme weather conditions. In this study, all years were weighted equally. The set of parameters consists of the 10th, the 50th and the 90th percentile courses (PC10, PC50, PC90), as well as the mean of the two values PC10 and PC90. The set of percentile courses describes the inter-year index variability of all years within the model period. All four courses can serve as the reference phenology time series for the subsequent detection of anomalies (Section 2.4). In this study, we use the mean of PC10 and PC90 as the main phenology course (MPC) for the forest disturbance analyses.

Finally, a set of phenology metrics are extracted from the MPC. Phenology metrics are measures that allow for a straightforward, rather intuitive interpretation of phenological characteristics. In this study, we computed the phenological metrics "start of vegetation period" (SVP), "end of vegetation period" (EVP) as well as the maximum value of the MPC (MAX_{MPC}) and the date when the MAX_{MPC} is reached (MPC_{max}). For all dates, the day-of-year (DOY) notation is used. The SVP is the date, where the day-to-day gradient of the MPC is a maximum considering all values between DOY 90 (i.e., end of March) and DOY 182 (i.e., end of June). The EVP is the date where the day-to-day gradient of the MPC is a minimum, considering all values between DOY 245 (i.e., beginning of September) and DOY

340 (i.e., beginning of December). Other metrics, such as the growing season length or the number of phenology peaks, are not considered as they are beyond the scope of this study.

2.4. Anomaly Detection

Phenological anomalies are significant deviations of the spectral index from the "expected" course (base line). The base line (BL) is the MPC shifted in the Euclidean space to allow for more distinct deviation patterns in periods with a naturally high phenological variation (e.g., in spring and fall). The result is an "inward-buffered" reference course. Thereby, the detection of anomalies is robust in terms of slight shifts of the course along the time axis (e.g., a shifted start of the vegetation period). Such shifts can occur, for example, due to varying weather conditions from year to year.

For anomaly detection, the cumulative sum of the daily difference between the BL and the smoothed detection period curve (DPTS; Section 2.2.5) is computed for each pixel. The cumulative sum of daily index deviations serves as a proxy for the emergence and manifestation of anomalies [45–47]. Periods with an insufficient number of data points (usually in winter) are ignored in the calculation. The beginning and the end of the considered period are not set to fixed dates but are chosen according to the available data. For this, all observations in the MP are pooled as if they were collected within one year. Then the day-of-year (DOY) values are determined that correspond to the 2nd and to the 98th percentile. These dates serve as the beginning and the end of the period for calculating the cumulative sum of deviations. Optionally, the last m data points at the end of the time series can be omitted in the cumulative sum calculation, usually with $0 < m < 2$ to avoid incorrect results induced by data points that could not be filtered and smoothed by subsequent observations. For ad hoc results, e.g., right after a storm event, however, $m = 0$ is recommended because otherwise the most recent deviations cannot be detected. In this study, $m = 1$ is used. The date of the last effective data point used for the TSA is called "last usable observation" (LUO).

For forest disturbance analysis, two pieces of information are important, i.e., the level of disturbance and the date of disturbance. Both pieces of information are deduced from our high-density times series.

The forest disturbance level (FDL) is measured by means of the cumulative sum of daily index deviations (Equation (1)) with FDL = 1 corresponding to a cumulative sum equal to 1 and so on.

$$\text{FDL} = \sum_{t=\text{FDD}}^{\text{LUO}} \max(0, BL(t) - \text{DPTS}(t)) \tag{1}$$

The FDL is a kind of measure to identify pixels that are affected by a certain level of disturbance. The FDL refers to the severity and the duration of the disturbance. The higher the FDL, the higher the level of manifestation of the disturbance. In this study, the threshold for the FDL (T_{FDL}) is set to a medium level of 7, meaning that all pixels with an FDL of 7 and higher are labelled as disturbance pixels. In addition to the information, whether or not a pixel is a disturbance pixel, the date when the specified FDL is reached is stored, called the Cumulative Deviation Date (CDD; Equation (2)).

$$\text{CDD} = \min_{n \in \mathbb{N}} \left(n \,\middle|\, \widetilde{\text{FDL}} \geq T_{FDL} \right) \text{ with } \widetilde{\text{FDL}} = \sum_{t=\text{FDD}}^{n} \max(0, BL(t) - \text{DPTS}(t)) \tag{2}$$

Beyond that, for each disturbance pixel, the date is reconstructed in a backwards direction when the identified disturbance shows up in the data the first time during the backwards reconstruction, called the Forest Disturbance Date (FDD). The FDD is a "theoretical" date estimated from the BL and the actual (non-modelled) index course. It is the date, when both curves intersect for the last time, previous to the corresponding CDD.

All features (CDD, FDD, etc.) are exported as grids with a spatial resolution of 10 m. So, as a result, we get maps that report for each Sentinel-2 pixel if it shows anomalies according to the specified level of disturbance and, if true, an estimated date corresponding to the starting point of the abnormal spectral behaviour.

2.5. Validation

The TSA's ability to detect disturbance events is assessed by checking the derived forest disturbance map at the positions of the in situ observations (Section 2.1.3). To account for spatial deviations between the Sentinel-2 data and the in situ data, the area within a radius of 20 m around each point is considered. The detection of the disturbance event is regarded as successful if there is at least one disturbance pixel within this area. Additionally, the minimum FDD of all disturbance pixels per point within the specified circle is computed. We chose the minimum FDD as a benchmark, because usually one aims at detecting forest disturbance events as early as possible, both in the field and by means of remote sensing.

Based on this dataset, the detection rate, i.e., the number of detected disturbance events divided by the total number of disturbance events in the in situ data, is computed. For the temporal evaluation, the date difference (in situ date minus FDD) is analysed.

To complete the validation, we manually check the forest disturbance map at randomly selected positions within the "No Disturbance" stratum (Section 2.1.3) based on visual interpretation of a series of Sentinel-2 images. For this inspection, cloud-reduced natural- and false-colour-composite mosaics of Sentinel-2 images from spring, summer, and fall of 2018 and 2019 are used. For each point, including a 3 by 3 pixel neighbourhood, it is visually checked if the spectral signature is stable (corresponding to the class "No Disturbance") or changing (corresponding to the class "Disturbance") over the deviation period. In the case of deciduous forest, mainly the summer images are considered to avoid misclassifications due to seasonal phenology changes.

2.6. Implementation

The entire workflow is implemented via the open source software "R" [48], benefitting from its comprehensive package libraries, primarily *raster* [49], *rgdal* [50], *gdalUtils* [51], *rgeos* [52], *doParallel* [53], *foreach* [54] and *signal* [55]. All processed data are saved on a local network-attached storage (NAS). The computation-intensive TSA approach highly relies on memory-optimised and parallelised computing: first during the parallelised batch-mode of Sen2Cor, second when copying the data from the NAS to the local environment and third when executing the per-pixel TSA itself. The implemented parallelisation allows for the full use of all CPU-power available.

3. Results

Following the structure of the method section, the results section presents the main findings concerning (1) the phenology modelling with Sentinel-2 time series applied to the entire forest area of Austria (Section 2.3), and (2) the multiyear forest disturbance mapping, focusing on damages by the bark beetle infestation in Northern Austria (Upper and Lower Austria) between 2018 and 2019 (Section 2.4).

3.1. Phenology Modelling with Sentinel-2 Time Series

The phenology modelling procedure analyses per-pixel time-series data, covering more than 40,000 km^2 of forest area in Austria, which results in around 400 million unique models per spectral index. The models comprise Sentinel-2 data from the years 2017 to 2019. In Figures 3 and 4 examples of phenology courses typical for deciduous and coniferous forest are plotted together with detailed additional information, such as phenology metrics, derived from the time-series data.

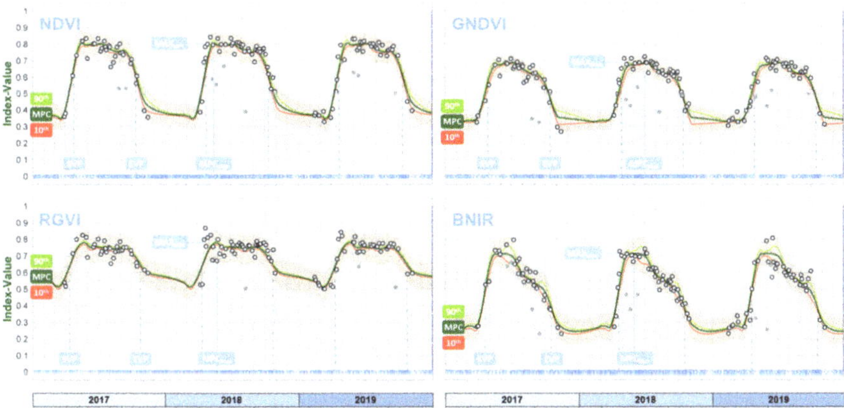

Figure 3. Phenology courses and metrics of the Normalised Difference Vegetation Index (NDVI), the Green-NDVI (GNDVI), the Red-Green Vegetation Index (RGVI) and the near infrared band (BNIR), based on data points from 2017 to 2019 for a deciduous forest pixel.

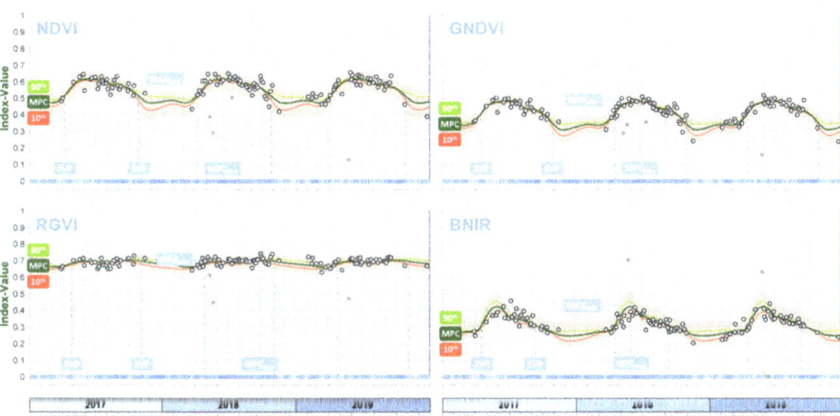

Figure 4. Phenology courses and metrics of the Normalised Difference Vegetation Index (NDVI), the Green-NDVI (GNDVI), the Red-Green Vegetation Index (RGVI) and the near infrared band (BNIR), based on data points from 2017 to 2019 for a coniferous forest pixel.

The white dots indicate valid data points. The grey dots are data points excluded by the outlier filtering procedure. The blue circles at the bottom of the plot show all available data points (from granules with more than 80% valid pixels), including observations that were eliminated, e.g., due to clouds or shadows. Each graph comprises the 10th percentile index course (thin red line), the 90th percentile index course (thin green line), and the resulting MPC (bold dark green line). The brownish ribbon represents the variability of the MPC. It is plotted just for illustration. The pixel plots show significant differences, depending on the forest type. Pixels representing deciduous forest (Figure 3) show generally more variation over the year than pixels representing coniferous forest (Figure 4).

The seasonal course patterns typical for deciduous and coniferous forest vary from index to index. In general, the average NDVI- and RGVI-values are higher, compared to the GNDVI and the BNIR, both for deciduous and coniferous forest. The seasonal course pattern of MPC is less distinct for GNDVI and RGVI than for NDVI and BNIR. The latter shows a notable peak in spring to early summer, highlighting the BNIR's higher sensitivity to depict vegetation productivity. These temporal differences in the MPCs underline that each spectral index has special characteristics.

The deviations of the data points from the MPC (noise) vary from index to index. For pixels of coniferous forest, the RGVI noise is clearly the lowest compared to the other indices, whereas for pixels of deciduous forest, RGVI noise is the highest.

The vertical lines in blue indicate selected phenology metrics. The first one (solid line) denotes the date when MPC reaches its maximum (MPC_{Max}). Deciduous forest pixels show basically higher maximum MPC values (MAX_{MPC}) than coniferous forest pixels (Figures 3 and 4).

Figure 5 shows the NDVI-based MAX_{MPC} map for Austria, limited to forest. The values range from about 0.5 to 1.0. The highest values of 0.8 and higher are found in areas covered by deciduous forest, such as in the north-eastern part of Austria. NDVI values around 0.65 indicate spruce-dominated areas, as existing in alpine regions or in the northern parts of Austria. Lowland pine stands and high-alpine dwarf pines, for example, show values below 0.55.

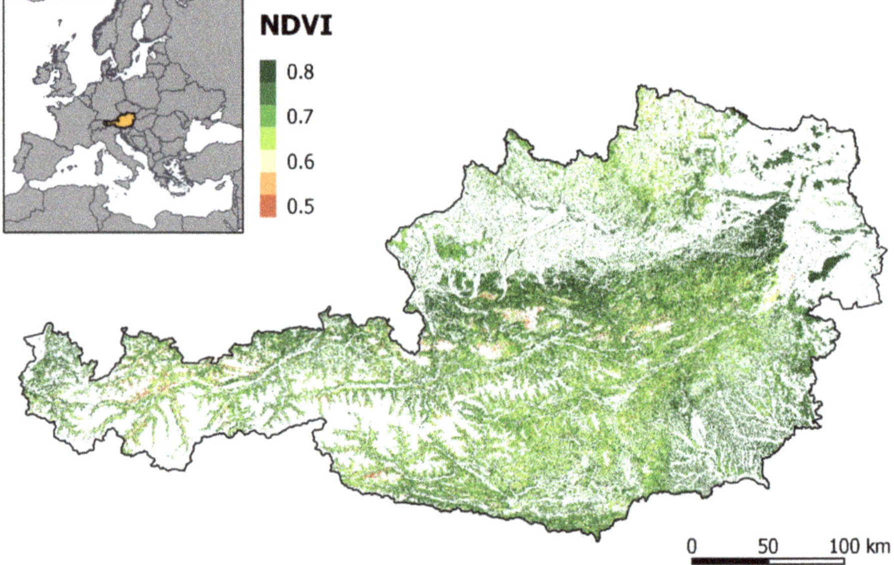

Figure 5. Maximum MPC value (MAX_{MPC}) in terms of NDVI between 2017 and 2019 for areas covered by forest in Austria.

The start date of the vegetation period (SVP) and the end date (EVP) are denoted by dashed lines. SVP and EVP slightly differ depending on the used index and can significantly vary between different forest types and locations. Figure 6 shows the SVP for Vorarlberg, the most western region of Austria, derived from the GNDVI model of the years 2017 to 2019. Early SVPs (about mid of April) mainly occur in areas of low to mid altitudes dominated by broadleaved tree species, as found in the western and northern part of Vorarlberg. Late SVPs are mainly found in areas covered by coniferous forest, such as in the alpine region in the south of Vorarlberg. At a closer look, one can also see heterogeneous spatial patterns and distinct differences in the SVP that can be explained by differences in the elevation and the tree species composition (Figure 6, right).

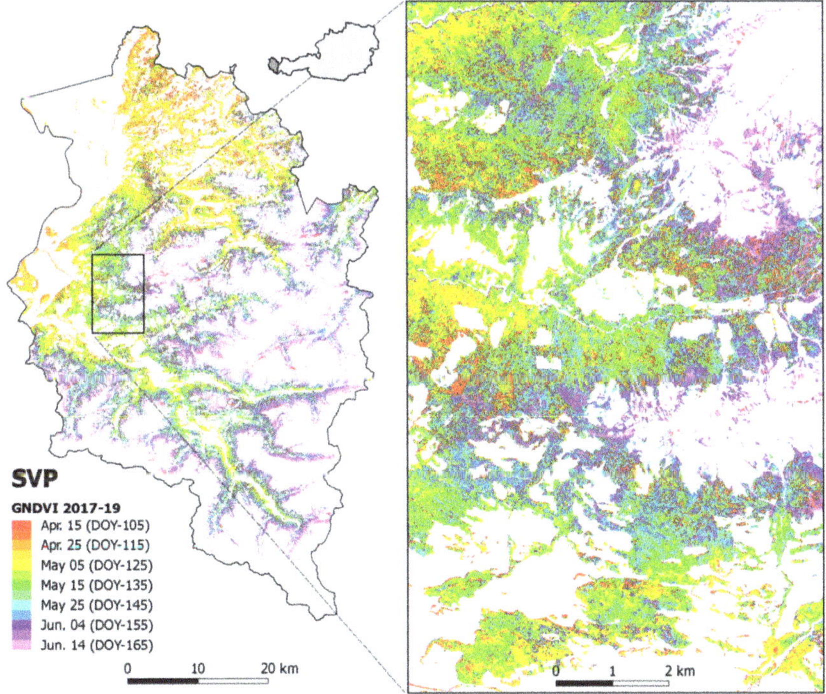

Figure 6. Start of vegetation period (SVP), derived from the modelled GNDVI course in the forests of the state of Vorarlberg.

3.2. Forest Disturbance Mapping in Northern Austria

The presented forest disturbance mapping results are based on the RGVI that proved to be the best index for negative deviation detection as it shows little noise and robust courses. This is especially true for coniferous forest (Figure 4). For the presentation of the forest disturbance results, we chose the northern region of Austria, which is currently a hotspot in terms of bark beetle infestation.

First, we exemplify the basic results of the per-pixel anomaly-detection procedure using four pixels (Figure 7, P1 to P4) selected from the study area that represent typical events when dealing with forest anomaly detection. Note that the last data point of each pixel time series is excluded from the TSA (Section 2.4). Excluded data points are highlighted by a grey dashed ellipse.

Example P1 (Figure 7, first row): In 2018, the first year of the detection period (DP), the index course shows the same stable and inconspicuous trend as in 2017. In 2019, however, we observe gradually lower values. In early June 2019, a strong disturbance occurs, finally reaching the deviation level FDL-7 on 30th June (CDD, orange marker). The red area below the baseline (BL, blue line) corresponds to the cumulative deviation of the time series. The date of origin of the disturbance (FDD-7, yellow marker) is estimated to be 9th February 2019. The data points after the CDD-7 label, all of them lying significantly below the baseline, confirm the detected disturbance.

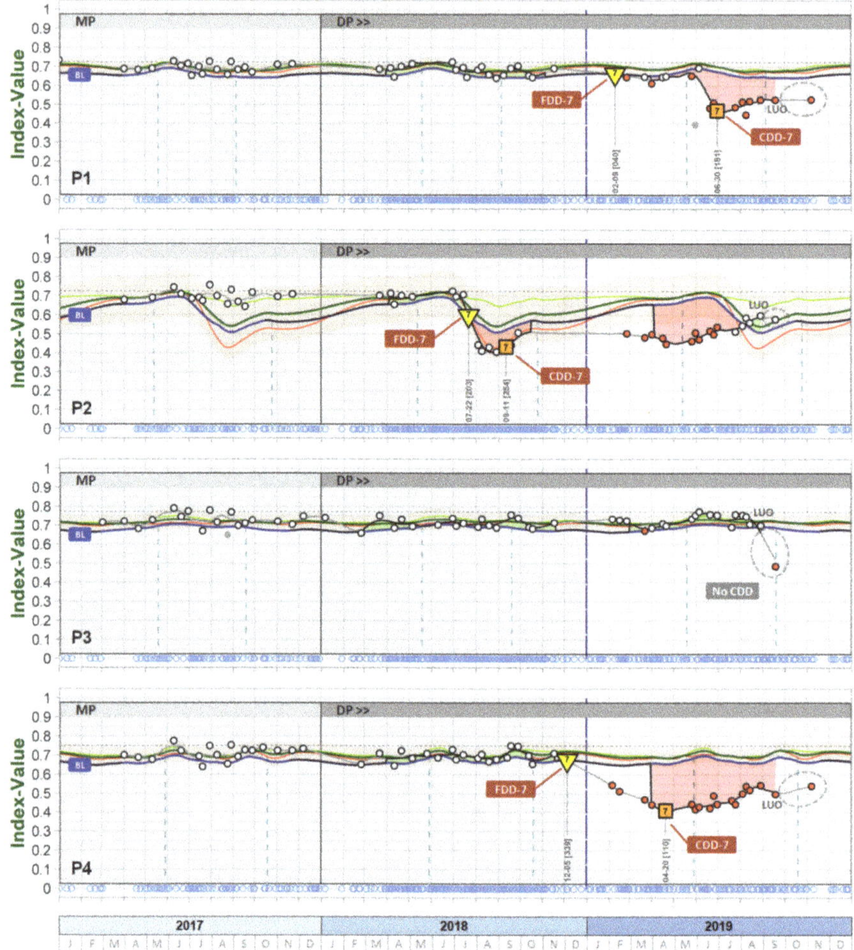

Figure 7. Forest disturbance detection 2018-2019 with medium detection sensitivity (T_{FDL} = 7). Compilation of single pixel courses, their individual deviation from the RGVI-model (2017-2018), and the identified CDD-7 and FDD-7. (light-green line: 90th-percentile index course (PC90), red line: 10th-percentile index course (PC10), dark-green line: main phenology index course (MPC, mean of PC10 and PC90), dark-blue line: reference index baseline (BL) for calculating the deviations from the actual index time series in grey).

Example P2 (Figure 7, second row): The pixel shows a disturbance occurring between 13th July 2018 and 4th August 2018. The medium damage level (FDL-7) manifests in early September (CDD-7 = 11th September) and the related FDD-7 is July 22nd 2018. In this example, the main deviation happens in the period, when MP and DP overlap. The index course and the resulting deviation area (red area) of 2019 clearly confirm the disturbance detected in 2018. Note that the winter period is excluded when computing the cumulative deviation sum (no red area between mid of October and beginning of April), because the number of data points is not sufficient (Section 2.4).

Example P3 (Figure 7, third row): The time series of this pixel follows the modelled course and no change is detected. Only the last data point possibly indicates a major deviation, but this data point was excluded in the truncation process described in Section 2.4. Further observations are required to confirm or discard this assumption.

Example P4 (Figure 7, fourth row): This pixel does not show any anomalies until November 2018, but after the excluded winter period, a severe change becomes evident, reaching FDL-7 on 20th April 2019. The corresponding FDD-7 is traced back to 5th December 2018. This example represents common winter dynamics, such as forest management activities (e.g., clear-cutting, selective timber extraction or thinning) or natural disturbances (e.g., snow and avalanche damages).

Figure 8 shows the FDD-7 map for a subset of the study area, including the pixels P1 to P4 presented in Figure 7. The selected area is heavily affected by recurring bark beetle infestations [56]. In the background, Sentinel-2 RGB-composites (10 m, Level L2A), acquired in Sep. 2017 (Figure 8a,b), Sep. 2018 (Figure 8c,d) and Sep. 2019 (Figure 8e,f) are shown.

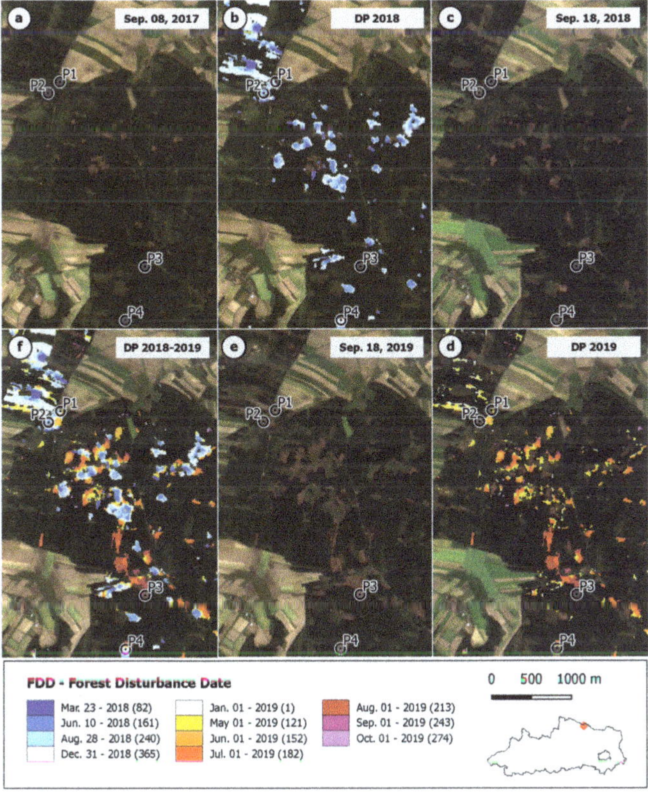

Figure 8. Forest disturbance maps (FDD-7) for a subset of the study area for the deviation periods September 2017–September 2018 (**b**), September 2018–September 2019 (**d**) and September 2017–September 2019 (**f**). The Sentinel-2 RGB-composites (10 m) in the background are from September 2017 (**a,b**), September 2018 (**c,d**) and September 2019 (**e,f**). The phenology courses of the pixels P1 to P4 are shown in Figure 7 and described in the text.

Figure 8b highlights the pixels where FDD-7 is in 2018 (dark-blue–blue–white). Figure 8d additionally highlights the pixels where FDD-7 is in 2019 (white–yellow–orange–red–pink). The FDD map shows rectangular to round patches of change, most of them with an FDD minimum (earliest date) close to the centre of the patch surrounded by continuously increasing FDD values, which is a characteristic pattern of spreading in the context of bark beetle infestation. Areas with a late FDD (near the end of the year 2019) are usually found in close proximity to areas with an earlier change (lower FDD).

The FDD maps can be spatially aggregated at any level. Figure 9 illustrates three FDD-mapping products, computed for the whole study area: (a) the original FDD-map—simplified to the categories "Disturbance" and "No Disturbance"—with a spatial resolution of 10 m, (b) the percentage of forest area affected by forest disturbance for hexagons of 100 hectares, and (c) the percentage of forest area affected by forest disturbance at municipality level.

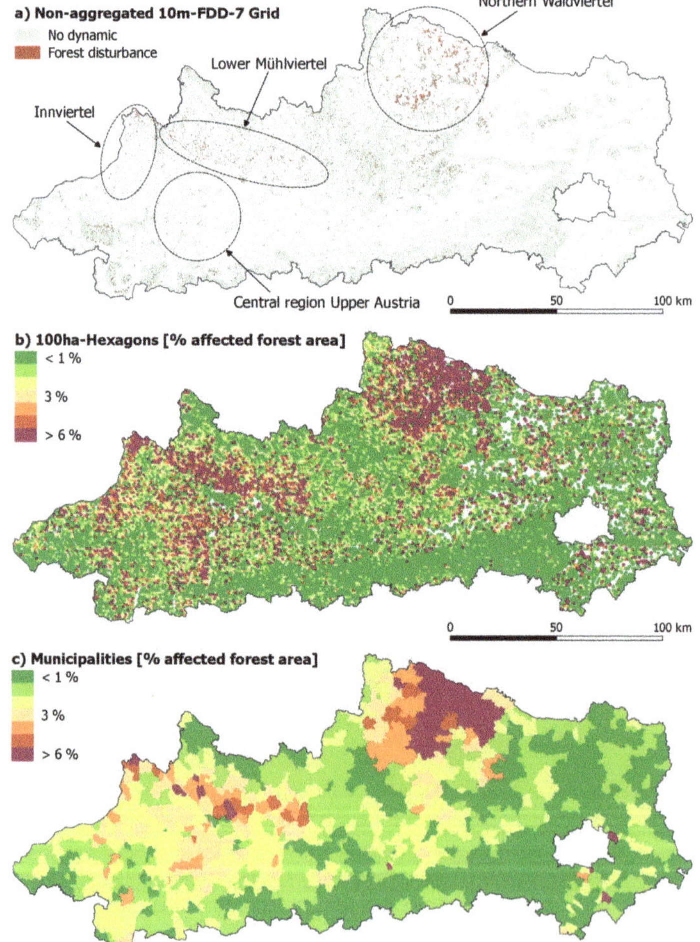

Figure 9. Forest disturbance maps based on a medium detection sensitivity (FDD-7) at three spatial levels; (**a**) non-aggregated 10m FDD-7 grid, (**b**) percentage affected forest area aggregated on 100ha-hexagons and (**c**) percentage affected forest area aggregated on the municipality level.

It was found that forests at higher altitudes show generally less disturbance than forests in lowland areas. In total, the disturbed forest area is 23,400 hectares, i.e., on average 2.8% of the forest area in the study area. The forest disturbance is not evenly distributed over the whole study area but concentrates on a few regions (Figure 9). Approximately, one quarter of all municipalities show an affected forest area of more than 4%, comprising primarily municipalities of the regions "Lower Mühlviertel" and "Innviertel", of the central region of Upper Austria and foremost of the region "Northern Waldviertel".

3.3. Validation

The validation results of the in situ dataset show that, in 1251 out of 1500 in situ cases, the TSA identified a disturbance. In 249 in situ cases, the Forest Disturbance Date grid (FDD-7) does not show a disturbance. So, the TSA identified 83.4% of the disturbances recorded by field data (Table 3).

Table 3. Validation results of the FDD grid.

Reference Dataset	Disturbance (Count)	No Disturbance (Count)	Disturbance (Fraction)	No Disturbance (Fraction)
In situ dataset for class "Disturbance"	1251	249	83.4%	16.6%
Random sample dataset for class "No Disturbance"	13	258	4.8%	95.2%

The validation results of the random sample dataset show that 258 out of 271 random points (95.2%) could be verified, by visual interpretation, to be not disturbed. In 13 cases, we observe phenological deviations, which are not detected by the TSA.

The histograms in Figures 10–12 show the temporal difference (in days) between the in situ date and the FDD. The bin width is 30 days. The black line in the centre denotes a difference of zero, which means that the FDD is equal to the recorded in situ date. Counts to the left of the line (negative date difference) indicate disturbance events, where the theoretical FDD of the TSA lies after the in situ date. Counts to the right of the line (positive date difference) correspond to cases, where the TSA detects disturbances earlier compared to the field observations.

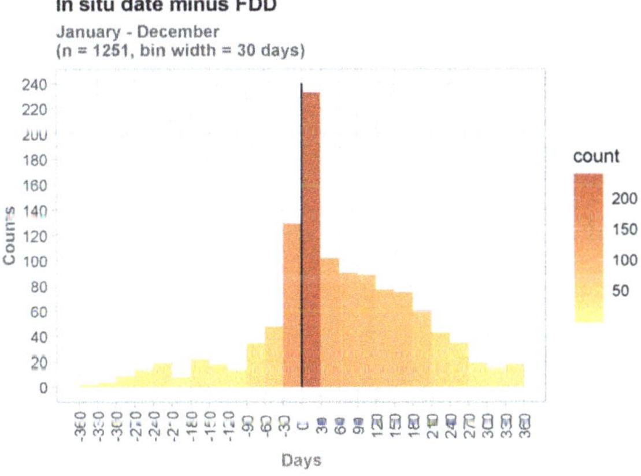

Figure 10. Histogram of in situ date minus FDD for all validation points detected as disturbances (January to December).

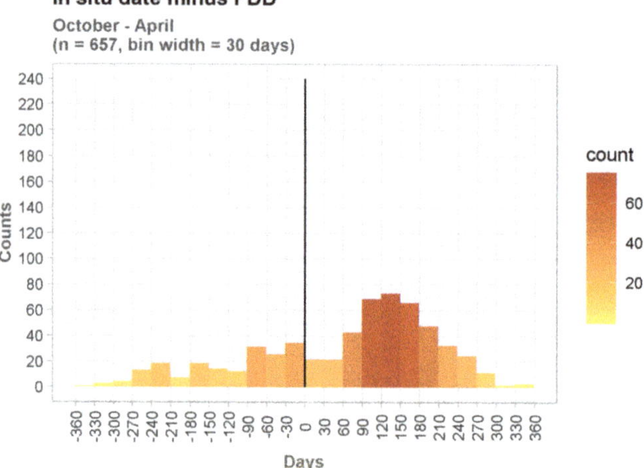

Figure 11. Histogram of in situ date minus FDD outside the vegetation period (October to April).

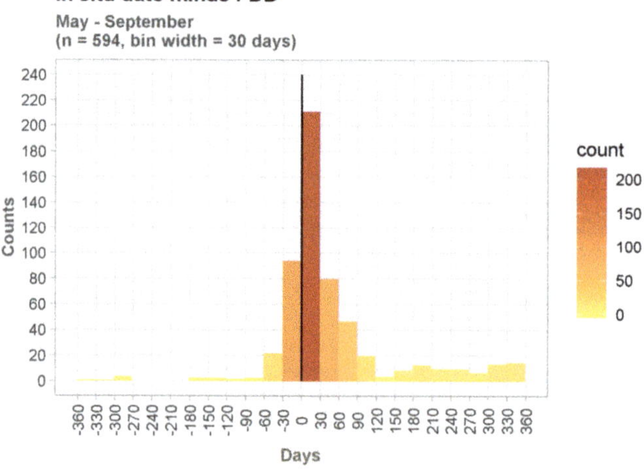

Figure 12. Histogram of in situ date minus FDD within the vegetation period (May to September).

The first histogram (Figure 10) comprises all 1251 cases with an FDD from January to December. There are 358 counts on the left and 893 counts on the right (29% and 71%). The column of the first bin on the right (0 to +30 days) is the highest of all and includes 233 cases (19%).

The second histogram (Figure 11) comprises a subset of Figure 10 including only counts outside the vegetation period with FDDs from October to April. In total, there are 657 cases with 218 counts (33%) on the left and 439 counts on the right (67%). The maximum on the right side indicates that the TSA detects disturbances with FDDs outside the vegetation period about 130 days earlier compared to the corresponding in situ date.

The third histogram (Figure 12) comprises a subset of Figure 10, including just counts within the vegetation period with FDDs from May to September. In total, there are 594 counts with 140 counts on the left (24%) and 454 counts on the right (76%). The graph indicates that, for disturbances with FDDs within the vegetation period, the FDD strongly correlates with the in situ date. The column of the first bin on the right side (0 to +30 days) is the highest of all and includes 211 cases (36%). About 51% of the cases have an FDD that deviates less than ±30 days from the in situ date, and about 69% of the cases have an FDD that deviates less than ±60 days from the in situ date.

4. Discussion

4.1. Phenology Modelling with Sentinel-2 Time Series

4.1.1. Take-Home Messages

The approach described in this article uses a new and advanced workflow to compile, preprocess and analyse dense Sentinel-2 (S2) time series. The presented TSA approach uses all S2 granules with less than 80% cloud cover available for a chosen period. The approach significantly benefits from the improved data availability due to the launch of the Sentinel-2B satellite in spring 2018. Thereby, very dense time series can be compiled, allowing for the application of advanced fitting methods. With such methods, intraseasonal variations can be preserved [25] and phenology developments can be traced with a high level of detail.

Recent studies stressed the capabilities of an updating S2 time series that predicts forest phenology using a recursive Kalman filter [31]. Unlike these studies, we use a Savitzky–Golay filtering (SGF) approach, as former studies showed the advantages of SGF to smooth out signal noise but retain temporal details. This holds true for dense time series analyses, such as S2-imagery time series [57].

The presented TSA is based on a dynamic SGF approach [35]. We use a dynamic window width, because a fixed window width can lead to insufficient or non-meaningful smoothing if the data is noisy [44]. So far, SGF has primarily been used with remote sensing data of medium resolution (e.g., MODIS with 250 m) and a fixed and rather wide window that results in a high degree of generalisation [57,58]. Most TSA studies based on Sentinel-2 or Landsat data, apply harmonic regressions (ordinary least square models) on the generated time series to characterise the seasonality of the vegetation canopy [24,25,47,59]. The periodic character of harmonic regression models, the fast computing time and the robust results are clear advantages of harmonic regressions and in the case of lower frequencies of data, they may be the only option for achieving robustness [25]. However, they become impractically complex when describing phenology courses with a more segmented type of the seasonal phenology dynamics, as it is in the case of forests in the mid-latitudes. Forest vegetation in the mid-latitudes goes through an inactive winter period, a sharp greening period in spring, followed by a slightly lower stable state in summer, and a constant defoliation process in fall, which is substantially different to vegetation in tropical regions with a more smoothed gradual phenology course.

The approach described in this article is based on TOA data, referred to as Level 1C (L1C). According to our experience, BOA data, referred to as Level 2A (L2A), produced with the Sen2Cor algorithm, provided by ESA, have considerable radiometric deficiencies, such as effects of overcorrection. Due to these problems, quite a few images cannot be used in the TSA, although the original images (L1C) are of good quality. Thus, we clearly get denser time series with L1C data than with L2A data. On the downside, L1C data are affected by mainly atmospherically induced noise, which is, however, successfully reduced by efficient outlier filtering and smoothing.

The innovative multiyear percentile modelling approach traces high courses (90th percentile) and low courses (10th percentile) of single year time series, whereas the mean of both provides robust multiyear courses (MPC). The MPC levels out extreme years, which further reduces distortions caused by possible outliers.

As a byproduct, the phenology modelling procedure delivers meaningful phenology metrics (e.g., Figure 6). Phenology metrics, such as the start and end of vegetation period [60–62], can be deduced for deciduous forest pixels quite easily, due to the typical seasonal characteristics. For coniferous forest pixels, it is more challenging. Here the dynamic SGF window width proves to be an appropriate mean to deduce reasonable metrics across various forest types and forest growing regions in Austria. Phenology metrics and the reliable MPCs can be used for various downstream analyses, such as for forest type classification [63] or habitat modelling.

4.1.2. Limitations

The accuracy of the TSA results is dependent on the data availability, which varies significantly from region to region. Primarily, the number of available observations per year is determined by the revisit time. This is at least 5 days and is halved in areas where the swaths overlap. In addition, local weather conditions, such as clouds and cloud shadows, and factors concerned with topography, such as topographic shadows and snow coverage, determine the actual data point density. In some regions of Austria, we can compile up to about 40 valid data points per year for the TSA (Figure 13). In swath overlapping areas, we can utilize 30 to 35 data points on average. In areas where the swaths do not overlap, the TSA can make use of about 15 to 20 data points on average. The single pixel courses, shown in Figures 3 and 4, lie in areas of overlapping swaths with significantly more observations than the single pixel courses, shown in Figure 7, that lie in areas without overlapping swaths.

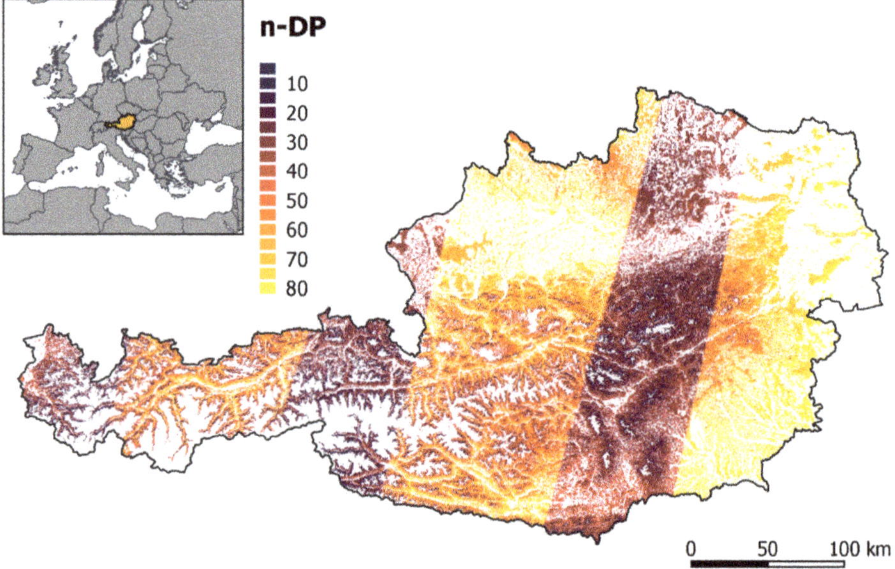

Figure 13. Number of valid data points per year (n-DP) across Austrian forests (2017–2018 average).

4.1.3. Implications

The TSA provides large area phenological information about the Earth's surface covered by vegetation. Although the approach was developed in the context of forestry, with a focus on forest applications, it also shows a high potential for interesting applications in other fields, such as conservation ecology, social ecology, and agriculture. The reconstructed phenology models provide an outstanding database not only for habitat modelling or wall-to-wall forest-type mapping, but also for models used for biomass and carbon stock estimation. Multitemporal and, in particular, phenological information, also play an increasing role in the analyses of nonforest environments. Crop type classification and monitoring [28,64–67], the reconstruction of the harvesting time of crops, cycle durations or the delineation of multiannual crops can benefit from the TSA and its outputs. For grassland management, the systematic capturing of cutting times would be highly relevant (e.g., for funding provided by the European Union) [68,69].

4.1.4. Recommendations

The described TSA is constantly adjusted, improved and extended. This comprises the used input data, parameter tuning, new analysis tools and downstream applications.

So far, the TSA relies on some Sen2Cor products, but the used indices are based on L1C values (no atmospheric correction), due to Sen2Cor-failures. In the long run, we aim to use atmospherically corrected input data. Other preprocessing procedures (e.g., ATCOR [70]), as alternatives to the Sen2Cor algorithm, still need to be tested. Consistent surface reflectance data would clearly be beneficial to further reduce signal noise effects.

The general data availability, the annual distribution of valid observations and the seasonal data variability constitute unique phenology courses on a per-pixel level. These unique time series require dynamic parameters (e.g., SGF-polynomial order, tree species dependent smoothing factor, data gap detection) for individual modelling in terms of outlier filtering, interpolation, smoothing, and multiyear data fusing. We plan to further optimise existing parameters and introduce new dynamic parameters.

In the next years, the TSA can be used to study long-term trends caused by climate change. Multiyear fusions of more than about five years will allow for investigating spatiotemporal shifts in forest phenology patterns. We are confident that the TSA will meet future demands of tracing altering site-specific forest phenology, including slightly changing tree species compositions or shifting growing periods.

4.2. Forest Disturbance Mapping in Northern Austria

4.2.1. Take-Home Messages

In the last decade, North America and Europe experienced massive bark beetle outbreaks with serious impacts on the landscape, forest industry, and ecosystem services. The extent and intensity of many recent outbreaks are widely believed to be unprecedented [71]. Therefore, there is an urgent need for operational tools to assess the affected area fast and reliably over large areas [11]. The presented TSA approach proves to be a proper forest monitoring tool for large-scale analysis as demonstrated in a test region located in Upper and Lower Austria. We found severe phenology anomalies, especially in the northern parts of this region, which corresponds well to recent reports about bark beetle calamities in Austria [56].

The main benefit of the described forest disturbance mapping approach is its ability to determine and map the date when an anomaly occurs with a high level of detail. The TSA can reconstruct a theoretical intra-annual forest disturbance date (FDD), expressed in the day-of-year format and with a spatial resolution of 10 m.

The FDD validation (Section 3.3) shows that 83.4% of the recorded field observations were successfully detected by the TSA. The results confirm that the anomaly detection procedure performs well. The error of omission of about 16% can be explained, to some degree, by the way the in situ data were collected, as discussed in Section 4.2.2. Furthermore, the validation results for the class "No Disturbance" with an agreement of 95.2% confirm that the TSA provides results with high accuracy.

Overall, we can reconstruct and map the forest disturbance date with a high level of detail on the time axis, as shown in Figure 8. Such maps compactly visualize the comprehensive spatiotemporal information contained in dense Sentinel-2 time series and can make a substantial contribution to the assessment and monitoring of forest disturbances.

The FDD maps (Figure 8) show patches of disturbance that are growing in a ring-like manner. These spatiotemporal disturbance patterns are typical for the spreading of bark beetle infestations [72]. They result from a temporal sequence of timber harvesting to counteract further bark beetle spreading. The detected patterns also indicate that the FDD maps are plausible. Otherwise, they would show a rather random distribution of disturbance patches.

The anomaly detection procedure is very flexible in terms of the disturbance level to be detected. This is realized by using the cumulative sum of deviation as a measure for disturbance. The sensitivity level threshold for the FDL (T_{FDL}) can be chosen, corresponding to different degrees of forest disturbance. The optimal level of sensitivity depends on the overall goal. FDL thresholds from 5 (very sensitive) to 10 (highly reliable) were found to be most reasonable. We recommend stepwise processing with a range of FDL thresholds and finally to choose the FDL threshold that is most appropriate. In this article, results for a T_{FDL} of 7 (CDD-7/FDD-7) are shown, corresponding to a medium detection sensitivity (i.e., minor anomalies are not considered).

The applied anomaly detection approach, based on dynamic SGF modelling, shows a high temporal sensitivity. In the vegetation period (May–September) we can detect more than half of the disturbances within at least ±30 days using the in situ date as a reference. Future studies need to investigate the strengths and weaknesses of the temporal TSA outputs, compared to similar information derived by approaches based on generalised harmonic model fitting and trajectory segmentation, which are widely established to detect temporal breakpoints in a time period of interest [45,46].

The TSA approach is expected to be suitable for different use cases, ranging from rapid disturbance mapping (e.g., after storm events) at local or regional scale to operational nation-wide disturbance mapping. Depending on the use case, different parameter configurations can be chosen.

For anomaly detection, we chose the vegetation index, RGVI. It was found to be useful particularly for detecting anomalies in coniferous forests which generally show index courses with little seasonal variation. Among all considered indices, the RGVI shows the lowest seasonal variation, which is preferable when it comes to anomaly detection. In this study, which concentrates on forest disturbances in coniferous forests, the RGVI index proved to be very efficient to detect distinct, as well as marginal vegetation, anomalies in the time series and can be recommended for studies on bark beetle infestation.

In general, some indices are more suitable to detect forest disturbances, others are more useful to derive phenology metrics (Figures 3 and 4). The GNDVI, for example, is probably a good candidate for analysing shifted spring greening due to seasonal drought stress. Here further research is needed.

4.2.2. Ground Truthing

A quantitative validation of TSA outcomes is generally difficult, as is the case for many monitoring applications based on remote sensing data [73,74]. Ground truth data that comprise temporal information on land cover changes are rare. Besides, it is generally difficult to obtain data on disturbances that are consistent over large areas, because how it is collected often varies with the responsible institution or person.

Being aware of all these challenges, the in situ validation dataset used in this study cannot be valued highly enough. Nevertheless, there are some limitations. First of all, reference data for the category "No Disturbance" is missing. Therefore, it cannot be used to estimate the rate of true negatives and false positives. In this study, this shortcoming is compensated by an extra reference dataset obtained by visual image interpretation. In future, in situ data for both classes, "Disturbance" and "No Disturbance", are desirable.

In addition, the in situ date, i.e., the date when the site was visited, does not necessarily correspond to the date when the disturbance event happened and became evident in the spectral signature the first time. This fact limits the possibilities of temporal evaluation.

The in situ data were not collected with the purpose of validating remote sensing-based products but with the purpose of documentation. Therefore, the data points are often placed close to but not within the affected groups of trees or forest stands. This fact may lead to an underestimation of the derived detection rate. Thus, the accuracy figures reported in this study are assumed to be rather conservative.

4.2.3. Limitations

The quality of the anomaly detection results is heavily dependent on the data availability, which can vary from pixel to pixel, as described in Section 4.1.2. The validation (Section 3.3) was done in an area where the Sentinel-2 swaths do not overlap with about 15 to 20 observations per year. Therefore, our results suggest that the number of data points usually available is sufficient to reliably detect forest disturbances. Only in areas where the data availability is additionally reduced by clouds, shadows or snow do we expect limitations. Therefore, in alpine regions a temporally precise reconstruction of forest disturbance dates remains challenging.

The exclusion of one or more data points at the end of the time series guarantees that only anomalies are considered that are affirmed by following data points. In this way, the risk of mapping false positives can be reduced. However, for "near-real-time" applications (e.g., the mapping of storm damages within a narrow time frame) the most recent data points are indispensable and consequently the risk of false positives has to be accepted.

The FDD-validation histogram of FDDs outside the vegetation period shows a skew distribution (Figure 11). The maximum of counts shifted from zero to the right, which can be explained by two reasons. First, compared to the vegetation period, there is a tendency that foresters note possible bark beetle infestations with a time lag. Second, the current reconstruction approach of the FDD shows some constraints in winter. If the disturbance lies in the winter period with insufficient data points and, therefore, the deviation calculation is deactivated, the TSA shows the tendency to result in too-early FDDs.

In this study, the modelling period and the deviation period partly overlap. This should be avoided in the future and was only accepted here because, when the study was carried out, the Sentinel-2 data archive comprised only three complete years (2017, 2018, and 2019) of consistent data. Alternatively, a baseline derived from Landsat data could be used. Phenology modelling could benefit from the much longer time-series compared to Sentinel-2. This is especially true if minor anomalies are to be detected. However, the integration of Landsat data also comes with some negative effects, primarily the lower spatial resolution of Landsat compared to Sentinel-2 as well as the differences in the spectral characteristics between Sentinel-2 and Landsat.

4.2.4. Implications

The spatiotemporal information provided by the FDD maps is highly relevant for the pest control management typically conducted by local forest authorities. At the same time, the approach can also be applied on larger scales, such as at the national level and providing information for stakeholders and policymakers. The presented TSA approach is not designed for the early detection of bark-beetle-attack—also referred to as "green-attack",—detection, which is currently a hot topic both in forest management and research. However, the FDD maps are a unique data source for entomological studies investigating the spreading behaviour of bark beetles. Beyond the assessment of natural disturbances, anomalies also caused by activities, such as illegal logging, are assumed to be detectable.

Further, the anomaly detection results deliver reliable information for a systematic large-scale assessment of forest disturbances with a spatial resolution of 10 m. For large parts of Northern Austria, the aggregated results (Section 3.2) reveal that the disturbed forest area is much too large for being consistent with sustainable forest management, according to the commonly used forest management model called "Normalwaldmodell" in German. The "Normalwaldmodell", according to Hundeshagen [75], is used to determine the annual allowable cut in the case where forest management is focused on even-aged, monospecific stands. The model is characterized by a specific production cycle (rotation period) and a uniform area distribution of the corresponding age-classes.

If we assume a relatively low average rotation period of 80 years for spruce stands [76], an annual harvest rate (including unscheduled timber extractions) of up to 1.3% is sustainable. Thus, from a forest management perspective, more than 4% (i.e., 2% per year) of harvested forest area in the period from 2018 to 2019, as it was found in some regions within the study area, clearly indicate unsustainable developments. At least a quarter of the investigated municipalities—for whatever reason—show such dynamics. This underlines that zonal statistics covering different scales, and different damage levels are of high relevance for the forestry sector. Zonal statistics provide aggregated overviews and comprehensively inform policy makers and stakeholders about the extent of forest disturbances.

4.2.5. Recommendations

The presented TSA is able to detect anomalies, but so far it cannot distinguish between different causes of anomaly, such as bark beetle infestations, storm events or timber harvesting activities. However, this information is very important, for example, to specifically assess the amount of damage caused by bark beetle infestation [11]. Previous studies have already tried to discriminate different categories of change by directly using various disturbance metrics or by using those metrics as input data for machine learning algorithms (e.g., Random Forest) [20,77–80]. First tests based on our data suggest that there are specific patterns both in the single pixel courses as well as in the FDD maps that could help to categorize disturbances by the cause of disturbance. The validation data source used in this study possibly provides appropriate training data to distinguish different forest disturbances. This issue will be addressed in follow-up studies.

So far, we have concentrated on coniferous forests as in Austria this forest type is most affected by natural disturbance processes. Thus, there is a need to also fine tune the TSA for deciduous forests.

5. Conclusions and Perspectives

In this study, we present the first nationwide operational forest phenology modelling and forest monitoring system optimised for 10 m Sentinel-2 time series data and based on a Savitzky–Golay modelling approach. The method was successfully tested in Austria and is expected to also be applicable in many other regions all over the world.

Overall, the study shows that, even with TOA data, instead of BOA data, the robust forest phenology modelling is feasible. Even so, further tests with atmospherically corrected data (e.g., from ATCOR [70] or an improved Sen2Cor version) are planned. The workflow is very flexible so that the TOA data can be replaced by BOA data without any effort. Besides, the TSA is extendable to additional input data. In a next step, the Sentinel-2 20 m bands will be included in the TSA.

The main benefit of the described approach, compared to Sentinel-2 approaches that exist so far, is its capability to derive meaningful phenology courses without eliminating intra-annual characteristics at the same time. Our TSA is more than a fixed sequence of single snapshots. It is capable of balancing between temporal sensitivity and certainty, as different applications need differently adjusted TSA-settings. Besides the basic index choice, we can define various parameters as BL-offset from MPC, smoothing degree, and many more.

Important outputs of the TSA are day-of-year spectral quantities (e.g., MAX_{MPC} for the NDVI) and seasonal metrics (e.g., start of vegetation period). These output features offer the opportunity to derive area-wide consistent wall-to-wall products. They are relevant to NFIs for many purposes.

Recent efforts of the Austrian NFI have aimed for operational implementation to use phenology modelling metrics to derive reliable nationwide forest type maps. Forest type classifications [81] will definitely benefit from such consistent input data. The resulting forest type maps can, for example, further improve the biomass estimations of NFIs [34].

The main added value of the presented TSA is the provision of novel temporal information about forest phenology anomalies. The TSA does not only map phenology anomalies with a high spatial resolution but also assigns a time stamp to each disturbed pixel with a high temporal resolution, indicating the estimated date when the anomaly is recognizable in the dataset the first time. The high sensitivity of the TSA's outcomes serves the forestry and forest ecosystem sciences' aim to monitor forests. Finally, the TSA also opens new fields for various applications on a forest-management level. Here the described nationwide wall to wall application, which focuses on bark beetle damages, demonstrates that the TSA is a useful monitoring system to scrutinise spatiotemporal patterns of forest disturbance. The results of this study show that it is possible to map the spreading of bark beetle infestations and other disturbances with a high accuracy.

The upcoming decades demand long-term analytic tools that focus on the incremental impact of climate change effects on forest ecosystems. Therefore, subsequent efforts should extend the TSA in such a way that also transannual anomalies can be captured.

Author Contributions: M.L. prepared the imagery data, developed and implemented the time-series algorithm and performed downstream analyses from the time series outputs. T.K. was a major contributor in interpreting the results and in writing the manuscript. Both authors have read and agreed to the published version of the manuscript.

Funding: This research received no external funding.

Acknowledgments: Thanks to the forest section of the Federal Government of Lower Austria for providing the in situ data, to Ambros Berger for revising the manuscript and three anonymous reviewers whose comments helped to improve this paper.

Conflicts of Interest: The authors declare no conflict of interest.

References

1. Kindermann, G.; McCallum, I.; Fritz, S.; Obersteiner, M. A Global Forest Growing Stock, Biomass and Carbon Map Based on FAO Statistics. *Silva Fenn.* **2008**, *42*, 387–396. [CrossRef]
2. Hansen, M.C.; Potapov, P.V.; Moore, R.; Hancher, M.; Turubanova, S.A.; Tyukavina, A.; Thau, D.; Stehman, S.V.; Goetz, S.J.; Loveland, T.R.; et al. High-Resolution Global Maps of 21st-Century Forest Cover Change. *Science* **2013**, *342*, 850–853. [CrossRef] [PubMed]
3. Mackey, B.; DellaSala, D.A.; Kormos, C.; Lindenmayer, D.; Kumpel, N.; Zimmerman, B.; Hugh, S.; Young, V.; Foley, S.; Arsenis, K.; et al. Policy Options for the World's Primary Forests in Multilateral Environmental Agreements. *Conserv. Lett.* **2015**, *8*, 139–147. [CrossRef]
4. FAO Forests, the Global Carbon Cycle and Climate Change. Available online: http://www.fao.org/3/XII/MS14-E.htm (accessed on 26 April 2020).
5. Henders, S.; Persson, U.M.; Kastner, T. Trading Forests: Land-Use Change and Carbon Emissions Embodied in Production and Exports of Forest-Risk Commodities. *Environ. Res. Lett.* **2015**, *10*, 125012. [CrossRef]
6. O'Hara, K.L. What Is Close-to-Nature Silviculture in a Changing World? *Forestry* **2016**, *89*, 1–6. [CrossRef]
7. McDowell, N.G.; Coops, N.C.; Beck, P.S.A.; Chambers, J.Q.; Gangodagamage, C.; Hicke, J.A.; Huang, C.; Kennedy, R.; Krofcheck, D.J.; Litvak, M.; et al. Global Satellite Monitoring of Climate-Induced Vegetation Disturbances. *Trends Plant Sci.* **2015**, *20*, 114–123. [CrossRef]
8. Birdsey, R.; Pan, Y. Trends in Management of the World's Forests and Impacts on Carbon Stocks. *For. Ecol. Manag.* **2015**, *355*, 83–90. [CrossRef]
9. Goetz, S.; Dubayah, R. Advances in Remote Sensing Technology and Implications for Measuring and Monitoring Forest Carbon Stocks and Change. *Carbon Manag.* **2011**, *2*, 231–244. [CrossRef]
10. Matthews, E.; Payne, R.; Rohweder, M.; Murray, S. *Pilot Analysis of Global Ecosystems: Forest Ecosystems*; World Resources Institute: Washington, DC, USA, 2000.
11. Senf, C.; Seidl, R.; Hostert, P. Remote Sensing of Forest Insect Disturbances: Current State and Future Directions. *Int. J. Appl. Earth Obs. Geoinf.* **2017**, *60*, 49–60. [CrossRef]
12. Bastin, J.-F.; Finegold, Y.; Garcia, C.; Mollicone, D.; Rezende, M.; Routh, D.; Zohner, C.M.; Crowther, T.W. The Global Tree Restoration Potential. *Science* **2019**, *365*, 76–79. [CrossRef]
13. Sebald, J.; Senf, C.; Heiser, M.; Scheidl, C.; Pflugmacher, D.; Seidl, R. The Effects of Forest Cover and Disturbance on Torrential Hazards: Large-Scale Evidence from the Eastern Alps. *Environ. Res. Lett.* **2019**, *14*, 114032. [CrossRef]
14. Erb, K.-H.; Kastner, T.; Plutzar, C.; Bais, A.L.S.; Carvalhais, N.; Fetzel, T.; Gingrich, S.; Haberl, H.; Lauk, C.; Niedertscheider, M.; et al. Unexpectedly Large Impact of Forest Management and Grazing on Global Vegetation Biomass. *Nature* **2018**, *553*, 73–76. [CrossRef]
15. Pugh, T.A.M.; Lindeskog, M.; Smith, B.; Poulter, B.; Arneth, A.; Haverd, V.; Calle, L. Role of Forest Regrowth in Global Carbon Sink Dynamics. *Proc. Natl. Acad. Sci. USA* **2019**, *116*, 4382–4387. [CrossRef] [PubMed]
16. Seidl, R.; Albrich, K.; Erb, K.; Formayer, H.; Leidinger, D.; Leitinger, G.; Tappeiner, U.; Tasser, E.; Rammer, W. What Drives the Future Supply of Regulating Ecosystem Services in a Mountain Forest Landscape? *For. Ecol. Manag.* **2019**, *445*, 37–47. [CrossRef]
17. Wulder, M.A.; Loveland, T.R.; Roy, D.P.; Crawford, C.J.; Masek, J.G.; Woodcock, C.E.; Allen, R.G.; Anderson, M.C.; Belward, A.S.; Cohen, W.B.; et al. Current Status of Landsat Program, Science, and Applications. *Remote Sens. Environ.* **2019**, *225*, 127–147. [CrossRef]

18. García-Mora, T.J.; Mas, J.-F.; Hinkley, E.A. Land Cover Mapping Applications with MODIS: A Literature Review. *Int. J. Digit. Earth* **2012**, *5*, 63–87. [CrossRef]
19. Nguyen, T.H.; Jones, S.D.; Soto-Berelov, M.; Haywood, A.; Hislop, S. A Spatial and Temporal Analysis of Forest Dynamics Using Landsat Time-Series. *Remote Sens. Environ.* **2018**, *217*, 461–475. [CrossRef]
20. Moisen, G.G.; Meyer, M.C.; Schroeder, T.A.; Liao, X.; Schleeweis, K.G.; Freeman, E.A.; Toney, C. Shape Selection in Landsat Time Series: A Tool for Monitoring Forest Dynamics. *Glob. Change Biol.* **2016**, *22*, 3518–3528. [CrossRef]
21. Bullock, E.L.; Woodcock, C.E.; Olofsson, P. Monitoring Tropical Forest Degradation Using Spectral Unmixing and Landsat Time Series Analysis. *Remote Sens. Environ.* **2020**, *238*, 110968. [CrossRef]
22. Hermosilla, T.; Wulder, M.A.; White, J.C.; Coops, N.C. Prevalence of Multiple Forest Disturbances and Impact on Vegetation Regrowth from Interannual Landsat Time Series (1985–2015). *Remote Sens. Environ.* **2019**, *233*, 111403. [CrossRef]
23. Pasquarella, V.; Bradley, B.; Woodcock, C. Near-Real-Time Monitoring of Insect Defoliation Using Landsat Time Series. *Forests* **2017**, *8*, 275. [CrossRef]
24. Zhu, Z.; Woodcock, C.E. Continuous Change Detection and Classification of Land Cover Using All Available Landsat Data. *Remote Sens. Environ.* **2014**, *144*, 152–171. [CrossRef]
25. Jönsson, P.; Cai, Z.; Melaas, E.; Friedl, M.; Eklundh, L. A Method for Robust Estimation of Vegetation Seasonality from Landsat and Sentinel-2 Time Series Data. *Remote Sens.* **2018**, *10*, 635. [CrossRef]
26. European Space Agency Copernicus Programme. Available online: https://www.esa.int/Applications/Observing_the_Earth/Copernicus (accessed on 26 April 2020).
27. European Space Agency Sentinel-2 Mission. Available online: https://sentinel.esa.int/web/sentinel/missions/sentinel-2 (accessed on 26 April 2020).
28. Belgiu, M.; Csillik, O. Sentinel-2 Cropland Mapping Using Pixel-Based and Object-Based Time-Weighted Dynamic Time Warping Analysis. *Remote Sens. Environ.* **2018**, *204*, 509–523. [CrossRef]
29. Kanjir, U.; Đurić, N.; Veljanovski, T. Sentinel-2 Based Temporal Detection of Agricultural Land Use Anomalies in Support of Common Agricultural Policy Monitoring. *IJGI* **2018**, *7*, 405. [CrossRef]
30. Vrieling, A.; Meroni, M.; Darvishzadeh, R.; Skidmore, A.K.; Wang, T.; Zurita-Milla, R.; Oosterbeek, K.; O'Connor, B.; Paganini, M. Vegetation Phenology from Sentinel-2 and Field Cameras for a Dutch Barrier Island. *Remote Sens. Environ.* **2018**, *215*, 517–529. [CrossRef]
31. Puhm, M.; Deutscher, J.; Hirschmugl, M.; Wimmer, A.; Schmitt, U.; Schardt, M. A Near Real-Time Method for Forest Change Detection Based on a Structural Time Series Model and the Kalman Filter. *Remote Sens.* **2020**, *12*, 3135. [CrossRef]
32. Gabler, K.; Schadauer, K. *Methods of the Austrian Forest Inventory 2000/02. Origins, Approaches, Design, Sampling, Data Models, Evaluation and Calculation of Standard Error*; BFW-Reports; Austrian Research Center for Forests (BFW): Wien, Austria, 2006; Volume 135, ISBN 1013-0713.
33. Breidenbach, J.; Waser, L.T.; Debella-Gilo, M.; Schumacher, J.; Hauglin, M.; Puliti, S.; Astrup, R. National Mapping and Estimation of Forest Area by Dominant Tree Species Using Sentinel-2 Data, 2020. Available online: https://arxiv.org/abs/2004.07503 (accessed on 28 October 2020).
34. Puliti, S.; Hauglin, M.; Breidenbach, J.; Montesano, P.; Neigh, C.S.R.; Rahlf, J.; Solberg, S.; Klingenberg, T.F.; Astrup, R. Modelling Above-Ground Biomass Stock over Norway Using National Forest Inventory Data with ArcticDEM and Sentinel-2 Data. *Remote Sens. Environ.* **2020**, *236*, 111501. [CrossRef]
35. Savitzky, A.; Golay, M.J.E. Smoothing and Differentiation of Data by Simplified Least Squares Procedures. *Anal. Chem.* **1964**, *36*, 1627–1639. [CrossRef]
36. European Space Agency. *Sen2Cor v2.5.5*; European Space Agency: Paris, France, 2020.
37. European Space Agency Open Access Hub. Available online: https://scihub.copernicus.eu/twiki/do/view/SciHubWebPortal/APIHubDescription (accessed on 26 April 2020).
38. Hauk, E.; Schadauer, K. *Instruktion für die Feldarbeit der Österreichischen Waldinventur 2007–2009*; BFW: Vienna, Austria, 2009.
39. Olofsson, P.; Foody, G.M.; Herold, M.; Stehman, S.V.; Woodcock, C.E.; Wulder, M.A. Good Practices for Estimating Area and Assessing Accuracy of Land Change. *Remote Sens. Environ.* **2014**, *148*, 42–57. [CrossRef]
40. Xue, J.; Su, B. Significant Remote Sensing Vegetation Indices: A Review of Developments and Applications. *J. Sens.* **2017**, *2017*, 1353691. [CrossRef]

41. Rouse, J.W.; Haas, R.H.; Scheel, J.A.; Deering, D.W. Monitoring Vegetation Systems in the Great Plains with ERTS. In Proceedings of the 3rd Earth Resource Technology Satellite (ERTS) Symposium, Washington, DC, USA, 10–14 December 1974; Volume 1, pp. 48–62.

42. Gitelson, A.; Kaufman, Y.; Merzlyak, M. Use of a Green Channel in Remote Sensing of Global Vegetation from EOS-MODIS. *Remote Sens. Environ.* **1996**, *58*, 289–298. [CrossRef]

43. Motohka, T.; Nasahara, K.N.; Oguma, H.; Tsuchida, S. Applicability of Green-Red Vegetation Index for Remote Sensing of Vegetation Phenology. *Remote Sens.* **2010**, *2*, 2369–2387. [CrossRef]

44. Jönsson, P.; Eklundh, L. TIMESAT—A Program for Analyzing Time-Series of Satellite Sensor Data. *Comput. Geosci.* **2004**, *30*, 833–845. [CrossRef]

45. Verbesselt, J.; Hyndman, R.; Newnham, G.; Culvenor, D. Detecting Trend and Seasonal Changes in Satellite Image Time Series. *Remote Sens. Environ.* **2010**, *114*, 106–115. [CrossRef]

46. Verbesselt, J.; Hyndman, R.; Zeileis, A.; Culvenor, D. Phenological Change Detection while Accounting for Abrupt and Gradual Trends in Satellite Image Time Series. *Remote Sens. Environ.* **2010**, *114*, 2970–2980. [CrossRef]

47. Deijns, A.A.J.; Bevington, A.R.; van Zadelhoff, F.; de Jong, S.M.; Geertsema, M.; McDougall, S. Semi-Automated Detection of Landslide Timing Using Harmonic Modelling of Satellite Imagery, Buckinghorse River, Canada. *Int. J. Appl. Earth Obs. Geoinf.* **2020**, *84*, 101943. [CrossRef]

48. R Core Team. *R: A Language and Environment for Statistical Computing*; R Foundation for Statistical Computing: Vienna, Austria, 2019.

49. Hijmans, R. Raster: Geographic Data Analysis and Modeling. 2020. Available online: https://cran.r-project.org/package=raster (accessed on 18 December 2020).

50. Bivand, R.; Keitt, T.; Rowlingson, B. rgdal: Bindings for the "Geospatial" Data Abstraction Library. 2019. Available online: https://cran.r-project.org/package=rgdal (accessed on 18 December 2020).

51. Greenberg, J.A.; Mattiuzzi, M. gdalUtils: Wrappers for the Geospatial Data Abstraction Library (GDAL) Utilities. 2020. Available online: https://cran.r-project.org/package=gdalUtils (accessed on 18 December 2020).

52. Bivand, R.; Rundel, C. rgeos: Interface to Geometry Engine - Open Source ('GEOS'). 2019. Available online: https://cran.r-project.org/package=rgeos (accessed on 18 December 2020).

53. Microsoft Cooperation; Weston, S. doParallel: Foreach Parallel Adaptor for the "parallel" Package. 2019. Available online: https://cran.r-project.org/package=doParallel (accessed on 18 December 2020).

54. Microsoft Cooperation; Weston, S. foreach: Provides Foreach Looping Construct. 2020. Available online: https://cran.r-project.org/package=foreach (accessed on 18 December 2020).

55. Signal Developers. Signal: Signal Processing. 2013. Available online: http://r-forge.r-project.org/projects/signal/ (accessed on 18 December 2020).

56. Hoch, G.; Perny, B. *BFW Praxisinformation*; BFW: Vienna, Austria, 2019; pp. 18–21.

57. Chen, J.; Jönsson, P.; Tamura, M.; Gu, Z.; Matsushita, B.; Eklundh, L. A Simple Method for Reconstructing a High-Quality NDVI Time-Series Data Set Based on the Savitzky–Golay Filter. *Remote Sens. Environ.* **2004**, *91*, 332–344. [CrossRef]

58. Cao, R.; Chen, Y.; Shen, M.; Chen, J.; Zhou, J.; Wang, C.; Yang, W. A Simple Method to Improve the Quality of NDVI Time-Series Data by Integrating Spatiotemporal Information with the Savitzky-Golay Filter. *Remote Sens. Environ.* **2018**, *217*, 244–257. [CrossRef]

59. Shimizu, K.; Ota, T.; Mizoue, N. Detecting Forest Changes Using Dense Landsat 8 and Sentinel-1 Time Series Data in Tropical Seasonal Forests. *Remote Sens.* **2019**, *11*, 1899. [CrossRef]

60. Kowalski, K.; Senf, C.; Hostert, P.; Pflugmacher, D. Characterizing Spring Phenology of Temperate Broadleaf Forests Using Landsat and Sentinel-2 Time Series. *Int. J. Appl. Earth Obs. Geoinf.* **2020**, *92*, 102172. [CrossRef]

61. Prabakaran, C.; Singh, C.P.; Panigrahy, S.; Parihar, J.S. Retrieval of Forest Phenological Parameters from Remote Sensing-Based NDVI Time-Series data. *Curr. Sci.* **2013**, *105*, 9.

62. Cai, Y.; Li, X.; Zhang, M.; Lin, H. Mapping Wetland Using the Object-Based Stacked Generalization Method Based on Multi-Temporal Optical and SAR Data. *Int. J. Appl. Earth Obs. Geoinf.* **2020**, *92*, 102164. [CrossRef]

63. Kollert, A.; Bremer, M.; Löw, M.; Rutzinger, M. Exploring the Potential of Land Surface Phenology and Seasonal Cloud Free Composites of One Year of Sentinel-2 Imagery for Tree Species Mapping in a Mountainous Region. *Int. J. Appl. Earth Obs. Geoinf.* **2021**, *94*, 102208. [CrossRef]

64. Csillik, O.; Belgiu, M.; Asner, G.P.; Kelly, M. Object-Based Time-Constrained Dynamic Time Warping Classification of Crops Using Sentinel-2. *Remote Sens.* **2019**, *11*, 1257. [CrossRef]

65. Rufin, P.; Frantz, D.; Ernst, S.; Rabe, A.; Griffiths, P.; Özdoğan, M.; Hostert, P. Mapping Cropping Practices on a National Scale Using Intra-Annual Landsat Time Series Binning. *Remote Sens.* **2019**, *11*, 232. [CrossRef]

66. Picoli, M.C.A.; Camara, G.; Sanches, I.; Simões, R.; Carvalho, A.; Maciel, A.; Coutinho, A.; Esquerdo, J.; Antunes, J.; Begotti, R.A.; et al. Big Earth Observation Time Series Analysis for Monitoring Brazilian Agriculture. *ISPRS J. Photogramm. Remote Sens.* **2018**, *145*, 328–339. [CrossRef]

67. Qiu, B.; Zou, F.; Chen, C.; Tang, Z.; Zhong, J.; Yan, X. Automatic Mapping Afforestation, Cropland Reclamation and Variations in Cropping Intensity in Central East China during 2001–2016. *Ecol. Indic.* **2018**, *91*, 490–502. [CrossRef]

68. Kolecka, N.; Ginzler, C.; Pazur, R.; Price, B.; Verburg, P.H. Regional Scale Mapping of Grassland Mowing Frequency with Sentinel-2 Time Series. *Remote Sens.* **2018**, *10*, 1221. [CrossRef]

69. Schwieder, M.; Buddeberg, M.; Kowalski, K.; Pfoch, K.; Bartsch, J.; Bach, H.; Pickert, J.; Hostert, P. Estimating Grassland Parameters from Sentinel-2: A Model Comparison Study. *PFG* **2020**. [CrossRef]

70. GEOSYSTEMS. *ATCOR Workflow for IMAGINE*; GEOSYSTEMS: Germering, Germany, 2020.

71. Morris, J.L.; Cottrell, S.; Fettig, C.J.; Hansen, W.D.; Sherriff, R.L.; Carter, V.A.; Clear, J.L.; Clement, J.; DeRose, R.J.; Hicke, J.A.; et al. Managing Bark Beetle Impacts on Ecosystems and Society: Priority Questions to Motivate Future Research. *J. Appl. Ecol.* **2017**, *54*, 750–760. [CrossRef]

72. Baier, P. Ausbreitung. In *Der Buchdrucker. Biologie, Ökologie, Management*; Hoch, G., Schopf, A., Weizer, G., Eds.; Austrian Research Centre for Forests (BFW): Vienna, Austria, 2019; pp. 57–71.

73. Mayr, S.; Kuenzer, C.; Gessner, U.; Klein, I.; Rutzinger, M. Validation of Earth Observation Time-Series: A Review for Large-Area and Temporally Dense Land Surface Products. *Remote Sens.* **2019**, *11*, 2616. [CrossRef]

74. Loew, A.; Bell, W.; Brocca, L.; Bulgin, C.E.; Burdanowitz, J.; Calbet, X.; Donner, R.V.; Ghent, D.; Gruber, A.; Kaminski, T.; et al. Validation Practices for Satellite-Based Earth Observation Data across Communities. *Rev. Geophys.* **2017**, *55*, 779–817. [CrossRef]

75. Hundeshagen, J.C. Die Forstabschätzung Auf Neuen, Wissenschaftlichen Grundlagen: Nebst Einer Charakteristik und Vergleichung Aller Bisher Bestandenen Forsttaxations-Methoden. In *2 Abtl.*; Laupp: Tübingen, Austria, 1826; Available online: http://mdz-nbn-resolving.de/urn:nbn:de:bvb:12-bsb10300397-0 (accessed on 27 October 2020).

76. Ledermann, T.; Rössler, G. Fichte - Klima - Umtriebszeit 2019. Available online: https://bfw.ac.at/cms_stamm/050/PDF/praxistag19/BFWPraxistag2019_fichte_umtriebszeit.pdf (accessed on 30 April 2020).

77. Senf, C.; Pflugmacher, D.; Hostert, P.; Seidl, R. Using Landsat Time Series for Characterizing Forest Disturbance Dynamics in the Coupled Human and Natural Systems of Central Europe. *ISPRS J. Photogramm. Remote Sens.* **2017**, *130*, 453–463. [CrossRef]

78. Huo, L.-Z.; Boschetti, L.; Sparks, A.M. Object-Based Classification of Forest Disturbance Types in the Conterminous United States. *Remote Sens.* **2019**, *11*, 477. [CrossRef]

79. Hais, M.; Jonášová, M.; Langhammer, J.; Kučera, T. Comparison of Two Types of Forest Disturbance Using Multitemporal Landsat TM/ETM+ Imagery and Field Vegetation Data. *Remote Sens. Environ.* **2009**, *113*, 835–845. [CrossRef]

80. Zhao, F.; Huang, C.; Zhu, Z. Use of Vegetation Change Tracker and Support Vector Machine to Map Disturbance Types in Greater Yellowstone Ecosystems in a 1984–2010 Landsat Time Series. *IEEE Geosci. Remote Sens. Lett.* **2015**, *12*, 1650–1654. [CrossRef]

81. Fassnacht, F.E.; Latifi, H.; Stereńczak, K.; Modzelewska, A.; Lefsky, M.; Waser, L.T.; Straub, C.; Ghosh, A. Review of Studies on Tree Species Classification from Remotely Sensed Data. *Remote Sens. Environ.* **2016**, *186*, 64–87. [CrossRef]

Publisher's Note: MDPI stays neutral with regard to jurisdictional claims in published maps and institutional affiliations.

Article

A Comparison of Two Machine Learning Classification Methods for Remote Sensing Predictive Modeling of the Forest Fire in the North-Eastern Siberia

Piotr Janiec [1,2] and Sébastien Gadal [1,2,*]

[1] Aix Marseille Univ, Université Côte d'Azur, Avignon Université, CNRS, ESPACE, UMR 7300, Avignon, 13545 Aix-en-Provence CEDEX 04, France; piotr.janiec@etu.univ-amu.fr
[2] Department of Ecology and Geography, Institute of Natural Sciences, North-Eastern Federal University, 67000 Yakutsk, Russia
*   Correspondence: sebastien.gadal@univ-amu.fr

Received: 27 November 2020; Accepted: 16 December 2020; Published: 18 December 2020

Abstract: The problem of forest fires in Yakutia is not as well studied as in other countries. Two methods of machine learning classifications were implemented to determine the risk of fire: MaxENT and random forest. The initial materials to define fire risk factors were satellite images and their products of various spatial and spectral resolution (Landsat TM, Modis TERRA, GMTED2010, VIIRS), vector data (OSM), and bioclimatic variables (WORLDCLIM). The results of the research showed a strong human influence on the risk in this region, despite the low population density. Anthropogenic factors showed a high correlation with the occurrence of wildfires, more than climatic or topographical factors. Other factors affect the risk of fires at the macroscale and microscale, which should be considered when modeling. The random forest method showed better results in the macroscale, however, the maximum entropy model was better in the microscale. The exclusion of variables that do not show a high correlation, does not always improve the modeling results. The random forest presence prediction model is a more accurate method and significantly reduces the risk territory. The reverse is the method of maximum entropy, which is not as accurate and classifies very large areas as endangered. Further study of this topic requires a clearer and conceptually developed approach to the application of remote sensing data. Therefore, this work makes sense to lay the foundations of the future, which is a completely automated fire risk assessment application in the Republic of Sakha. The results can be used in fire prophylactics and planning fire prevention. In the future, to determine the risk well, it is necessary to combine the obtained maps with the seasonal risk determined using indices (for example, the Nesterov index 1949) and the periodic dynamics of forest fires, which Isaev and Utkin studied in 1963. Such actions can help to build an application, with which it will be possible to determine the risk of wildfire and the spread of fire during extreme events.

Keywords: wildfires; MaxENT; random forest; risk modeling; GIS; multi-scale analysis; Yakutia; Arctic; Siberia

1. Introduction

The disturbance regime in the boreal forest is extremely variable. Every year in Siberia, millions of hectares of forest are burned. Forest fires are one of the main factors causing not only long-term, harmful changes in plant ecosystems, but also contribute to the deterioration of living conditions in society, especially in the event of wildfires. Taiga fire is a natural phenomenon. Fires determine the normal, ecological functioning of the forest in this region. Forest fires are an inseparable part of the natural cycle. After the fire, there are favorable conditions for the young generation of trees.

Wildfires are important for the indigenous peoples of Siberia. The territory after the wildfire turns into a pasture and people harvest berries for winter stocks. Fires have deep consequences and issues to the transformation of the landscape in the specific context of permafrost. Fire in forest areas where permafrost is spread is considered as an important factor in modeling the surface and influencing geomorphological processes.

Boreal forest plays a big role in the circulation of carbon dioxide in the regional and global scale. Changes in fire regime and climate in this region have already started, and they have an impact on the carbon releasing dynamic not only during the fire, but also years after the fire when the permafrost is thawing [1,2]. Forest fires in Russia, in general, and in the Republic of Sakha (Yakutia) are one of the main and most common natural threats. Currently, the area covered by forest fires has increased significantly. According to the results of scientific research, not the least role in the spread of fires is played by the human factor. Almost half of all forest fires that have occurred in Yakutia are caused by humans [2]. The occurrence and spread of forest fires are influenced by a complex of different factors that mutually reinforce each other and create conditions that contribute to the forest fire.

Geoinformation systems and modern remote sensing methods are a popular and effective means of identifying the most important factors affecting fires. GIS technologies are effective for building the necessary long-term fire safety models in countries such as Canada, the USA, and most of the European countries, where sustainable forest management is wide. GIS forest fire risk models are commonly used, but most of them do not include the human factor [3]. In Russia, the Nesterov index, from the 1940s is still in use [4]. There is large interest in the development of an integrated fire hazard assessment system integrated with GIS, remote sensing, meteorological, and government data. A system like that, available for forestry, firefighters, and local authorities, can improve the fight against fires and reduce the damage, thanks to faster detection of fire ignition and focusing on the most endangered places. The purpose of this work was geoinformation modeling the long-term wildfire risk in the Republic of Sakha (Yakutia) on a macro- and microscale. To achieve this goal, the following tasks were defined: (1) study structure and regime of the wildfires in the boreal forest; (2) define the most important factors that cause forest fires in the region of the studies; (3) develop and implement statistical methods to determine the relationships between the wildfires and its factors; (4) develop spatial and statistical analysis methods for building a model of wildfire risk in macro- and microscale; and (5) validation of the built model and create wildfire risk maps at the macro-and microscale.

The methodology of the study of wildfire risk is based on forest management and geographical developments. The field of studies is built on Canadian, U.S., and European approaches to deal with natural hazards as in the works of San-Miguel-Ayanz [3], Tedim [5], Oliveira [6], Cardille [7], Martinez [8], and others. Knowledge on the boreal forests in Yakutia was taken from the publications of such authors as A.P. Isaev [9], E.I. Troeva [2], M.M. Cherosov [10], and others.

Work was divided into two main stages. The first stage presents the methodology validating the impact of individual fire factors on fire occurrence. The result of the analysis is a determination of dependencies between factors and exclusion from the further classification of factors that do not affect fires. In the second step, based on the literature analysis [11–13] two methods of machine learning classifications were implemented to determine the risk of fire: MaxENT and random forest. The initial materials to define fire risk factors were satellite images and their products of various spatial and spectral resolution (Landsat TM, Modis TERRA, GMTED2010, VIIRS), vector data (OSM), and bioclimatic variables (WORLDCLIM).

The results in the form of long-term fire risk maps can be used for fire prevention and planning fire-fighting measures in the territory of the republic. The results can help create an application that can be used to determine the risk of fire and the spread of fire during a disaster. This work has the potential to lay the foundations for the future of a fully automated application of fire risk assessment in the Sakha Republic.

2. Material and Methods

2.1. Yakutia

To correctly understand fire regimes in Yakutia, it is important to know the features of their origin. The geographical, geological, climatological, and ecological position in the landscape make it possible to explain the complexity of the fire phenomenon. Yakutia is a very specific region, with occurrence extremely high and low temperatures, a thick layer of permafrost, specific geological structures, and the occurrence of light taiga forests dependent on fire regimes. Due to the high complexity of the territory, it was decided to study wildfires on two scales to examine the dependencies between fires and their factors on the regional and global scale. At the macroscale, the territory of research was the territory of the Sakha Republic; on a microscale, one of the regions of the republic was chosen, which was the Nyurbinskii district (Figure 1). This region was chosen due to the high number of fire incidents in recent years.

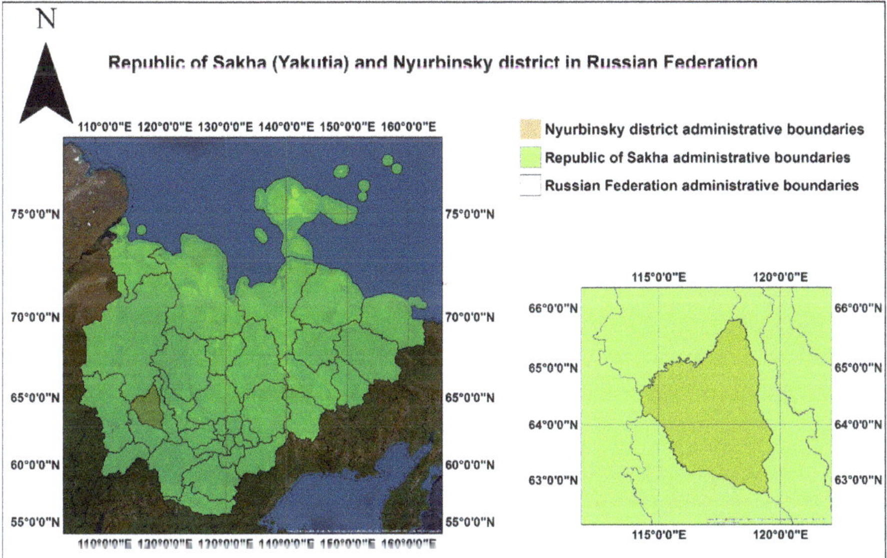

Figure 1. The geographical extent of the study area (source: Earthstar Geographics, low resolution 15 m Imagery).

The Republic of Sakha (Yakutia) is in northeastern Siberia and occupies 1/5 of the territory of the Russian Federation. The territory of the republic from the east and south is closed by mountain ranges; in the north, it has access to the Arctic Ocean. The relief and geological structure are distinguished by a complex and diverse structure. The orography of the territory determines the characteristics of a sharply continental climate, permafrost, soil, vegetation, wildlife, grasslands, and grazing land, and influences the nature of human economic activity [14]. A characteristic feature of the climate of Yakutia is a sharp continentality, which is manifested in large annual fluctuations in temperature and a relatively small amount of precipitation. The main factor of this state is the Siberian anticyclone. In the study area, the average air temperature during the winter months is −30 °C, and the average temperature of the coldest month (January) is −35 °C. The average annual precipitation is 200–400 mm [15]. The permafrost thickness in the territory is sometimes more than 100 m. Seasonal thawing varies between 0.8 and 3.3 m, depending on the landscape and the type of soil. Permafrost in the conditions of Yakutia has an impact on all soil processes [16]. Sakha can be divided into three great vegetation belts. About 40% of Yakutia is located above the Arctic Circle, and all of this is covered with permafrost, which greatly affects the region's ecology and limits the forests in the southern region. The arctic and subarctic tundra define a middle region where lichens and mosses grow like big green carpets and are favorite pastures for deer. In the southern part of the tundra belt along the rivers, scattered dwarf Siberian pines and larches grow. Below the tundra is a vast region of the taiga zone [2].

According to the state report "On the state and environmental protection of the Republic of Sakha (Yakutia) in 2018, the forested area occupies 82% of the territory of Yakutia, but the forest is 54% of the territory. The area of the forest zone is 252,820.0 million hectares. The forest cover is very diverse, from 11.5% in the Verkhoyansk district to 91.7% in South Yakutia. In the Nyurbinsky district, the forest area is 4416.3 thousand hectares and is more than 84% of the total area. The timber reserved is about 9.2 milliard m^3 and 96% of it is coniferous. The average timber reserve per one hectare is 58 m^3. The biggest average timber reserves are of *Pinus Sibirica* stands (188 m^3/ha) and *Picea* spp. (130 m^3/ha). The average timber reserves of *Pinus sylvestris* is 104 m^3/ha, *Larix* spp. is 62 m^3/ha, and *Betula Pendula* 41 m^3/ha.

A total of 95.6% of forest species in Yakutia are coniferous. The major species are *Larix* (4 main types: *Larix cajanderi, L. gmelinii, L. sibirica,* and *L. czekanowskii (L. sibirica x L. gmelinii)*), which represent 77.5% of the total forest resources. The second main species is *Pinus sylvestris* (6.5% of the total area) and *Pinus obovata* (0.24% of the total area). In southwest Yakutia, *Pinus sibirica* and *Abies sibirica* occur. *Picea ajanensis* is characteristic of the south mountainous regions. Other mountainous regions are occupied by *Pinus pumilo* (4.6% of the total forest area). The main deciduous tree species are *Betula pendula, B. pubescens, B. ermannii, Populus staveolens, P. tremula, Chosenia arbutifolia,* and *Salix* spp. Deciduous trees occupy only two million hectares, which are 1.24% of the total forest area.

Tree stands are well adapted to growing under extremely hard conditions of a dry climate and occurrence of permafrost. Forests are also very well adapted to the recovery process after fires. The process is determined by several factors. The most important factors are high seed production, good seed germination, and highly adaptive potential. *Larix cajanderi* is highly adapted to the existence with frequent fires. This species is pouring seeds in late summer in the ripening year. Seeds are being covered by needle litter and snow, where it creates the right conditions for sprouting and uses the spring moisture. The best conditions for forest growth are during the first years after a fire. Litter is destroyed, and the soil is enriched with ashy elements. The upper soil horizon's moisture is increased due to inflow from lower horizons [10].

Nyurbinsky ulus (region) is in the middle of the Vilyi River and occupies the territory adjacent to both Vilyi and its main tributary Marha. The region has an area of approximately 52.4 thousand sq km. Nyurbinsky region is part of the West Yakutia natural zone. It is characterized by plains and plateaus and the key elements are soil and vegetation on which relief has a big influence. According to Kurzhuev, the region is located on the Viluy Plateau, which is part of the Central Yakut Plateau. There is a characteristic occurrence of cryolithozone relief forms such as alases, yuryakhs, and bulgunyakhs. The number of days with snow cover is 210–225 days. The average annual temperature in the region is negative and in Markha it is −11.1 °C. The coldest month is January with an absolute minimum of −60 °C, and the warmest is July (the absolute maximum is +37 °C). The average amount of precipitation per year is 260–280 mm [14]. The maximum is in July. In the geobotanical regionalization of Yakutia, the Nyurbinsky region is situated in the boreal region in the taiga zone—a subzone of the middle taiga forests—sub-province, Central Yakutian middle taiga. The area of the region is dominated by Larix forests and Pinus Sylvestris forests. The river's valleys are characterized by rich meadows as well as common steppe and forest-steppe landscapes. Dry belts of alas vegetation are represented by the *Carex duriuscula* steppes [14].

2.2. Spatial Monitoring Factors for Fire Forest Monitoring in Yakutia

2.2.1. Fires Data

In the Republic of Sakha (Yakutia), open data on forest fires are not available. There is a database of forest fires at the Ministry of Environmental Protection (https://minpriroda.sakha.gov.ru), but these data are tabular, and without georeferencing, it is impossible to create a GIS database necessary for this type of research. These data are understated due to the addition to the database of only those fires in which there was action taken to extinguish the fire. Fires that are far from human activity often do not extinguish, so are not classified in the database. The most popular global data source on fire data is the Fire Information for Resource Management System (FIRMS). Data available in the service are collected from the VIIRS 375 m, 750 m, and MODIS Collection 6 Active Fire Product. The data are collected from 2002 until now. For the studies, we chose data from the MODIS Collection 6 sensor because of their longer availability and enough spatial resolution. We used data between 2001 and 2018 from the FIRMS fire archive. The shapefiles were in the Geographic WGS84 projection. The confidence values ranged from 0% to 100% and ranged in one of three fire classes (low-confidence fire, nominal-confidence fire, or high-confidence fire) [17].

2.2.2. Factors Data

Forest fire regimes are extremely diverse and vary due to the spread of fires and climate changes [8]. In difficult to manage forest areas like in Yakutia, it is necessary to properly characterize the factors that are causing forest fires. Researchers have used several different variables to assess wildfire risk [18–21]. Due to the lack of data, it was decided to investigate only long-term fire factors data like constant, long-term factors such as slope, aspect, fuel, climate, NDVI, etc. Constant factors are those factors that do not change rapidly, but gradually, in the long perspective. Constant factors can be calculated in medium- or long-term periods before the fire season [3]. In Yakutia, there is a very big data shortage. If some data are already collected, it is very difficult to access them. For these reasons, mostly global datasets were used.

Variables were divided into four groups: (1) meteorological (precipitations, temperature, maximum summer temperature, radiation); (2) NDVI; (3) landform (elevation, slope, slope direction); and (4) human activity (distance from roads, distance from settlements, distance from rivers). All data types were in other formats and had different spatial resolutions (Table 1).

Table 1. Technical description of the data.

Factor	Source	Format	Resolution	Preprocessing
Radiation Precipitations Temperature Max temperature	WorldClim	GeoTiff	30 s	ROI
NDVI	MODIS/LANDSAT	Hdf/GeoTiff	250 m/30 m	Mosaic, resampling, atmospheric correction ROI
Elevation Slope Slope direction	GMTED2010	GeoTiff	7.5 s	Mosaic, resampling ROI
Distance from settelments Distance from roads Distance from water lines	OSM	Shape		Eulicdean distance/ROI

To collect bioclimatic variables, the WorldClim dataset of global climate layers was used. The WorldClim dataset has a spatial resolution of about 1 km^2 [22]. WorldClim is a set of global climate layers (gridded climate data), specifically developed for ecological modeling on GIS. Currently, WorldClim provides several datasets for different temporal scenarios (past, current, and future conditions). In this work, data for the current condition scenario (1970–2000) were used. WordClim bioclimatic variables are analysis-ready data so the preprocessing was not necessary.

NDVI is greatly used in the evaluation of the phenology and productivity of the vegetation. Onigemo et al. claims that the values of NDVI obtained through images at the peak of drought were related to the content moisture and fresh phytomass, showing its potential to estimate fire risk [23]. Illaera et al. found a good correlation between the NDVI values and the location of wildfires [24]. To assess NDVI in macro-and microscale, it was decided to use NDVI from two sources: MODIS and LANDSAT images. The Landsat data were used for the determination of NDVI in the Nyurbinsky region. As the region of interest, the following paths and rows were defined: 129-15; 129-16; 130-14; 130-15; 130-16; 131-15. There were selected available images from the year with the smallest cloud cover. Images without cloud cover at the peak of the growing season (July) were available only for Landsat 5 in 2009. Before creating the mosaic, atmospheric correction was necessary. It allowed for improving images from level 1 to level 2. For NDVI calculations, bands 3 and 4 were used by the formula described by J.W. Rouse in 1973 [25]. To determine the average NDVI in the vegetation season for Yakutia between 2001 and 2018 with MODIS, dataset MOD13A2 Version 6 was used. MOD13A2 provides NDVI values with a resolution of 1 km. The product is derived with a monthly interval. There were selected images at the peak of the growing season (July) from each year. At the territory of Yakutia, there are the following tiles, which were defined as region of interests: h21v01, h22v01, h23v01, h21v02, h22v02, h23v02, h24v02, h23v03, h24v03, h25v03. After the selection of the data yearly, an NDVI mosaic was created and the mean NDVI calculated between 2001 and 2015.

To model landform factor data, it was decided to use one of the most used in these types of works, digital elevation models: the Global Multi-resolution Terrain Elevation Data 2010 (GMTED2010). It incorporates the current best available global elevation data. GMTED2010 is commonly used for radiometric and geometric correction, cover mapping, and extraction of drainage features for hydrologic modeling [26]. There was a necessity for mosaicking. For the GMTED, the following entities were used: GMTED2010N50E120, GMTED2010N50E150, GMTED2010N70E120, and GMTED2010N70E150. From the DEM, the information about elevation, slope, and slope direction was derived.

To investigate the relationship between the fire and the presence of a human, it was necessary to obtain information on the distance of fires from roads, buildings, and rivers. The data in shape format were downloaded from Open Street Map (OSM). According to previous works on this subject [27,28], to achieve this goal, the Euclidean distance was used. Euclidean distance is a straight line between two points in Euclidean space. The locations were converted to raster. The resolution of the raster was defined by the shortest of the width or height of the extent of the input feature, in the input spatial reference, divided by 250.

After the preprocessing, two regions of interest (ROI) were defined for each obtained raster. Each of them was clipped using the forest cover of the Nyurbinsky region and Republic of Sakha (Yakutia) forest cover. The forest cover was made available by the Institute of Biological Problems of Cryolithozone in Yakutia. The same ROI was made for fire points between 2001–2015 from the FIRMS dataset.

2.3. Modeling the Risk Assessment of Forest Fires

2.3.1. Fire Factors Analysis

The first step after the preprocessing (Figure 2) was the classification of the values in each raster, for ordering data and eliminating individual pixels strongly deviating from the average pixels were probably deviating due to measurement error or instrument inaccuracy. Due to a large amount of data, the surface to cover (3,084,000 km^2), and computing capabilities, it was necessary to simplify the data to some extent but retain their characteristics. Jenks's optimization method was used. The Jenks classification is designed to identify data clusters as well as maintain a representation of all data in the set. The optimization lasts as long as the limits of the intervals are obtained, respecting the principle of the smallest possible differentiation of the observations contained in them, with the greatest distance between the intervals at the same time. This method aims to reduce the variance within classes and maximize the variance between classes. This is done by striving to minimize the average deviations of each class from the middle class while maximizing the deviations of each class from the means of other groups [29]. New values in rasters have been extracted to each fire point from the FIRMS dataset of over 17 years in the territory of Yakutia and the Nyurbinsky region.

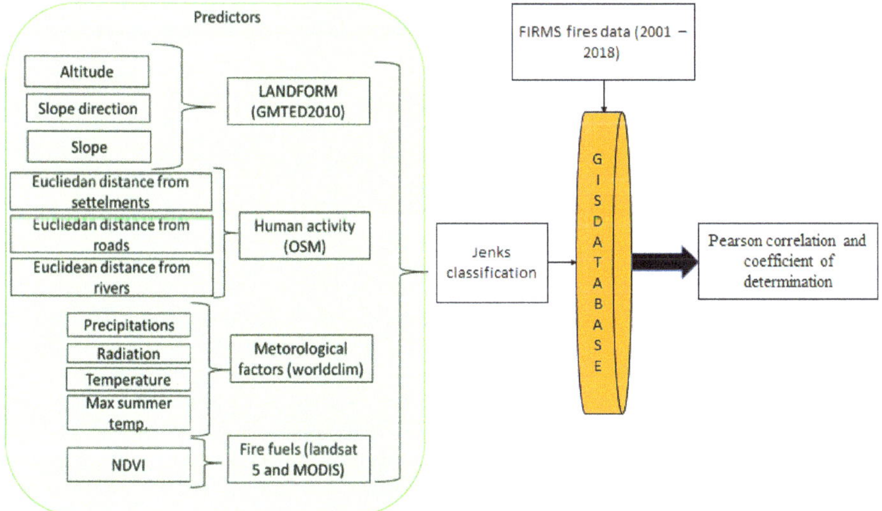

Figure 2. Schema of the analysis of the relationship between the fires and their factors.

The next step was to calculate the area of each class in each factor. Then, all fires between 2001–2018 were summed, separately for each category in each obtained earlier categorized raster. With these data, it was possible to determine the regression, and correlation between the dependent variables and explanatory variables. The same calculation scheme was repeated for each variable. In the last step, Pearson correlation and coefficient of determination between the fire data and factors data were calculated for each model.

2.3.2. Long-Term Fire Risk Modeling

The next step was to build risk models using two machine learning methods (Figure 3). The training dataset was built from the FIRMS fire data between 2001–2015. All registered fire points at the territory in the Republic of Sakha Yakutia and Nyurbinsky region in this period were used. The variables were divided, according to the previously obtained correlation coefficient. As the predictors, two types of datasets were examined. A dataset with a coefficient of correlation equal or higher than satisfactory (higher than 0.6) and a dataset with all 11 previously selected fire factors. All predictor rasters have been resampled and converted to grid format. Such a set of training data and predictor data were created, separately for the territory of Yakutia and the territory of the Nyurbinsky region.

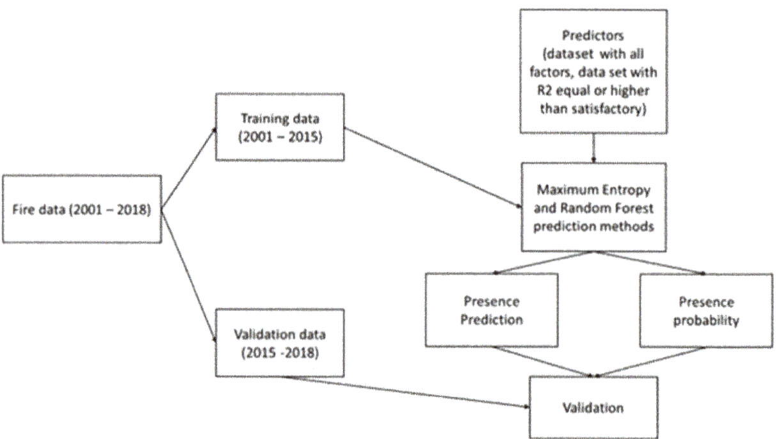

Figure 3. Long-term fire risk modeling methodology.

In the simulation, two types of prediction methods were selected, referring to the works of authors such as Peters [11], Parisien [30] (maximum entropy prediction model) and Oliveira [6] (random forest prediction model). In their studies, the authors showed a good correlation between the risk models obtained when using machine learning and forest fires. The authors demonstrated the superiority of methods using machine learning over traditional ones.

The maximum entropy prediction model is a widely used and accepted statistical method to produce predicting probability distributions. The model is adapted in diverse topics such as thermodynamics, economics, forensics, imaging technologies, and recently, ecology. Maximum entropy can provide accurate predictions of patterns in macroecology and help identify the mechanisms that matter most [31]. The algorithm is widely used for mapping species distributions [32] and conservation planning [31]. In recent years, the model of maximum entropy began to be used in the ecology of fires. Maximum entropy is a density estimation method based on a probability distribution. It is a presence-only machine learning algorithm that iteratively contrasts environmental predictor values at occurrence locations with those of a large background sample taken across the study area [33]. Maximum entropy has proved to be an enormously powerful tool for reconstructing images from many types of data [34].

The biggest advantage of a random forest is that it is a very flexible method, and it can be used in different types of problems. The random forest algorithm is a fully nonparametric machine learning method for data analysis. Classification and regression random forest is competitive with the best available methods and superior to most methods in common use [35]. The application of random forest can be an effective methodology to predict fire occurrence in different sites. The random forest algorithm is a technique developed by Breiman (2001) [35]. It combines a large set of decision trees, which is the biggest advantage compared to a simple decision tree algorithm. Each tree is trained by a set of variables, which are randomly selected from the training dataset.

All classifications were carried out using SAGA—6.4.0 and its modules "Maximum Entropy Presence Prediction", "Random Forest Presence Prediction (ViGrA) Classification". After the processing, two types of models were obtained: presence prediction and presence probability maps. Presence prediction maps are only meant to determine if there is a possibility of a fire. All raster pixels are classified into two classes:

- Absence—there is the possibility of a forest fire.
- Presence—there is no possibility of the forest fire.

Presence probability maps are meant to determine how high the possibility is of the fire. All raster pixels are classified into six classes:

- Very low—very low possibility of the fire.
- Low—low possibility of fire.
- Moderate—moderate possibility of fire.
- High—high possibility of fire.
- Very high—very high possibility of fire.
- Extreme—extreme possibility of fire.

To obtain only six possibility classes, it was necessary to reclassify the models. For reclassification, the natural breaks method was used.

For the validation of the results, raw fire points data between 2015 and 2018 in the Republic of Sakha (Yakutia) and the Nyurbinsky region were used. To carry out statistical analysis pixel count in each probability class in the presence probability maps and each class at the presence prediction maps were summed. Fire points in each class were summed and divided by the pixel sum in each class. This process was carried out to consider the surface of each class when verifying models. Next, the percentage share of fires in each class was calculated. The last step was regression and correlation analysis.

3. Results

3.1. Fire Factors

In the first part of the studies, attempts were made to uncover the main factors affecting the possibility of a wildfire. Of the 11 preselected factors, not all of them showed a good correlation with fire points (Table 2).

Table 2. Coefficient of correlation for each variable in the Republic of Sakha (Yakutia) and the Nyurbinsky region.

Factor	Yakutia	Nyurbinksy
	R	R
Radiation	0.97	0.39
Precipitations	−0.33	−0.99
Temperature	−0.41	0.20
Max temperature	0.93	0.47
NDVI	0.91	0.97
DEM	−0.76	0.28
Slope	−0.97	−0.73
Slope direction	−0.43	0.15
Settlements	0.97	0.86
Roads	−0.80	−0.82
Water lines	0.91	−0.54

In Yakutia (macroscale), the correlation above 70% was shown by factors such as solar radiation, maximum summer temperature, NDVI, elevation, slope, distance from roads, distance from settlements, and distance from rivers.

The problem was studied in the scale of the region using the example of the Nyurbinsky. In this example, only five factors out of 11 showed a good correlation. A very strong relationship between fires and precipitation was demonstrated. As the amount of precipitation falls, the number of points of ignition increases, which was not observed on a macroscale. The distribution of NDVI in microscale is almost the same as for the Yakutia. With an increase of the slope, fewer fires are observed, as in the macroscale.

3.2. Fire Risk

The next step was to create fire hazard models using two types of modeling methods. Fire risk models in the territory of Yakutia are shown in Figures 4 and 5. Results of the modeling differed, but both methods gave satisfying results. For the territory of the Republic of Sakha (Yakutia), high coefficients of correlation for each prediction method can be observed (Tables 3 and 4). Coefficients were slightly higher for the random forest prediction method. In this method, the model with 11 variables did not give considerably better results than the model with nine variables.

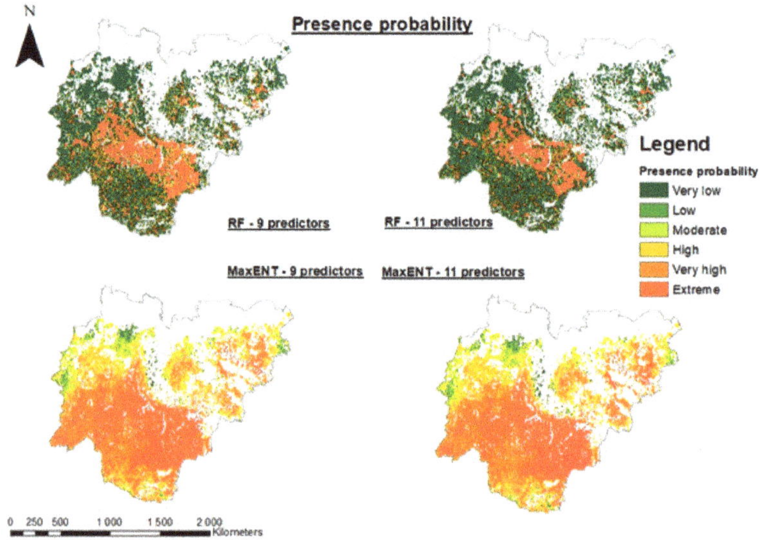

Figure 4. Results of the long-term wildfire presence probability in Yakutia (macroscale).

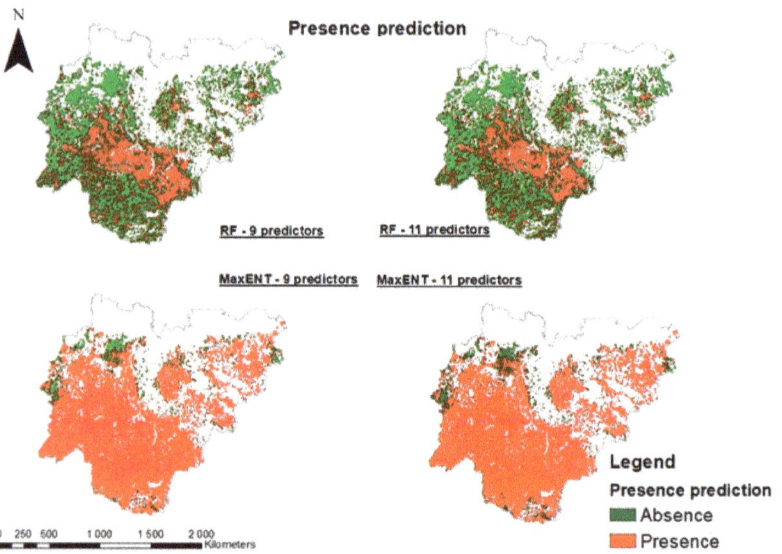

Figure 5. Results of long-term wildfire presence prediction modeling in Yakutia (macroscale).

Table 3. Comparison of the different types of long-term wildfire presence probability modeling methods in Yakutia (macroscale).

Fire Probability	Percentage of Fire Points in Each Probability Class			
	Random Forest		MaxENT	
	9 Predictors	11 Predictors	9 Predictors	11 Predictors
Very Low	0.07	0.07	0.69	0.95
Low	1.74	1.65	1.57	2.30
Moderate	8.96	9.89	9.19	9.59
High	19.82	18.53	10.64	10.58
Very High	25.52	28.02	27.01	25.37
Extreme	43.89	41.84	50.89	51.21
R	0.97	0.98	0.91	0.90

Table 4. Comparison of the different types of long-term wildfire presence prediction modeling methods in Yakutia (macroscale).

Presence Prediction	Percentage of Fire Points in Presence and Absence Class			
	Random Forest		MaxENT	
	9 Predictors	11 Predictors	9 Predictors	11 Predictors
Absence	0.20	0.14	12.50	14.13
Presence	99.80	99.86	87.50	85.87

When modeling using the maximum entropy prediction model, better results gave a dataset with nine variables. In both models, the highest number of fires between 2015–2018 is in the high, very high, and extreme probability classes. In the very low-risk class, there were less than 1% of fire points. Presence prediction models showed a superior random forest method over the maximum entropy method. In the random forest method, almost 100% of the fire points were in the presence class. The maximum entropy method was characterized by worse results. In the absence, the class located more than 10% of the fire points from the validation dataset.

Modeling carried out at the territory of the Nyurbinsky region showed slightly different results than in Yakutia (Figures 6 and 7). Correlation coefficients were not as high and did not exceed 0.9 (Table 5). According to the coefficient of correlation, the smallest error was characterized by a model of maximum entropy using six predictor variables. The coefficient of correlation for 11 predictors was only slightly lower and was equal to 0.89.

Table 5. Comparison of the different types of long-term wildfire presence probability modeling methods in Nyurbinsky (microscale).

Fire Probability	Percentage of Fire Points in Each Probability Class			
	Random Forest		MaxENT	
	6 Predictors	11 Predictors	6 Predictors	11 Predictors
Very Low	498	4.97	0.00	0.00
Low	15.09	17.04	0.00	0.00
Moderate	20.40	16.06	1.92	12.81
High	15.60	20.25	32.28	31.23
Very High	24.94	21. 56	31.73	29.02
Extreme	19.00	20.13	34.07	26.94
R	0.75	0.82	0.90	0.89

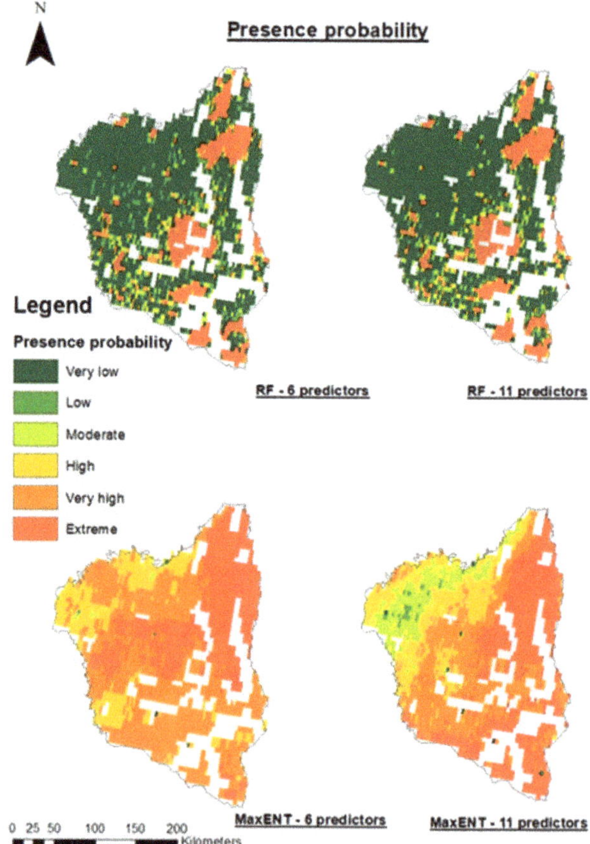

Figure 6. Results of the long-term wildfire presence probability modeling in Nyurbinksy (microscale).

It was observed that in the low and very low fire risk classes, there were no fire points. Significantly worse results were shown by the random forest prediction method. In both cases of the random forest method (six and 11 predictors), the coefficient of correlation did not exceed 0.82. Over 20% of all fire points were at low-risk classes. Similar results were observed by analyzing the presence prediction maps (Table 6). The random forest prediction method was characterized by more than 30% of fire points in the absence class. The results were distributed in a similar way in the maximum entropy prediction model, which is characterized by 11 predictors. The best results were observed in the maximum entropy classification using six predictive variables. In this model, more than 90% of the points were in the presence class.

Table 6. Comparison of the different types of long-term wildfire presence prediction modeling methods in Nyurbinsky (microscale).

Presence Prediction	Percentage of Fire Points in Presence and Absence Class			
	Random Forest		MaxENT	
	6 Predictors	11 Predictors	6 Predictors	11 Predictors
Absence	32.73	31.22	8.34	32.66
Presence	67.27	68.78	91.66	67.34

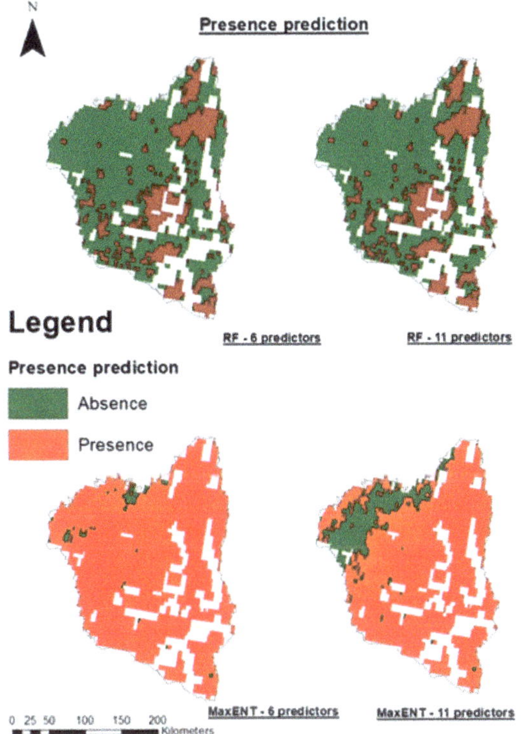

Figure 7. Results of the long-term wildfire presence prediction modeling in Nyurbinsky (macroscale).

It was observed that in both regions (Yakutia and Nyurbinsky), the random forest presence prediction method gave more accurate results. Even though the Nyurbinsky region incorrectly classified a larger number of fire points, it can be unambiguously stated that it is wrong. This is because (Figures 5 and 7) in the presence maximum entropy, presence prediction method, almost all of the territory of Yakutia and Nyurbinsky was classified as territory with the presence of fires. The results of the random forest presence prediction method showed a much narrower territory of the possibility of the occurrence of fires and yet it classified more points correctly in the territory of Yakutia.

4. Discussion

In fire studies, fire risk is one of the major topics and there are many different approaches to this subject. Blanchi et al., in their work about a methodological approach in fire risk studies, collected more than 50 works connected with fire probability cartography. Risk mapping has less than twenty years. Previously, there were rather preferred descriptive approaches. Since the 1990s, there has been an increasing interest in this field [36]

The study presented the possibilities of using different types of GIS and remote sensing data in modeling the wildfire risk. Results allow us to reflect various aspects of fire studies. The difficulties in fire risk assessment were to point out and clarify possibilities to define wildfire risk. There are many different approaches in hazard mapping based on different datasets, scales, and algorithms. The multi-approach is relevant in fire management studies [37,38]. Gai et al. [39] developed a spatially weighted index model for fire risk assessment. You et al. [40] integrated a Forest Resource Inventory Database based on four aspects of topographical, human activity, climate, and forest characteristic factors. Goldarag et al. [41] used neural networks and logistic regression for fire risk assessment.

The biggest challenge in the study was to collect the necessary data due to poor exploration of the area of studies. GIS technologies in Yakutia are under development [42]. For this purpose, we chose a semi-probabilistic mode of modeling, which gave possibilities of combining historical fire data and physical fire mechanisms [43].

The results of the analysis showed that the fire risk assessment in the Republic of Sakha (Yakutia) is not a problem that can be easily solved. The specificity of the studied territory is significantly different from other parts of the world that are facing the problem of wildfires. Boreal forests of Yakutia are perfectly adapted to extremely severe climatic conditions. Fires have been affecting the natural functioning of the forests of the region for centuries. Fires not only affect the climatic conditions, but also the formation of the terrain [36].

Results of the correlation analysis in macroscale showed that with increasing radiation, the risk of fires increases, and the same happened with increasing maximum summer temperature. Lim et al., in their study, highlighted that fire data showed a high correlation with climate factors [44]. NDVI was highly correlated with fires (R = 0.91). With increasing NDVI, the number of points of ignition increased. This situation was associated with the accumulation of combustible materials along with an increase in forest biomass. Elevation and slope correlated inversely proportional. According to Pourghasemi et al., slope and aspect are some of the most important factors controlling forest fire occurrence [45]. Kurbatsky et al. wrote that long drought conditions in the Yakutian valleys caused intense droughts and accumulation of fire fuels, which caused strong fires [46]. As the slope increased, we observed a decreasing number of wildfires. This situation may be due to the strongly inclined slopes that act as a barrier to the spread of fire. Additionally, on inclined slopes, many combustible materials cannot accumulate. Ghorbanzadeh et al. claimed that environmental variables are not the only reason for fire susceptibility and risk [47]. A very strong connection between the increase in the number of fire points and the human factors was observed. In places with proximity to transport routes, the number of fires increases, and the same thing happens if we consider the settlements. This situation indicates a strong anthropogenic impact on the risk of fire. There are not so many fires near water lines, which can be caused by the presence of moist areas close to the rivers.

The largest correlation at the microscale was observed when taking rainfall into account. As the rainfall increased, the number of fires increased. The other climatic factors did not show a strong correlation. NDVI, similar to the macroscale, strongly influences the number of fires. The properties of combustible materials relate to type, phytomass, condition, and moisture, among which the moisture content is the most important for fire protection [48]. In a completely different way, anthropogenic factors were distributed on the microscale than on the macroscale. The Nyurbinsky region has a relatively high population density compared to the entire territory of Yakutia, which may affect the differences between the results. Most fires continue to occur along roads, but the closer the settlements are, it occurs less frequently. This situation may be because fires that approach the villages pose a threat to people and fire-fighting starts quickly (fires do not reach large sizes), while those that are far from populated areas remain without any action as they do not pose a great threat. Along roads that are far from villages, fires can reach serious sizes. Additionally, villages in this region are usually not surrounded by forests, but by meadows called "alas", with lower susceptibility to ignition.

Human activity has a greater influence on the fire regimes and differs in the macro- and microscale. The results showed that distance from settlements (R = −0.97, and R = 0.86), distance from rivers (R = 0.91, and R = −0.54), and distance from roads (R = −0.80, and R = −0.82) had a great importance in fire risk assessment. Ajin et al. [49] showed a high correlation between fire occurrence and distance from roads and settlements. The study by Werf showed that in residential areas and near roads, more human activities are witnessed, and human activity is the most significant factor in the fire outbreak [50]. Studies of the territory of Yakutia in residential areas and near roads show that more human activities are witnessed, and the human activity is the most significant factor in fire outbreaks.

Very large differences were observed between the results of the research on the macro- and microscale. Often the results were quite the opposite. It is probably related to the specificity of the

Yakutia region. The Niyurbinsky region is quite highly urbanized compared to other regions of Yakutia, which may affect the results. Additionally, the regional climatic conditions differed from the average for the entire region. The Nyuribnsky region is characterized by flat terrain, and most of Yakutia is mountainous, which can also affect the conditions for fires. To analyze the macroscale, it is necessary to take into account a wide variety of conditions: climatic, geographical, and human influence. However, on a microscale, these conditions are more homogeneous.

The analysis of the different methods of wildfire risk assessment allows for the identification of the high danger areas in Nyurbinsky and the entire Yakutia. The maximum accuracy was demonstrated by the random forest method with 11 predictors (R = 0.98) for the entire territory of Yakutia. The analysis showed that the random forest method gave more accurate results and a much narrower area of the possibility of fires than the MaxENT method, which allows us to propose that this model is preferable. A model created using the maximum entropy method has a small differentiation into zones, which does not allow for its use in practice as uninformative. Parisien used the MaxENT and random forest model to predict fire ignition across the USA [30] with satisfactory results. The MaxENT model was used to assess fire risk in India's Ghats Mountains [34]. In their studies, the authors showed a good correlation between the risk models obtained when using machine learning and forest fires. Due to their power, versatility, and ease of use, random forests are quickly becoming one of the most popular machine learning methods [48]. The studies confirmed that the use of GIS and remote sensing technologies can be used to assess the long-term risk and probability of fires in Yakutia. The results showed that reducing the number of factors during modeling did not show a significant impact on the results in both methods. In the few last years, an increase in the number and severity of fires has been observed as has more years with extreme fire seasons [9]. This situation may also be affected by climate change. On the other hand, a larger number of fires may have an impact on the climate, permafrost ecosystems, and the functioning of indigenous people. In connection with this situation, it is necessary to work on new fire risk assessment methods in the Arctic and subarctic zone of Yakutia. Methods should use global databases of fire, climate, and human activities. For building new models that include all these factors, it is necessary to use different types of datasets, like OSM data, climatic datasets, and remote sensing data from different sensors (MODIS, Landsat, Sentinel). According to Gigović et al., the random forest model could be used at the regional level for forest fire susceptibility mapping [51]. Banerjee et al. claim that the MaxENT method can be used as a decision support tool for stakeholders of forest resources [52]. Pham et al. claim that models that consider climate variables, vegetation, and human influences can explain fire risk better than those that only account for some of these factors [53].

The objective of work on modeling long-term wildfire risk in the Nyurbinsky region of Yakutia was accomplished. The accuracy of the results depends on the data used, and here available. The results were valid within the limits caused by the data used, the approach was chosen as was the research objective. Despite these limitations, it was possible to obtain interesting results that enabled us to answer the main questions posed by the problem.

5. Conclusions

The problem of the forest fires in Yakutia is not as well studied as in other countries. The results of the research have shown a strong human influence on the risk in this region despite the low population density. Anthropogenic factors showed a high correlation with the occurrence of wildfires more than climatic or topographical factors. Other factors affect the risk of fires at the macroscale, and others the microscale, which should be considered when modeling.

The random forest method showed better results in the macroscale, however, the maximum entropy model was better in the microscale. The exclusion of variables that do not show high correlation does not always improve the modeling results. The random forest presence prediction model is a more accurate method and significantly reduces the risk territory. The reverse is the method of maximum entropy, which is not as accurate and classifies very large areas as endangered.

Further study of this topic requires a clearer and conceptually developed approach to the application of remote sensing data. Therefore, this work makes sense to lay the foundations for the future of a completely automated fire risk assessment application in the Republic of Sakha. The results can be used for fire prophylactics and planning fire prevention. In the future, to determine the risk well, it is necessary to combine the obtained maps with the seasonal risk determined using indices (for example, the Nesterov index 1949) and the periodic dynamics of forest fires, which Isaev and Utkin studied in 1963. Such actions can help build an application with which it will be possible to determine the risk of wildfire and the spreading of fire during extreme events.

Author Contributions: S.G. and P.J. participated in the conceptualization of the manuscript. S.G. and P.J. participated in the methodology development and software development. P.J. conceived the designed the experiments. S.G. and P.J. wrote the manuscript. S.G. provided supervision on the design and implementation of the research. All authors contributed to the review and improvement of the manuscript. All authors have read and agreed to the published version of the manuscript.

Funding: This research was funded by the French ANR PUR (Polar Urban Centers) in the Work Package 4 focusing on the question of settlements and urban centers risk exposure modeling. The present research received support from Campus France and the CNRS PEPS RICOCHET.

Conflicts of Interest: The authors declare no conflict of interest.

References

1. Kasischke, E.S.; Stocks, B.J. *Fire, Climate Change, and Carbon Cycling in the Boreal Forest*; Springer Science & Business Media: Berlin/Heidelberg, Germany, 2012; pp. 203–204, ISBN 978-0-387-21629-4.
2. Troeva, E.I.; Isaev, A.P.; Cherosov, M.M.; Karpov, N.S. *The Far North: Plant Biodiversity and Ecology of Yakutia*; Springer Science & Business Media: Berlin/Heidelberg, Germany, 2010; pp. 203–204, ISBN 978-90-481-3774-9.
3. San-Miguel-Ayanz, J.; Carlson, J.D.; Alexander, M.; Tolhurst, K.; Morgan, G.; Sneeuwjagt, R.; Dudley, M. Current Methods to Assess Fire Danger Potential. In *Wildland Fire Danger Estimation and Mapping*; Series in Remote Sensing; World Scientific: Singapore, 2003; Volume 4, pp. 21–61, ISBN 978-981-238-569-7.
4. Жданко, В.А.; Гриценко, М.В. Метод анализа лесопожарных сезонов: Практические рекомендации. Л. ЛНИИЛХ **1980**. (In Russian)
5. Tedim, F.; Leone, V.; Amraoui, M.; Bouillon, C.; Coughlan, M.R.; Delogu, G.M.; Fernandes, P.M.; Ferreira, C.; McCaffrey, S.; McGee, T.K.; et al. Defining Extreme Wildfire Events: Difficulties, Challenges, and Impacts. *Fire* **2018**, *1*, 9. [CrossRef]
6. Oliveira, S.; Oehler, F.; San-Miguel-Ayanz, J.; Camia, A.; Pereira, J.M.C. Modeling spatial patterns of fire occurrence in Mediterranean Europe using Multiple Regression and Random Forest. *For. Ecol. Manag.* **2012**, *275*, 117–129. [CrossRef]
7. Cardille, J.A.; Ventura, S.J.; Monica, G.T. Environmental and social factors influencing wildfires in the Upper Midwest, United States. *Ecol. Appl.* **2001**, *11*, 111–127. [CrossRef]
8. González-Cabán, A. *Proceedings of the Second International Symposium on Fire Economics, Planning, and Policy: A Global View*; U.S. Department of Agriculture, Forest Service, Pacific Southwest Research Station: Albany, CA, USA, 2008; 720p.
9. Isaev, A.P.; Protopopov, A.V.; Protopopova, V.V.; Egorova, A.A.; Timofeyev, P.A.; Nikolaev, A.N.; Shurduk, I.F.; Lytkina, L.P.; Ermakov, N.B.; Nikitina, N.V.; et al. Vegetation of Yakutia: Elements of Ecology and Plant Sociology. In *The Far North: Plant Biodiversity and Ecology of Yakutia*; Troeva, E.I., Isaev, A.P., Cherosov, M.M., Karpov, N.S., Eds.; Plant and Vegetation; Springer: Dordrecht, The Netherlands, 2010; pp. 143–260, ISBN 978-90-481-3774-9.
10. Cherosov, M.M.; Isaev, A.P.; Mironova, S.I.; Lytkina, L.P.; Gavrilyeva, L.D.; Sofronov, R.R.; Arzhakova, A.P.; Barashkova, N.V.; Ivanov, I.A.; Shurduk, I.F.; et al. Vegetation and Human Activity. In *The Far North: Plant Biodiversity and Ecology of Yakutia*; Troeva, E.I., Isaev, A.P., Cherosov, M.M., Karpov, N.S., Eds.; Plant and Vegetation; Springer: Dordrecht, The Netherlands, 2010; pp. 261–295, ISBN 978-90-481-3774-9.
11. Peters, M.P.; Iverson, L.R.; Matthews, S.N.; Prasad, A.M. Wildfire hazard mapping: Exploring site conditions in eastern US wildland–urban interfaces. *Int. J. Wildland Fire* **2013**, *22*, 567–578. [CrossRef]
12. Phillips, S.J.; Dudík, M. Modeling of species distributions with Maxent: New extensions and a comprehensive evaluation. *Ecography* **2008**, *31*, 161–175. [CrossRef]

13. Gull, S.F.; Skilling, J. Maximum entropy method in image processing. In *IEE Proceedings F (Communications, Radar and Signal Processing)*; IET Digital Library: London, UK, 1984; Volume 131, pp. 646–659.

14. Атлас сельского хозяйства ЯАССР, А. под редакциейАГ Гущиной. *М. Главное УправлениеГеодезии и Картографии приСоветеМинистровСССР* **1989**. (In Russian)

15. Воробьев, К.А.; Ким, А.Ю. ГеографическийАтлас *Республика Саха (Якутия)*; Роскартография: Якутск, Russia, 2000. (In Russian)

16. Еловская, Л.Г. Классификация и диагностика мерзлотных почв Якутии. Якутский филиалСОАНСССР. 1987. Available online: https://search.rsl.ru/ru/record/01001419857 (accessed on 27 November 2020). (In Russian).

17. Giglio, L. MODIS Collection 6 Active Fire Product User's Guide Revision A. Available online: /paper/MODIS-Collection-6-Active-Fire-Product-User%27s-Guide-Giglio/4aacae34ad3bcd557591067399ebc38580eb8286 (accessed on 14 September 2020).

18. Prichard, S.J.; Kennedy, M.C. Fuel treatments and landform modify landscape patterns of burn severity in an extreme fire event. *Ecol. Appl.* **2014**, *24*, 571–590. [CrossRef]

19. Martínez, J.; Vega-Garcia, C.; Chuvieco, E. Human-caused wildfire risk rating for prevention planning in Spain. *J. Environ. Manag.* **2009**, *90*, 1241–1252. [CrossRef]

20. Syphard, A.D.; Radeloff, V.C.; Keuler, N.S.; Taylor, R.S.; Hawbaker, T.J.; Stewart, S.I.; Clayton, M.K. Predicting spatial patterns of fire on a southern California landscape. *Int. J. Wildland Fire* **2008**, *17*, 602–613. [CrossRef]

21. Kwak, H.; Lee, W.-K.; Saborowski, J.; Lee, S.-Y.; Won, M.-S.; Koo, K.-S.; Lee, M.-B.; Kim, S.-N. Estimating the spatial pattern of human-caused forest fires using a generalized linear mixed model with spatial autocorrelation in South Korea. *Int. J. Geogr. Inf. Sci.* **2012**, *26*, 1589–1602. [CrossRef]

22. Fick, S.E.; Hijmans, R.J. WorldClim 2: New 1-km spatial resolution climate surfaces for global land areas. *Int. J. Climatol.* **2017**, *37*, 4302–4315. [CrossRef]

23. Onigemo, A.E.; Santos, S.A.; Pellegrin, L.A.; Abreu, U.G.P.; Silva, E.T.J.; Soriano, B.M.A.; Crispim, S.M.A. Application of vegetation index to assess fire risk in open grasslands with predominance of cespitous grasses in the Nhecolândia sub-region of the Pantanal. *Simpósio Bras. Sens. Remoto* **2007**, *13*, 4493–4500.

24. Illera, P.; Fernández, A.; Delgado, J.A. Temporal evolution of the NDVI as an indicator of forest fire danger. *Int. J. Remote Sens.* **1996**, *17*, 1093–1105. [CrossRef]

25. Goward, C.R.; Murphy, J.P.; Atkinson, T.; Barstow, D.A. Expression and purification of a truncated recombinant streptococcal protein G. *Biochem. J.* **1990**, *267*, 171–177. [CrossRef]

26. Danielson, J.J.; Gesch, D.B. *Global Multi-Resolution Terrain Elevation Data 2010 (GMTED2010)*; US Department of the Interior, US Geological Survey: Newton, MA, USA, 2011.

27. Pew, K.L.; Larsen, C.P.S. GIS analysis of spatial and temporal patterns of human-caused wildfires in the temperate rain forest of Vancouver Island, Canada. *For. Ecol. Manag.* **2001**, *140*, 1–18. [CrossRef]

28. Zhang, Z.X.; Zhang, H.Y.; Zhou, D.W. Using GIS spatial analysis and logistic regression to predict the probabilities of human-caused grassland fires. *J. Arid Environ.* **2010**, *74*, 386–393.

29. Jenks, G.F. The Data Model Concept in Statistical Mapping. *Int. Yearb. Cartogr.* **1967**, *7*, 186–190.

30. Parisien, M.-A.; Snetsinger, S.; Greenberg, J.A.; Nelson, C.R.; Schoennagel, T.; Dobrowski, S.Z.; Moritz, M.A. Spatial variability in wildfire probability across the western United States. *Int. J. Wildland Fire* **2012**, *21*, 313–327. [CrossRef]

31. Harte, J.; Newman, E.A. Maximum information entropy: A foundation for ecological theory. *Trends Ecol. Evol.* **2014**, *29*, 384–389. [CrossRef]

32. Elith, J.; Graham, C.H.; Anderson, R.P.; Dudík, M.; Ferrier, S.; Guisan, A.; Hijmans, R.J.; Huettmann, F.; Leathwick, J.R.; Lehmann, A. Novel methods improve prediction of species' distributions from occurrence data. *Ecography* **2006**, *29*, 129–151. [CrossRef]

33. Franklin, J. *Mapping Species Distributions: Spatial Inference and Prediction*; Cambridge University Press: Cambridge, UK, 2010; ISBN 978-1-139-48529-6.

34. Renard, Q.; Pélissier, R.; Ramesh, B.R.; Kodandapani, N. Environmental susceptibility model for predicting forest fire occurrence in the Western Ghats of India. *Int. J. Wildland Fire* **2012**, *21*, 368–379. [CrossRef]

35. Cutler, D.R.; Edwards, T.C., Jr.; Beard, K.H.; Cutler, A.; Hess, K.T.; Gibson, J.; Lawler, J.J. Random forests for classification in ecology. *Ecology* **2007**, *88*, 2783–2792. [CrossRef]

36. Blanchi, R.; Jappiot, M.; Alexandrian, D. Forest fire risk assessment and cartography. A methodological approach. In Proceedings of the IV International Conference on Forest Fire Research, Luso, Portugal, 18–23 November 2002.

37. Tehrany, M.S.; Jones, S.; Shabani, F.; Martínez-Álvarez, F.; Bui, D.T. A novel ensemble modeling approach for the spatial prediction of tropical forest fire susceptibility using logitboost machine learning classifier and multi-source geospatial data. *Theor. Appl. Climatol.* **2019**, *137*, 637–653. [CrossRef]

38. Koetz, B.; Morsdorf, F.; van der Linden, S.; Curt, T.; Allgöwer, B. Multi-Source land cover classification for forest fire management based on imaging spectrometry and LiDAR data. *For. Ecol. Manag.* **2008**, *256*, 263–327. [CrossRef]

39. Gai, C.; Weng, W.; Yuan, H. GIS-Based forest fire risk assessment and mapping. In Proceedings of the 2011 Fourth International Joint Conference on Computational Sciences and Optimization, Kunming/Lijiang, China, 5–19 April 2011; pp. 1240–1244.

40. You, W.; Lin, L.; Wu, L.; Ji, Z.; Yu, J.; Zhu, J.; Fan, Y.; He, D. Geographical information system-based forest fire risk assessment integrating national forest inventory data and analysis of its spatiotemporal variability. *Ecol. Indic.* **2017**, *77*, 176–184. [CrossRef]

41. Goldarag, Y.J.; Mohammadzadeh, A.; Ardakani, A.S. Fire risk assessment using neural network and logistic regression. *J. Indian Soc. Remote Sens.* **2016**, *44*, 885–894. [CrossRef]

42. Andreev, D.V.; Makarova, M.E. GIS technologies application in useful fossils search in the territory of the Republic of Sakha (Yakutia). In Proceedings of the IOP Conference Series: Earth and Environmental Science, Changsha, China, 18–20 September 2020; IOP Publishing: Bristol, UK, 2020; Volume 548, p. 032006.

43. Boike, J.; Grau, T.; Heim, B.; Günther, F.; Langer, M.; Muster, S.; Gouttevin, I.; Lange, S. Satellite-derived changes in the permafrost landscape of central Yakutia, 2000–2011: Wetting, drying, and fires. *Glob. Planet. Chang.* **2016**, *139*, 116–127. [CrossRef]

44. Lim, C.-H.; Kim, Y.S.; Won, M.; Kim, S.J.; Lee, W.-K. Can satellite-based data substitute for surveyed data to predict the spatial probability of forest fire? A geostatistical approach to forest fire in the Republic of Korea. *Geomat. Nat. Hazards Risk* **2019**, *10*, 719–739. [CrossRef]

45. Pourghasemi, H.R.; Kariminejad, N.; Amiri, M.; Edalat, M.; Zarafshar, M.; Blaschke, T.; Cerda, A. Assessing and mapping multi-hazard risk susceptibility using a machine learning technique. *Sci. Rep.* **2020**, *10*, 1–11. [CrossRef]

46. Курбатский, Н.П. Техника и тактика тушения лесных пожаров. М. Гослесбумиздат **1962**. (In Russian)

47. Ghorbanzadeh, O.; Blaschke, T.; Gholamnia, K.; Aryal, J. Forest fire susceptibility and risk mapping using social/infrastructural vulnerability and environmental variables. *Fire* **2019**, *2*, 50. [CrossRef]

48. Dennison, P.E.; Roberts, D.A.; Reggelbrugge, J.C. Characterizing chaparral fuels using combined hyperspectral and synthetic aperture radar data. In Proceedings of the 9th AVIRIS Earth Science Workshop, Pasadena, CA, USA, 23–25 February 2000; Jet Propulsion Lab: Pasadena, CA, USA, 2000; Volume 6, pp. 23–25.

49. Ajin, R.S.; Loghin, A.-M.; Jacob, M.K.; Vinod, P.G.; Krishnamurthy, R.R. The risk assessment study of potential forest fire in Idukki Wildlife Sanctuary using RS and GIS techniques. *Int. J. Adv. Earth Sci. Eng.* **2016**, *5*, 308–318.

50. Werf, G.R.; Randerson, J.T.; Giglio, L.; Collatz, G.J.; Mu, M.; Kasibhatla, P.S.; Morton, D.C.; DeFries, R.S.; Jin, Y.; van Leeuwen, T.T. Global fire emissions and the contribution of deforestation, savanna, forest, agricultural, and peat fires (1997–2009). *Atmos. Chem. Phys.* **2010**, *10*, 11707–11735. [CrossRef]

51. Gigović, L.; Pourghasemi, H.R.; Drobnjak, S.; Bai, S. Testing a new ensemble model based on SVM and random forest in forest fire susceptibility assessment and its mapping in Serbia's Tara National Park. *Forests* **2019**, *10*, 408. [CrossRef]

52. Banerjee, P. Drivers and distribution of forest fires in Sikkim Himalaya: A maximum entropy-based approach to spatial modelling. **2020**, In Review.

53. Pham, B.T.; Jaafari, A.; Avand, M.; Al-Ansari, N.; Dinh Du, T.; Yen, H.P.H.; Phong, T.V.; Nguyen, D.H.; Le, H.V.; Mafi-Gholami, D. Performance evaluation of machine learning methods for forest fire modeling and prediction. *Symmetry* **2020**, *12*, 1022. [CrossRef]

Publisher's Note: MDPI stays neutral with regard to jurisdictional claims in published maps and institutional affiliations.

Article

Tree Species Classification in Mixed Deciduous Forests Using Very High Spatial Resolution Satellite Imagery and Machine Learning Methods

Martina Deur [1], Mateo Gašparović [2,*] and Ivan Balenović [3]

[1] Institute for spatial planning of Šibenik-Knin County, Vladimira Nazora 1/IV, 22000 Šibenik, Croatia; martina.deur@zpu-skz.hr

[2] Chair of Photogrammetry and Remote Sensing, Faculty of Geodesy, University of Zagreb, 10000 Zagreb, Croatia

[3] Division for Forest Management and Forestry Economics, Croatian Forest Research Institute, Trnjanska cesta 35, HR-10000 Zagreb, Croatia; ivanb@sumins.hr

* Correspondence: mgasparovic@geof.unizg.hr; Tel.: +385-1-4639-223

Received: 10 November 2020; Accepted: 27 November 2020; Published: 30 November 2020

Abstract: Spatially explicit information on tree species composition is important for both the forest management and conservation sectors. In combination with machine learning algorithms, very high-resolution satellite imagery may provide an effective solution to reduce the need for labor-intensive and time-consuming field-based surveys. In this study, we evaluated the possibility of using multispectral WorldView-3 (WV-3) satellite imagery for the classification of three main tree species (*Quercus robur* L., *Carpinus betulus* L., and *Alnus glutinosa* (L.) Geartn.) in a lowland, mixed deciduous forest in central Croatia. The pixel-based supervised classification was performed using two machine learning algorithms: random forest (RF) and support vector machine (SVM). Additionally, the contribution of gray level cooccurrence matrix (GLCM) texture features from WV-3 imagery in tree species classification was evaluated. Principal component analysis confirmed GLCM variance to be the most significant texture feature. Of the 373 visually interpreted reference polygons, 237 were used as training polygons and 136 were used as validation polygons. The validation results show relatively high overall accuracy (85%) for tree species classification based solely on WV-3 spectral characteristics and the RF classification approach. As expected, an improvement in classification accuracy was achieved by a combination of spectral and textural features. With the additional use of GLCM variance, the overall accuracy improved by 10% and 7% for RF and SVM classification approaches, respectively.

Keywords: pixel-based supervised classification; random forest; support vector machine; gray level cooccurrence matrix (GLCM); principal component analysis (PCA); WorldView-3

---

1. Introduction

To reduce labor-intensive and time-consuming field-based forest surveys, the potential of remote sensing for forestry applications has long been investigated by both researchers and practicing foresters [1,2]. Over the last few decades, the rapid development of different remote sensing sensors and instruments has resulted in the development of various methods and techniques for retrieval of various types of forest information from remote sensing data [3,4]. As a result, remote sensing has been experimentally and practically used for diverse forestry tasks (e.g., inventory, management, modeling, ecology, protection, health, etc.) [4].

The sensors most commonly used in remote sensing are optical (multispectral or hyperspectral) [4,5]. Moderate (e.g., Landsat) [6–8] or high spatial (e.g., SPOT—fra. Satellite Pour l'Observation de la Terre, RapidEye) [9–12] resolution satellite imagery have been proven to be very efficient for land use and land cover mapping of large areas, which is one of the most common applications for remote sensing [4]. Very high resolution (VHR) satellite imagery (e.g., IKONOS, Pleidas, and WorldView) provide much more detailed information on the observed object of interest at both local and regional scales. In combination with constantly evolving automatic classification approaches, such as contemporary machine learning algorithms, multispectral VHR satellite imagery presents an effective tool for detailed assessment of large areas. Therefore, the possibility of their application for individual tree species classification has been examined by a number of studies in the last few decades [5].

Regardless of the forest or vegetation type (managed, natural, and urban forest, and plantations), Fassnacht et al. [5] stressed the importance of spatially explicit information on tree species composition for a wide variety of applications in forest management and conservation sectors. These applications include various types of resource inventories, biodiversity assessment and monitoring, hazard and stress assessment, etc.

Immitzer et al. [13] classified tree species in temperate Austrian forests dominated by Norway spruce, Scots pine, European beech, and Pedunculate oak using VHR WorldView-2 (WV-2) satellite imagery. To test the classification accuracy, a total of 1465 individual tree samples were manually determined for 10 tree species. Tree species were classified with a random forest (RF) algorithm and by applying object- and pixel-based approaches. The results showed that the object-based approach outperformed the pixel-based approach, which was also reported by Ghosh et al. [14]. Immitzer et al. [13] reported the overall accuracy for classifying 10 tree species as being around 82% for the object-based approach and all eight bands. The species-specific producer's accuracies ranged between 33% and 94%, while the user's accuracies ranged between 57% and 92%. Immitzer et al. [13] further showed that the use of four additional channels (coastal, yellow, red edge, and near infrared 2) of the WV-2 satellite imagery only had a limited impact on classification accuracy for the four main tree species but led to significant improvements in classification accuracy when the other six secondary tree species were included.

Similar promising results for tree species classification with overall accuracy values ranging from 68% to 89% were obtained in several subsequent studies based solely on the classification of spectral bands of WV-2 satellite imagery [15–18]. Perrbhay et al. [15] used partial least squares discriminant analysis to classify six commercial tree species in a plantation in South Africa, with an overall accuracy of 85.4%. In another study from South Africa, Omer et al. [17] compared support vector machine (SVM) and artificial neural network (ANN) algorithms for pixel-based classification of six tree species in a mixed indigenous coastal forest. Both algorithms provided similar classification results; the overall accuracies for SVM and ANN were 77% and 75%, respectively. In the same study area (mixed indigenous coastal forest, South Africa), Cho et al. [16] used an SVM algorithm to compare pixel- and object-based approaches for pixel-based classification of three dominant tree species. The overall accuracies values were 85% and 89% for pixel- and object-based approaches, respectively. Karlson et al. [18] applied an RF algorithm for object-based classification of five dominant tree species in parkland landscape in Burkina Faso, West Africa, achieving an overall accuracy of 78.4%.

Besides the commonly used spectral characteristics, multispectral VHR imagery provides additional features that can be used to improve tree species classification. Waser et al. [19] reported that the additional use of vegetation indices in combination with spectral characteristics significantly improved the classification accuracy of different levels of damaged ash trees but only slightly improved tree species classification. Several studies [14,20,21] reported that the addition of texture features to spectral characteristics can considerably improve tree species classification accuracy. For example, Ferreira et al. [20] reported an increase of more than 25% in average producer's accuracies after combining pan-sharpened WorldView-3 (WV-3) imagery with gray level cooccurrence matrix (GLCM) texture features for eight tree species in tropical, semi-deciduous, and old-growth forests in Brazil. The Haraclick's [22] GLCM texture is the most used texture measure in remote sensing applications [20,23].

The abovementioned studies have confirmed the usefulness of VHR satellite imagery for tree species classification. However, previous studies were predominantly based on WV-2 imagery and covered a limited area, mostly focused on Africa. Only several studies dealt with tree species classification using more advanced satellite imagery, such as WV-3 [20,24,25]. Therefore, further research over different regions and forest types is needed to prove the applicability of VHR satellite imagery (e.g., WV-3) to tree species classification. According to the results of previous studies [14,20,21], additional metrics such as texture features can considerably improve the classification accuracy and therefore have to be considered as an additional contributive input in further studies. Furthermore, a number of classification algorithms have been evaluated, but mostly individually. A limited number of studies have evaluated and compared two [13,14,17,24] or more [21] algorithms using the same imagery, study area, and field reference data. Such studies are essential for identifying which algorithm provides the best results for certain forest areas using imagery and applying imagery characteristics (spectral, textural, etc.).

The main goal of this study was to assess the possibility of tree species classification in a lowland, mixed deciduous forest using the pixel based supervised classification of WV-3 satellite imagery with two machine learning algorithms (RF and SVM). In addition to spectral characteristics, the contribution of various GLCM texture features from WV-3 imagery to tree species classification was evaluated. To the best of our knowledge, no similar studies have been conducted in this or similar forest conditions; the vast majority of previous studies were conducted in less complex forest conditions, and only a few studies dealt with the more advanced satellite imagery such as WV-3.

2. Materials

2.1. Study Area

The study area is located in the Jastrebarski lugovi management unit in central Croatia, near the city of Jastrebarsko, 35 km southwest of Zagreb (Figure 1). It covers an area of 2128.77 ha of lowland deciduous forests and is a part of the Pokupsko Basin forest complex ($\approx$12,000 ha). The main forest type (management class) in the study area is even-aged pedunculate oak (*Quercus robur* L.) forests of different age classes, covering 77% of the study area. The oak stands are commonly mixed with secondary tree species, such as common hornbeam (*Carpinus betulus* L.), black alder (*Alnus glutinosa* (L.) Geartn.), and narrow-leaved ash (*Fraxinus angustifolia* Vahl.). Other secondary tree species that occur sporadically throughout the study area are European white elm (*Ulmus laevis* Pall.), silver (white) birch (*Betula pendula* Roth.), lime (*Tilia* sp.), and poplars (*Populus* sp.). In addition to oak management class, two other forest types present in the study area are even-aged narrow-leaved ash and even aged common hornbeam management classes, covering 17% and 6% of the study area, respectively. Unlike the oak stands, ash and hornbeam stands are more homogeneous and less mixed with secondary tree species. The forests of the study area are state-owned, and they are actively managed for sustained timber based on 140- (oak stands), 80- (ash stands), and 70-year (hornbeam stands) rotation cycles, with two or three regeneration fellings during the last 10 years of the rotation. The terrain is mostly flat with ground elevations ranging from 105 to 118 m above sea level (a.s.l.)

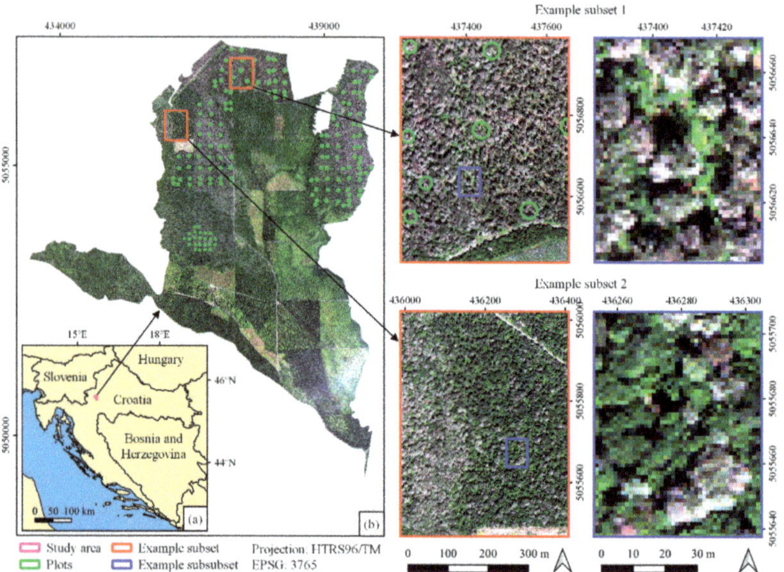

Figure 1. (**a**) Location of the study area in Croatia and (**b**) the study area with two example subset locations (red squares) and sublocations (blue squares) for visual assessment (background: true-color composite of WorldView-3 imagery bands: red-green-blue).

2.2. Field Data

Field data (tree species and tree locations) were collected from a total of 164 circular samples between March and June 2017. Sample plots were systematically distributed (100 × 100 m, 100 m × 200 m, 200 × 100 m, and 200 × 200 m) throughout the 30 stands (subcompartments) within the study area. Of 164 sampled plots, 156 plots were located in 28 oak stands of different ages ranging from 33 to 163 years, while 8 plots were located in two 78-year-old ash stands. The sample plots had radii of 8, 15, or 20 m depending on stand age and stand density. To obtain the precisely measured locations of the plot centers, the Global Navigation Satellite System Real-Time Kinematic (GNSS RTK) method was applied using the Stonex S9IIIN receiver (Stonex, Milan, Italy) connected to the Croatian network of GNSS reference stations [26,27]. The plot centers were measured during leaf-off conditions in the beginning of March 2017. The applied method and service provided fixed solutions for 53 plot centers with an average positioning precision (standard deviation reported by the receiver) of 0.038 m; for 111 plot centers, float solutions were obtained with an average positioning precision of 0.155 m. Within each plot, tree species were determined and recorded, and tree positions were measured for all trees with a diameter at breast height (dbh) above 10 cm. The position of each tree in the plot was recorded by measuring the distance and azimuth from plot center to each tree using a Vertex III hypsometer (Haglöf, Långsele, Sweden) and Haglöf compass (Haglöf, Långsele, Sweden), respectively. In total, species were recorded and positions were measured for 4953 trees within the 164 sample plots. Tree density in sampled plots ranged from 56 to 1840 trees·ha^{-1}, with an average of 526 trees·ha^{-1} and standard deviation (SD) of 303 trees·ha^{-1}. Table 1 shows the summary statistics of tree density at the plot level for the entire set of 164 field sample plots and for the plots grouped into age classes. Sample plots were not equally distributed across the different age classes due to the irregular distribution of age classes in the study area.

Table 1. Summary statistics (mean; Min, minimum; Max, maximum; and SD, standard deviation) of tree density (trees·ha^{-1}) at the plot level for the entire set of 164 field sample plots and for the plots grouped into age classes.

Age Class	Age (Years)	No. of Stands	No. of Sampled Plots	No. of Sampled Trees	Tree Density (Trees·ha^{-1})			
					Mean	Min	Max	SD
II	21–40	1	3	142	670	340	835	286
III	41–60	7	44	1230	808	311	1840	369
IV	61–80	9	45	1594	503	311	821	111
V	81–100	4	30	984	464	311	764	122
VI	101–120	-	-	-	-	-	-	-
VII	>121	9	42	1003	289	56	806	219
Total		30	164	4953	526	56	1840	303

2.3. WorldView-3 Satellite Imagery

The WV-3 multispectral imagery used in this study was acquired over the study area in June 2017 with a 2-m spatial resolution at a mean off nadir view angle of 29.2°; with mean in-track and cross-track view angles of −29.2° and −0.5°, respectively; and with mean Sun azimuth and Sun elevation angle of 158.6° and 66.6°, respectively. The WorldView-3 sensor provides eight multispectral bands: coastal blue (B1, 400–450 nm), blue (B2, 450–510 nm), green (B3, 510–580 nm), yellow (B4, 585–625 nm), red (B5, 630–690 nm), red-edge (B6, 705–745 nm), near-infrared 1 (NIR1; B7, 770–895 nm), and near-infrared 2 (NIR2; B8, 860–1040 nm). The satellite also has a panchromatic sensor (Pan, 450–800 nm) that provides imagery with a 0.5-m spatial resolution. However, in this study, only multispectral imagery was used.

3. Methods

The research workflow in this study was divided into four phases (Figure 2). The entire research was conducted using open source software, e.g., Orfeo ToolBox (6.6.1), Quantum GIS (3.10.1), SAGA GIS (7.0.0), SNAP (7.0.), and Grass GIS (7.8.1).

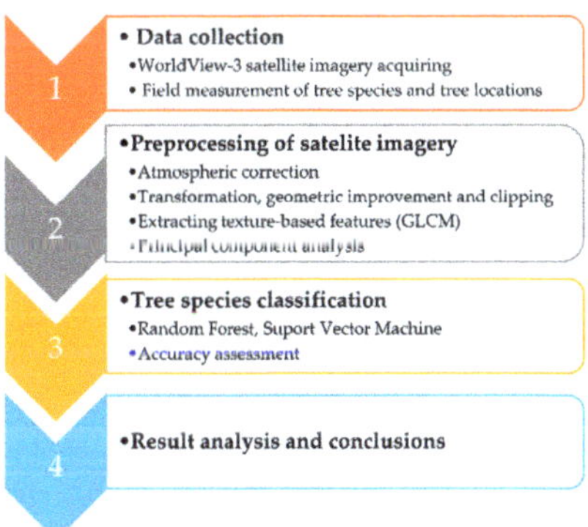

Figure 2. Research workflow of tree species classification in mixed deciduous forest using very high spatial resolution satellite imagery and machine learning algorithms.

3.1. Preprocessing of the WorldView-3 Satellite Imagery

To classify tree species using WV-3 imagery, several preprocessing steps were required. As a first step, atmospheric correction was performed. Imagery was transformed into the value of the reflection at the top of the atmosphere (top of atmosphere (TOA) reflectance). Atmospheric correction of the WV-3 satellite imagery was performed according to the second simulation of satellite signal in the solar spectrum (6S) algorithm using the i.atcorr module in GRASS GIS (version 7.8.1.). Based on information about the surface reflectance and atmospheric conditions, this model provides the reflectance of objects at the TOA [28].

Since the field data and satellite imagery did not use the same coordinate system, WV-3 was re-projected from WGS84 UTM 33N to the HTRS96/TM reference coordinate system. The geometric accuracy of WV-3 was improved in two steps: sensor orientation and orthorectification [29]. Sensor orientation was based on rational polynomial coefficients with a shift or zero-order (RPC0) bias correction using seven ground control points (GCPs). The two-dimensional positions of GPCs were recorded using a digital orthophoto map (DOF5), which was an official state map with a scale of 1:5000. Gašparović et al. [30] assessed the accuracy of the digital orthophoto map (DOF5) and WV-2 satellite imagery. The root mean square error (RMSE) for the tested DOF5 ranged between 0.37 and 0.46 m and was more than three times more accurate, on average, compared to orthorectified WV-2 satellite imagery. Since we used MS WV-3 imagery with a 2-m spatial resolution and the accuracy of DOF5 is below 0.5 m, the two-dimensional positions of GPCs recorded from DOF5 have sufficient accuracy. GPCs were acquired in an open-source program, Quantum GIS (QGIS) version 3.10.1, in the HTRS96/TM reference coordinate system. After sensor orientation, WV-3 was orthorectified using a global Shuttle Radar Topography Mission (SRTM) digital elevation model (DEM) with open-source software Orfeo ToolBox (OTB) version 6.6.1. The OTB algorithm for orthorectification was assessed using Monteverdi. Finally, in the last step of preprocessing, satellite imagery was cropped along the border of the Jastrebarski lugovi management unit.

3.2. Visual Interpretation of Reference Polygons for Image Classification

Reference polygons, i.e., polygons used for training and validation of tree species classification, were defined by visual interpretation of WV-3 imagery (B8 = NIR2, B3 = green, and B2 = blue) and by using field data on tree species and tree locations (Figure 3). Due to the complex forest structure in terms of the number of tree species and stand density, visual interpretation of tree species and creation of reference polygons were demanding. Therefore, reference polygons were defined only for the most common tree species (*Q. robur*, *C. betulus*, and *A. glutinosa*) within the field plots as well as for the classes "bare land", "low vegetation", and "shadow". Interpretation was conducted using different imagery compositions for which "true" and "false" color compositions were the most commonly used. One reference polygon mostly contained several neighboring trees of the same species in which crowns were merged or overlapped. A total of 306 reference polygons was collected for tree species, and 67 polygon samples were collected for other classes. All collected reference polygons were randomly divided into training (64%) and validation (36%) datasets (Table 2).

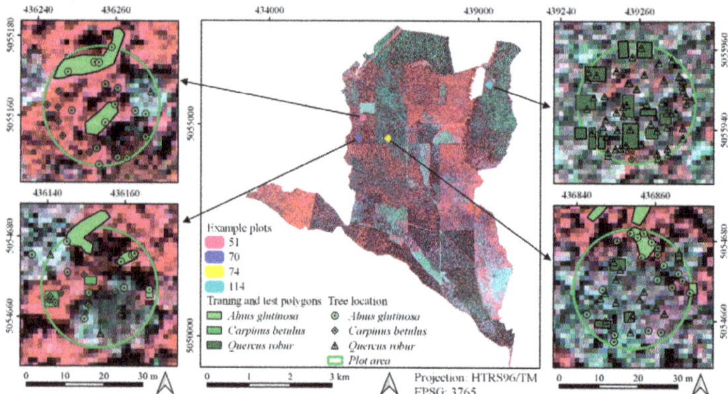

Figure 3. An example of a visual interpretation of reference polygons on four exemplary field plots (51, 70, 74, and 114) (background: false-color composite of WorldView-3 from 2017, bands: red-green-blue): the measured tree locations from field survey are shown on enlarged maps.

Table 2. Number of visually interpreted reference polygons (training and validation polygons) for observed tree species and other observed classes (bare land, low vegetation, and shadow).

Class	Training Polygons	Validation Polygons	Field Measured Trees Included in Polygon
Alnus glutinosa	24	15	61
Carpinus betulus	46	27	120
Quercus robur	124	70	219
Bare land	11	7	-
Low vegetation	18	11	-
Shadow	14	6	-
Total	237	136	400

3.3. Machine Learning Image Classification

In this study, pixel-based supervised image classification was performed using two machine learning algorithms: random forest (RF) and support vector machine (SVM).

RF is an automatic learning algorithm introduced by Breiman [31] and was improved by Adele Cutler [32]. Supervised RF pixel-based classification has been confirmed as useful in separating classes by spectral properties and has therefore been applied in a number of tree species classification studies [13,24,33,34]. RF represents a combination of tree predictors, where each tree depends on the values of a random vector sampled independently. All trees in the forest have the same distribution [31]. The algorithm input is a vector with every single decision tree. The result is a classification label with the most "votes". The RF classifier uses the Gini index [35]. The Gini index, as an attribute selection criterion, measures the impurity of an attribute in relation to the classes. A comprehensive overview of RF can be found in other studies [31,32,35,36]. To run a RF classifier, open-source software SAGA GIS (7.0.0) was used. Immtzer et al. [33] found that the default values of the Orfeo ToolBox (OTB) parameters for training and classification processes provide optimal results. According to Belgiu and Dragut [36], many studies have investigated the influence of nTree and Mtry parameters on the accuracy of RF classifiers. The most common recommendation, which was applied in this study as well, is to set the nTree parameter to 500 and Mtry to the square root of the number of input variables. In this study, a wide range of values for other parameters were tested, e.g., maximum tree depth of 1 to 1000 and minimum sample count of 1 to 100. Finally, the most accurate results for the RF algorithm were obtained when the maximum depth of the tree was set to 10, while the minimum sample count was set to 2. Regression accuracy was set to 0.01.

Another machine learning algorithm used in this study, the SVM algorithm, was developed by Cortes and Vapnik [37]. In general, SVMs are based on statistical learning theory and define the optimal hyperplane as a linear decision function with a maximal margin between the vectors of two classes. The support vector defines the margin of the largest separation between the two classes. SVM has been shown to be insensitive to high data dimensionality and to be robust in terms of small training sample sizes [38,39]. As in preview studies [17,21,24], the radial basis function (RBF) kernel was used in this study. Cost value and gamma are two user-defined parameters that influence the classification accuracy [40]. Cost value is used to fit the classification errors in the training data set [41] and gamma. The parameters of RBF were set as cost value = 1 and gamma = 0.0001, and probability thresholds were set to zero to avoid unclassified pixels. As well as for RF, to run an SVM classifier, open-source software SAGA GIS (7.0.0) was used.

3.4. Texture Features for Image Classification Improvement

Texture features are one of the important characteristics used for identifying objects from satellite imagery [23]. The most commonly used textural measure is the GLCM [22]. The GLCM is a record of how often different combinations of pixel brightness values (gray levels) occur in imagery [42]. According to Hall-Beyer [42], GLCM features can be categorized into three groups: contrast (contrast, dissimilarity, and homogeneity), orderliness (angular second moment and entropy), and statistics (mean, variance, and correlation). Several studies reported improved classification accuracy when combining GLCM texture and spectral features [43,44]. According to previous research [45,46], to reduce data redundancy, GLCM calculations were performed only for the red channel, which showed the highest entropy among of all bands.

Within this study, we used the GLCM implementation provided with the ESA Sentinel Application Platform SNAP (7.0.). It calculates the 10 Haralick measures: mean, variance, homogeneity, contrast, dissimilarity, energy, entropy, angular second moment, maximum probability, and correlation. Here, we combined the GLCM variance as a measure of the dispersion of the values around the mean with spectral features. The variance was measured according to Equation (1) [42]:

$$\sigma_i^2 = \sum_{i,j=0}^{N-1} P_{i,j}(i - \mu_i)^2 \quad \sigma_j^2 = \sum_{i,j=0}^{N-1} P_{i,j}(j - \mu_j)^2 \tag{1}$$

where i is the row number, j is the column number, $p(i, j)$ is the normalized value in the cell i,j, and N is the number of rows or columns. According to Chen et al. [47], who found that for spectrally homogenous classes smaller window sizes improve classification accuracy, GLCM features were computed with a small 5×5 pixel window size over all directions, a pixel displacement of 2, and a 32-level quantization.

To extract the most useful texture information from imagery, principal component analysis (PCA) was further applied. PCA is a dimension-reduction technique that can be used to reduce a large set of variables to a small set that still contains most of the information in the original set. PCA finds common factors in a given dataset and ranks them in order of importance [48]. In recent years, PCA has been used extensively in remote sensing classification problems, especially for hyperspectral imagery [49–51]. As with GLCM texture extraction, PCA was performed in SNAP (7.0.). The PCA of 10 GLCM measures was performed for WV-3 imagery.

3.5. Accuracy Assessment

For image classification, three different methods were applied and tested in this study:

1. Image classification using eight multispectral bands of WV-3 only and RF classifier (RF$_{MS}$),
2. Image classification using eight multispectral bands of WV-3 combined with texture features extracted from WV-3 and the RF classifier (RF$_{MS\text{-}GLCM}$), and
3. Image classification using eight multispectral bands of WV-3 combined with texture features extracted from WV-3 and the SVM classifier (SVM$_{MS\text{-}GLCM}$).

The accuracy assessment of image classification and, particularly, the accuracy assessment of tree species classification was conducted using a validation polygon dataset (Table 2).

To evaluate the results of the RF and SVM algorithms, a confusion matrix was produced and the producer's accuracies (PAs) and user's accuracies (UAs) for each class were calculated, as were the overall accuracy (OA) [52] and Kappa coefficient (k) [53]. The confusion matrix provided an overview of the classification errors between species. The diagonal represents correctly classified pixels according to reference data, while off-diagonals were misclassified. PA and UA were calculated by dividing the number of correctly classified pixels in each category by the total number of pixels in the corresponding column (PA) or row (UA). OA was computed by dividing the total number of correctly classified pixels by the total number of reference pixels. Besides the OA, within the confusion matrix, omission (O) and commission (C) errors were analyzed.

Developed by Cohen [53], k measures the proportion of agreement after chance agreements were removed from consideration and is calculated using the following equation [53]:

$$k = \frac{p_0 - p_e}{1 - p_e} \tag{2}$$

where p_0 is the relative observed agreement among raters and p_e is the hypothetical probability of chance agreement. Since the Kappa coefficient has certain limitations identified by Pontius and Millones [54], the figure of merit (FoM) was also calculated as an additional statistical measure (Equation (3)):

$$FoM = \frac{OA}{OA + O + C} \tag{3}$$

where OA represents overall accuracy, O is the number of omissions, and C is the number of commissions.

4. Results

For GLCM texture information extraction, various window sizes were tested (5×5, 7×7, and 9×9). The GLCM feature computed with a 5×5-pixel window size produced the best results (as confirmed by Chen et al. [47]), and it was used further in our research. Based on the extracted GLCM features and the conducted PCA, the GLCM variance was confirmed as the most important texture measure, with an eigenvalue considerably greater than that for other GLCM features (Table 3). Therefore, only GLCM variance was included in image classification ($RF_{MS+GLCM}$ and $SVM_{MS+GLCM}$).

Table 3. Eigenvalues of principal component analysis (PCA) on 10 gray level cooccurrence matrix (GLCM) texture features.

Component	Eigenvalue
GLCM Variance	1473.08
GLCM Mean	54.02
Contrast	17.77
Entropy	5.01
Dissimilarity	4.49
GLCM Correlation	0.99
Homogeneity	0.74
Energy	0.47
Angular Second Moment (ASM)	0.26
MAX	0.24

After the most important GLCM feature was determined, the GLCM variance for all eight MS bands was calculated. Prior to tree species classification, additional analysis (experimental classifications) was conducted for the study area to determine for which band or combination of bands GLCM variance provides the best results. In experimental classifications, GLCM variance was calculated for each band separately as well as for different combinations of all eight bands (Figure 4). The obtained results demonstrated that the highest effectiveness of GLCM variance was calculated for the first four bands (B1, B2, B3, and B4). The addition of the GLCM variance of the other bands did not improve the classification accuracy.

The classification of the WV-3 imagery using three different approaches (RF_{MS}, $RF_{MS-GLCM}$, and $SVM_{MS-GLCM}$) and visual analysis of the classification results was conducted for the entire study area; a detailed statistical accuracy assessment was conducted using validation polygons only (Table 2). The results of image classification for the entire study area are shown in Figure 5 (WV-3 true-color composite), whereas Figure 6 presents more detailed classification results for two example subsets.

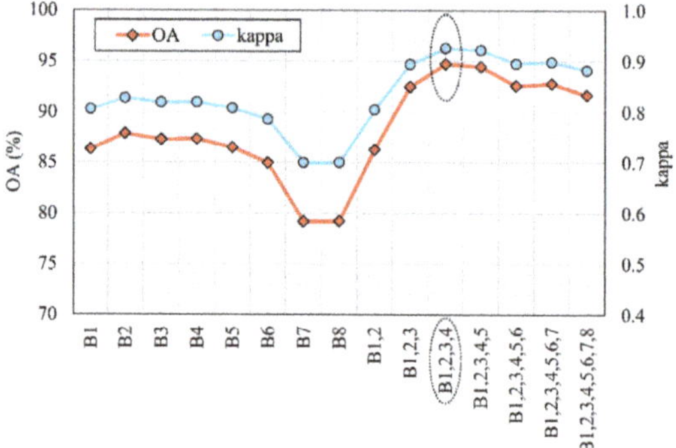

Figure 4. Overall accuracy (OA) and Kappa values for random forest (RF) experimental classifications based on spectral characteristics of all 8 bands and additional GLCM variance calculated for each band separately as well as for different combinations of bands: labels on the *x*-axis represent the band or bands for which GLCM variance was calculated, and the dashed oval emphasizes the combination with the highest OA and Kappa value.

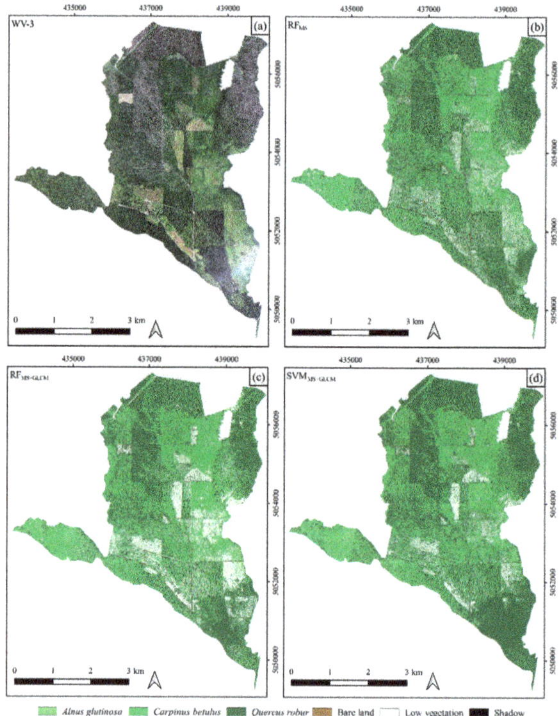

Figure 5. Study area shown as WV-3 true-color composite (**a**) and image classification maps derived by three different approaches: (**b**) using eight bands of WV-3 imagery and an RF classifier (RF_MS); (**c**) using eight bands of WV-3 combined with texture features extracted from WV-3 imagery and an RF classifier (RF_MS+GLCM); and (**d**) using eight bands of WV-3 combined with texture features extracted from WV-3 and a support vector machine (SVM) classifier (SVM_MS+GLCM).

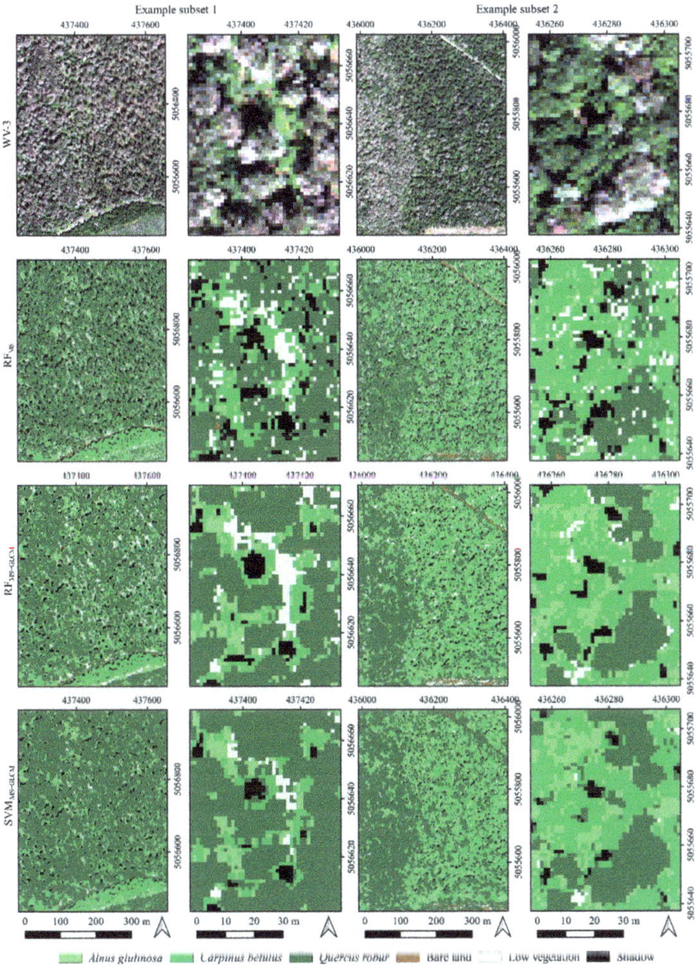

Figure 6. Two example subsets in the study area: (**first row**) WV-3 true-color composite; (**second row**) classification map derived using eight bands of WV-3 imagery and an RF classifier (RF$_{MS}$); (**third row**) classification map derived using eight bands of WV-3 combined with texture features extracted from WV-3 imagery and an RF classifier (RF$_{MS+GLCM}$); and (**fourth row**) classification map using eight bands of WV-3 combined with texture features extracted from WV-3 imagery and an SVM classifier (SVM$_{MS+GLCM}$).

An initial visual analysis of the performed classification for the entire study area showed that similar results were obtained using both algorithms (RF and SVM) when eight bands of WV-3 imagery were used in combination with its texture features. This can also be observed in Figure 7, which shows the proportion of observed tree species for the entire study. In other words, the initial visual assessment showed that the RF$_{MS+GLCM}$ and SVM$_{MS+GLCM}$ approaches (methods) provided similar results in automatic classification of mixed deciduous forests for the present study area whereas, when the RF$_{MS}$ approach was applied, i.e., when only multispectral bands of WV-3 imagery were used, different and slightly worse results were obtained. As shown in Figure 5, the classification of WV-3 using only spectral bands (RF$_{MS}$) produced noisy distribution of tree species classes. This is particularly evident for *Alnus glutinosa*, which has similar spectral properties to *Carpinus betulus*.

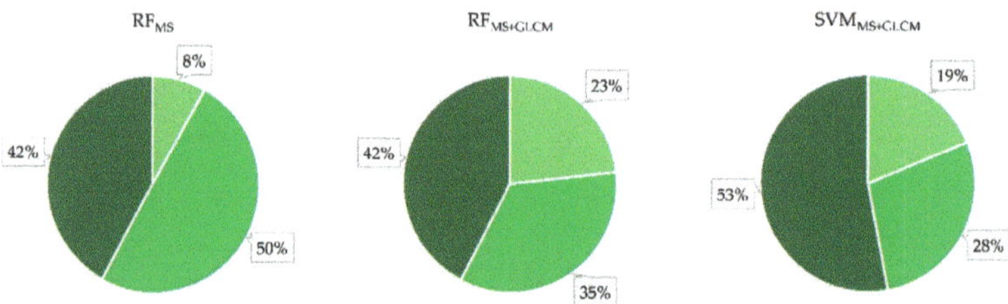

Figure 7. The share of observed tree species (*Alnus glutinosa*, *Carpinus betulus*, and *Quercus robur*) for the entire study area obtained with three different approaches (RF$_{MS}$, RF$_{MS-GLCM}$, and SVM$_{MS-GLCM}$).

Besides visual analysis, the classification accuracy was assessed and evaluated using a detailed statistical analysis performed on 136 validation polygons (Table 4). The obtained results showed that, for both algorithms (RF and SVM), the classification accuracy was considerably improved when the texture features (GLCM variance) of WV-3 imagery were considered in addition to the spectral characteristics. Namely, compared to the RF$_{MS}$ classification approach (based solely on WV-3 spectral characteristics), the OA values for RF$_{MS-GLCM}$ and SVM$_{MS-GLCM}$ approaches (both based on WV-3 spectral and texture characteristics) increased by 10% and 7%, respectively. Compared to the RF$_{MS}$ approach, the *k* values for the RF$_{MS-GLCM}$ and SVM$_{MS-GLCM}$ approaches increased by 0.13 and 0.09, respectively.

Table 4. Results of accuracy of WV-3 imagery classification using three different approaches (RF$_{MS}$, RF$_{MS-GLCM}$, and SVM$_{MS-GLCM}$) performed on 136 validation polygons. UA, user accuracy; PA, producer accuracy.

RF$_{MS}$								
Class (Latin Name)	A. glutinosa	C. betulus	Q. robur	Bare Land	Low Vegetation	Shadow	Total	UA
A. glutinosa	26	27	2	0	5	0	60	43%
C. betulus	8	433	0	0	2	0	443	98%
Q. robur	0	6	913	0	23	0	942	97%
Bare land	0	0	1	195	0	0	196	99%
Low vegetation	3	25	225	8	343	0	604	57%
Shadow	0	0	0	0	0	38	38	100%
Total	37	491	1141	203	373	38	OA = 85%	
PA	70%	88%	80%	96%	92%	100%	*k* = 0.79	

RF$_{MS+GLLCM}$								
Class (Latin name)	A. glutinosa	C. betulus	Q. robur	Bare land	Low vegetation	Shadow	Total	UA
A. glutinosa	50	10	0	0	0	0	60	83%
C. betulus	20	423	0	0	0	0	443	95%
Q. robur	1	0	936	2	3	0	942	99%
Bare land	0	0	1	195	0	0	196	99%
Low vegetation	0	0	68	15	521	0	604	86%
Shadow	0	0	0	0	0	38	38	100%
Total	71	433	1005	212	524	38	OA = 95%	
PA	70%	98%	93%	92%	99%	100%	*k* = 0.92	

SVM$_{MS+GLCM}$								
Class (Latin name)	A. glutinosa	C. betulus	Q. robur	Bare land	Low vegetation	Shadow	Total	UA
A. glutinosa	52	7	1	0	0	0	60	87%
C. betulus	28	415	0	0	0	0	443	94%
Q. robur	0	0	942	0	0	0	942	100%
Bare land	0	0	41	155	0	0	196	79%
Low vegetation	1	0	107	8	488	0	604	81%
Shadow	0	0	0	0	0	38	38	100%
Total	81	422	1091	163	488	38	OA = 92%	
PA	64%	98%	86%	95%	100%	100%	*k* = 0.88	

Observed by tree species and regardless of the applied classification approach, the lowest classification accuracy was obtained for *Alnus glutinosa*, with a UA ranging from 43% to 87%. Considerably higher accuracy was obtained for *Carpinus betulus* (UA = 94–98%), while the highest classification accuracy was obtained for *Quercus robur* (UA = 97–100%). The classification accuracy for *A. glutinosa* was considerably improved when the texture feature was added to classification; compared to RF_{MS} approach, the UA values for the $RF_{MS\text{-}GLCM}$ and $SVM_{MS\text{-}GLCM}$ approaches increased by 40% and 44%, respectively. Slight improvements in classification accuracy were also observed for *Q. robur*, where UA increased by 2% for $RF_{MS\text{-}GLCM}$ and by 3% for the $SVM_{MS\text{-}GLCM}$ approach compared to RF_{MS}. Inclusion of the texture feature in classification slightly deteriorated the accuracy for *C. betulus* compared to the RF_{MS} approach, by 3% and 4% for $RF_{MS\text{-}GLCM}$ and $SVM_{MS\text{-}GLCM}$, respectively.

All of the abovementioned were confirmed by an additional accuracy assessment (Table 5) based on FoM, O, C, and A measures. As well as PA and UA, FoM accuracy metrics showed that, for both algorithms (RF and SVM), the classification accuracy was improved by adding texture feature (GLCM variance) to the spectral characteristics.

Table 5. Additional accuracy assessment of WV-3 imagery classification using three different approaches (RF_{MS}, $RF_{MS\text{-}GLCM}$, and $SVM_{MS\text{-}GLCM}$). FoM, Figure of merit; O, omission; C, commission; A, overall agreement.

Class (Latin Name)	RF_{MS}			$RF_{MS+GLCM}$			$SVM_{MS+GLCM}$		
	FoM (%)	O (%)	C (%)	FoM (%)	O (%)	C (%)	FoM (%)	O (%)	C (%)
A. glutinosa	36.62	0.48	1.49	61.73	0.92	0.44	58.43	1.27	0.35
C. betulus	86.43	2.54	0.44	93.38	0.44	0.88	92.22	0.31	1.23
Q. robur	78.03	9.99	1.27	92.58	3.02	0.26	86.34	6.53	0.00
Bare land	95.59	0.35	0.04	91.55	0.74	0.04	75.98	0.35	1.80
Low vegetation	54.10	1.31	11.43	85.83	0.13	3.64	80.79	0.00	5.08
Shadow	100.00	0.00	0.00	100.00	0.00	0.00	100.00	0.00	0.00
A		85			95			92	

5. Discussion

In this study, we evaluated the possibility of tree species classification in a mixed deciduous forest using supervised pixel-based classification of WV-3 satellite imagery with two machine learning algorithms (RF and SVM). Additionally, the contribution of GLCM texture features from WV-3 imagery in tree species classification was evaluated. In general, our findings agree with a number of recent studies [13,15,17,19], which confirmed the potential of very high spatial resolution satellite imagery for tree species classification. However, compared to most of the previous studies conducted in a less complex forest environment [14,18,24,45,55], this research focused on a mixed deciduous forest with pedunculate oak as the main tree species and with large shares of other deciduous tree species. The proportion of tree species varied among the 164 plots distributed within 30 stands of different ages ranging from 33 to 163 years.

Within this study, we evaluated the contribution of 10 GLCM texture features for improvements of tree species classification using PCA. According to Hall-Breyer [56], PCA loadings show that contrast, dissimilarity, entropy, and GLCM variance are generally associated with visual edges of land-cover patches and that homogeneity, GLCM mean, GLCM correlation, and angular second moment are associated with patch interiors. The results obtained in this study confirmed that GLCM variance is the most (and only) important texture feature (Table 3) and, therefore, was the only texture feature included in further analysis.

Window size is an important component of texture analysis. Small windows can increase the differences and can increase the noise content, whereas larger windows cannot effectively extract texture information [57]. Therefore, the effect of window size on classification accuracy was additionally explored by calculating the GLCM with pixel window sizes of 5×5, 7×7, and 9×9. As in Chen et al. [47], the GLCM feature computed with a small 5×5 pixel window size produced the most accurate results. In the next step, the GLCM variance that was calculated for each of the eight bands of WV-3 imagery was tested in various combinations (individually for each band or grouped bands) with RF or SVM classifications based on all eight bands (Figure 4). The results showed that GLCM variance calculated for the first four bands (B1, B2, B3, and B4) in combination with the RF or SVM spectral classifications of all eight bands produced the most accurate results.

We also found that, for the applied RF algorithm, the addition of the texture feature (GLCM variance) to the spectral characteristics considerably improved the image classification accuracy (Tables 4 and 5). The SVM algorithm also produced very accurate results. Almost all accuracy parameters, including FoM, showed improvements for all tree species. These results are in line with other similar studies that also reported an increased overall tree species classification accuracy when spectral and texture features of very high resolution satellite imagery were combined [14,20,21,58,59]. However, the type and number of selected and used texture features differed among studies. For example, eight GLCM texture features (mean, variance, homogeneity, contrast, dissimilarity, entropy, angular second moment, and correlation) were successfully used to improve the species classification in two studies [20,58]. Wang et al. [59] used two texture features (mean, Angular Second Moment (ASM), and entropy) to improve the classification of coastal wetland vegetation from high spatial resolution Pleiades imagery. In several studies that used various remote sensing data for different aims [60–62], contrast, entropy, and GLCM mean were chosen as the most contributive texture features to predict forest structural parameters from WV-2 imagery [60], to identify old-growth forests from Sentinel-2 data [61], and to improve the classification in urban areas from SPOT imagery [62]. In this study, GLCM mean and contrast features were the second and third most important variables, respectively, but in comparison to GLCM variance, their contribution to tree species classification accuracy was nonsignificant (Table 3). In contrast to the abovementioned findings and those in this study, Yang et al. [63] reported that GLCM textural features did not improve the classification accuracy of tree species in two observed sites (homogeneous park forest and heterogeneous management forest) in China when combined with spectral features.

In this research, by combining spectral and textural features, the accuracy of tree species classification in mixed deciduous forest was improved for 10% (RF classification approach).

One of the crucial preprocessing steps is the generation of reference polygons. Reference polygons need to fulfill a number of requirements: training and validation polygons must be statistically independent, class-balanced, and representative of the target classes, and the training polygons needs to be large enough to accommodate the increased number of data dimensions [36]. According to Sabat-Tpomala et al. [64], the accuracy of any machine learning procedure is directly related to the quality of the reference polygons used for training and validation of a given classifier. Since, in this study, the research area consisted of natural mixed lowland forest, generation of reference polygons was complex and a time-intensive task. Compared to the other tree species, *Alnus glutinosa* had the smallest number of reference polygons (Table 2) and, according to the obtained UA and PA values, had the lowest classification accuracy (Table 4). The SVM$_{MS\text{-}GLCM}$ approach improved *Alnus glutinosa* classification accuracy by 4% and 10% in terms of UA and PA, respectively. The SVM algorithm is more resistant to smaller numbers of training patterns compared to the RF algorithm [64].

We demonstrated the potential of both RF and SVM to integrate spectral and textural features for the management of remotely sensed complex data [65–67]. Both algorithms classified tree species and other classes within the study area with high accuracy; OAs were 92% and 95% for SVM$_{MS\text{-}GLCM}$ and RF$_{MS\text{-}GLCM}$, respectively. Kupidura et al. [67] compared the efficacy of several texture analysis methods as tools for improving land use/land cover classification in satellite imagery and concluded that the choice of the classifier is often less important than adequate data preprocessing to obtain accurate results. Future research will consider other classification algorithms too, especially deep learning convolutional neural networks [68–71].

6. Conclusions

We confirmed the considerable potential of very high-resolution satellite imagery (WorldView-3) for tree species classification, even in areas with complex, natural, and mixed deciduous forest stands. A total of 373 polygon samples was collected by visual interpretation for six classes (*Quercus robur*, *Carpinus betulus*, *Alnus glutinosa*, bare land, low vegetation, and shadow). Image classification was based on 237 training polygons using pixel-based supervised classification conducted with two machine learning algorithms (RF and SVM). For accuracy assessment, 136 validation polygons were used. The validation results showed the relatively high overall accuracy (85%) for tree species classification based solely on WorldView-3 spectral characteristics and the RF classification approach. As expected, the classification accuracy was improved by a combination of spectral and textural features. With the additional use of GLCM variance calculated for the first four bands, overall accuracy improved by 10% and 7% using the RF and SVM classification approaches, respectively. Principal component analysis confirmed that GLCM variance was the most significant texture feature, whereas additional analysis where GLCM variance was calculated for various combinations of spectral bands demonstrated the greatest effectiveness of GLCM variance when calculated for the first four bands (B1, B2, B3, and B4). The findings of this research should serve as a basis for further studies that should test object-based imagery analysis as well as the influence of the use of pansharpened imagery on the classification accuracy.

Author Contributions: Conceptualization, M.D. and M.G.; methodology, M.D.; software, M.D.; validation, M.D., M.G., and I.B.; formal analysis, M.D. and M.G.; investigation, M.D.; resources, M.D. and M.G.; data curation, M.D., M.G., and I.B.; writing—original draft preparation, M.D.; writing—review and editing, M.D., M.G., and I.B.; visualization, M.D. and M.G.; supervision, M.G. and I.B.; project administration, M.G. and I.B.; funding acquisition, M.G. and I.B. All authors have read and agreed to the published version of the manuscript.

Funding: This research has been supported by projects: Retrieval of Information from Different Optical 3D Remote Sensing Sources for Use in Forest Inventory (3D-FORINVENT), funded by the Croatian Science Foundation (HRZZ IP-2016-06-7686); Operational Sustainable Forestry with Satellite-Based Remote Sensing (MySustainableForest), funded by the European Union's Horizon 2020 research and innovation program under grant agreement No 776045; and Advanced photogrammetry and remote sensing methods for environmental change monitoring (grant No. RS4ENVIRO), funded by the University of Zagreb.

Conflicts of Interest: The authors declare no conflict of interest.

References

1. Gering, L.R.; May, D.M.; Teuber, K.B. The use of aerial photographs and angle-gauge sampling of tree crown diameters for forest inventories. In Proceedings of the State-Of-The-Art Methodology of Forest Inventory, Syracuse, NY, USA, 30 July–5 August 1989; pp. 286–289.
2. Balenović, I.; Milas, A.S.; Marjanović, H. A Comparison of Stand-Level Volume Estimates from Image-Based Canopy Height Models of Different Spatial Resolutions. *Remote Sens.* **2017**, *9*, 205. [CrossRef]
3. McRoberts, R.E.; Cohen, W.B.; Næsset, E.; Stehman, S.V.; Tomppo, E.O. Using remotely sensed data to construct and assess forest attribute maps and related spatial products. *Scand. J. For. Res.* **2010**, *25*, 340–367. [CrossRef]
4. Lechner, A.M.; Foody, G.M.; Doreen, S.; Boyd, D.S. Applications in Remote Sensing to Forest Ecology and Management. *One Earth* **2020**, *2*, 405–412. [CrossRef]
5. Fassnacht, F.E.; Latifi, H.; Sterenczak, K.; Modzelewska, A.; Lefsky, M.; Waser, L.T.; Straub, C.; Ghosh, A. Review of studies on tree species classification from remotely sensed data. *Remote Sens. Environ.* **2016**, *186*, 64–87. [CrossRef]
6. Knorn, J.; Rabe, A.; Radeloff, V.C.; Kuemmerle, T.; Kozak, J.; Hostert, P. Land cover mapping of large areas using chain classification of neighboring Landsat satellite images. *Remote Sens. Environ.* **2009**, *113*, 957–964. [CrossRef]
7. Griffiths, P.; Linden, S.; Kuemmerle, T.; Hostert, P. A pixel-based Landsat compositing algorithm for large area land cover mapping. *IEEE J. Sel. Top. Appl. Earth Obs. Remote Sens.* **2013**, *6*, 2088–2101. [CrossRef]
8. Chaves, M.E.D.; Picoli, C.A.M.; Sanches, I.D. Recent Applications of Landsat 8/OLI and Sentinel-2/MSI for Land Use and Land Cover Mapping: A Systematic Review. *Remote Sens.* **2020**, *12*, 3062. [CrossRef]
9. Li, X.; Chen, W.; Cheng, X.; Liao, Y.; Chen, G. Comparison and integration of feature reduction methods for land cover classification with RapidEye imagery. *Multimed. Tools Appl.* **2017**, *76*, 23041–23057. [CrossRef]
10. Mas, J.F.; Lemoine-Rodríguez, R.; González-López, R.; López-Sánchez, J.; Piña-Garduño, A.; Herrera-Flores, E. Land use/land cover change detection combining automatic processing and visual interpretation. *Eur. J. Remote Sens.* **2017**, *50*, 626–635. [CrossRef]

11. Nampak, H.; Pradhan, B.; Rizeei, H.M.; Park, H.J. Assessment of land cover and land use change impact on soil loss in a tropical catchment by using multitemporal SPOT-5 satellite images and Revised Universal Soil Loss Equation model. *Land Degrad. Dev.* **2018**, *29*, 3440–3455. [CrossRef]

12. Saini, R.; Ghosh, S.K. Analyzing the impact of red-edge band on land use land cover classification using multispectral RapidEye imagery and machine learning techniques. *J. Appl. Remote Sens.* **2019**, *13*, 044511. [CrossRef]

13. Immitzer, M.; Atzberger, C.; Koukal, T. Tree Species Classification with Random Forest Using Very High Spatial Resolution 8-Band WorldView-2 Satellite Data. *Remote Sens.* **2012**, *4*, 2661–2693. [CrossRef]

14. Ghosh, A.; Joshi, P.K. A comparison of selected classification algorithms for mapping bamboo patches in lower Gangetic plains using very high resolution WorldView 2 imagery. *Int. J. Appl. Earth Obs. Geoinf.* **2014**, *26*, 298–311. [CrossRef]

15. Peerbhay, K.Y.; Mutanga, O.; Ismail, R. Investigating the capability of few strategically placed WorldView-2 multispectral bands to discriminate forest species in KwaZulu-Natal, South Africa. *IEEE J. Sel. Top. Appl. Earth Obs. Remote Sens.* **2013**, *7*, 307–316. [CrossRef]

16. Cho, M.A.; Malahlela, O.; Ramoelo, A. Assessing the utility worldview-2 imagery for tree species mapping in south african subtropical humid forest and the conservation implications: Dukuduku forest patch as case study. *Int. J. Appl. Earth Obs. Geoinf.* **2015**, *38*, 349–357. [CrossRef]

17. Omer, G.; Mutanga, O.; Abdel-Rahman, E.M.; Adam, E. Performance of Support Vector Machines and Artificial Neural Network for Mapping Endangered Tree Species Using WorldView-2 Data in Dukuduku Forest, South Africa. *IEEE J. Sel. Top. Appl. Earth Obs. Remote Sens.* **2015**, *8*, 4825–4840. [CrossRef]

18. Karlson, M.; Ostwald, M.; Reese, H.; Roméo, B.; Boalidioa, T. Assessing the potential of multi-seasonal WorldView-2 imagery for mapping West African agroforestry tree species. *Int. J. Appl. Earth Obs. Geoinf.* **2016**, *50*, 80–88. [CrossRef]

19. Waser, L.T.; Küchler, M.; Jütte, K.; Stampfer, T. Evaluating the Potential of WorldView-2 Data to Classify Tree Species and Different Levels of Ash Mortality. *Remote Sens.* **2014**, *6*, 4515–4545. [CrossRef]

20. Ferreira, M.P.; Wagner, F.H.; Aragão, L.E.; Shimabukuro, Y.E.; de Souza, C.R.F. Tree species classification in tropical forests using visible to shortwave infrared WorldView-3 images and texture analysis. *ISPRS J. Photogramm. Remote Sens.* **2019**, *149*, 119–131. [CrossRef]

21. Xie, Z.; Chen, Y.; Lu, D.; Li, G.; Chen, E. Classification of Land Cover, Forest, and Tree Species Classes with ZiYuan-3 Multispectral and Stereo Data. *Remote Sens.* **2019**, *11*, 164. [CrossRef]

22. Haralick, R.M. Statistical and structural approaches to texture. *Proc. IEEE* **1979**, *67*, 786–804. [CrossRef]

23. Gašparović, M.; Dobrinić, D. Comparative Assessment of Machine Learning Methods for Urban Vegetation Mapping Using Multitemporal Sentinel-1 Imagery. *Remote Sens.* **2020**, *12*, 1952. [CrossRef]

24. Li, D.; Ke, Y.; Gong, H.; Li, X. Object-Based Urban Tree Species Classification Using Bi-Temporal WorldView-2 and WorldView-3 Images. *Remote Sens.* **2015**, *7*, 16917–16937. [CrossRef]

25. Majid, I.A.; Latif, Z.A.; Adnan, N.A. Tree Species Classification Using WorldView-3 Data. In Proceedings of the IEEE 7th Control and System Graduate Research Colloquium, Shah Alam, Malaysia, 8 August 2016; pp. 73–76.

26. Dragčević, D.; Pavasović, M.; Bašić, T. Accuracy validation of official Croatian geoid solutions over the area of City of Zagreb. *Geofizika* **2016**, *33*, 183–206. [CrossRef]

27. Jurjević, L.; Gašparović, M.; Milas, A.S.; Balenović, I. Impact of UAS Image Orientation on Accuracy of Forest Inventory Attributes. *Remote Sens.* **2020**, *12*, 404. [CrossRef]

28. Vermote, E.F.; Tanre, D.; Deuze, J.L.; Herman, M.; Morcrette, J.J. Second simulation of the satellite signal in the solar spectrum, 6S: An overview. *IEEE Trans. Geosci. Remote Sens.* **1997**, *35*, 675–686. [CrossRef]

29. Gašparović, M.; Dobrinić, D.; Medak, D. Geometric accuracy improvement of WorldView-2 imagery using freely available DEM data. *Photogramm. Rec.* **2019**, *34*, 266–281. [CrossRef]

30. Gašparović, M.; Dobrinić, D.; Medak, D. Spatial accuracy analysis of aerial and satellite imagery of Zagreb. *Geodetski List.* **2018**, *72*, 1–14.

31. Breiman, L. Random Forests. *Mach. Learn.* **2001**, *45*, 5–32. [CrossRef]

32. Cutler, D.R.; Edwards, T.C.; Beard, K.H.; Cutler, A.; Hess, K.T.; Gibson, J.; Lawler, J.J. Random Forests For Classification In Ecology. *Ecology* **2007**, *88*, 2783–2792. [CrossRef]

33. Immitzer, M.; Vuolo, F.; Atzberger, C. First Experience with Sentinel-2 Data for Crop and Tree Species Classifications in Central Europe. *Remote Sens.* **2016**, *8*, 166. [CrossRef]

34. Huesca, M.; Roth, K.L.; García, M.; Ustin, S.L. Discrimination of Canopy Structural Types in the Sierra Nevada Mountains in Central California. *Remote Sens.* **2019**, *11*, 1100. [CrossRef]

35. Pal, M. Random forest classifier for remote sensing classification. *Int. J. Remote Sens.* **2005**, *26*, 217–222. [CrossRef]

36. Belgiu, M.; Drăguţ, L. Random forest in remote sensing: A review of applications and future directions. *ISPRS J. Photogramm. Remote Sens.* **2016**, *114*, 24–31. [CrossRef]

37. Cortes, C.; Vapnik, V. Support vector network. *Mach. Learn.* **1995**, *3*, 273–297. [CrossRef]

38. Van der Linden, S. Classifying segmented hyperspectral data from a heterogeneous urban environment using support vector machines. *J. Appl. Remote Sens.* **2007**, *1*, 013543. [CrossRef]

39. Brown, M.; Gunn, S.R.; Lewis, H.G. Support vector machines for optimal classification and spectral unmixing. *Ecol. Model.* **1999**, *120*, 167–179. [CrossRef]

40. Burges, C.J. A tutorial on support vector machines for pattern recognition. *Data Min. Knowl. Discov.* **1998**, *2*, 121–167. [CrossRef]

41. Adam, E.; Mutanga, O.; Odindi, J.; Abdel-Rahman, E.M. Land-use/cover classification in a heterogeneous coastal landscape using RapidEye imagery: Evaluating the performance of random forest and support vector machines classifiers. *Int. J. Remote Sens.* **2014**, *35*, 3440–3458. [CrossRef]

42. Hall-Beyer, M. *GLCM Texture: A Tutorial v. 3.0 March 2017*; University of Calgary: Calgary, Canada, 2017.

43. Johansen, K.; Phinn, S. Mapping Structural Parameters and Species Composition of Riparian Vegetation Using IKONOS and Landsat ETM+ Data in Australian Tropical Savannahs. *Photogramm. Eng. Remote Sens.* **2006**, *72*, 71–80. [CrossRef]

44. Mallinis, G.; Koutsias, N.; Tsakiri, M.; Karteris, M. Object-based classification using Quickbird imagery for delineating forest vegetation polygons in a Mediterranean test site. *ISPRS J. Photogramm. Remote Sens.* **2008**, *63*, 237–250. [CrossRef]

45. Pu, R.; Landry, S. A comparative analysis of high spatial resolution IKONOS and WorldView-2 imagery for mapping urban tree species. *Remote Sens. Environ.* **2012**, *124*, 516–533. [CrossRef]

46. Dorigo, W.; Lucieer, A.; Podobnikar, T.; Carni, A. Mapping invasive Fallopia japonica by combined spectral, spatial, and temporal analysis of digital orthophotos. *Int. J. Appl. Earth Obs. Geoinf.* **2012**, *19*, 185–195. [CrossRef]

47. Chen, D.; Stow, D.A.; Gong, P. Examining the effect of spatial resolution and texture window size on classification accuracy: An urban environment case. *Int. J. Remote Sens.* **2004**, *25*, 2177–2192. [CrossRef]

48. Snap Documentation. Available online: https://docs-snaplogic.atlassian.net/wiki/home (accessed on 7 May 2020).

49. Lee, J.; Cai, X.; Lellmann, J.; Dalponte, M.; Malhi, Y.; Butt, N.; Morecroft, M.; Schönlieb, C.B.; Coomes, D. Individual Tree Species Classification From Airborne Multisensor Imagery Using Robust PCA. *IEEE J. Sel. Top. Appl. Earth Obs. Remote Sens.* **2016**, *9*, 2554–2567. [CrossRef]

50. Deng, J.S.; Wang, K.; Deng, Y.H.; Qi, G.J. PCA-based land-use change detection and analysis using multitemporal and multisensor satellite data. *Int. J. Remote Sens.* **2008**, *29*, 4823–4838. [CrossRef]

51. Knauer, U.; von Rekowski, C.S.; Stecklina, M.; Krokotsch, T.; Pham Minh, T.; Hauffe, V.; Kilias, D.; Ehrhardt, I.; Sagischewski, H.; Chmara, S.; et al. Tree Species Classification Based on Hybrid Ensembles of a Convolutional Neural Network (CNN) and Random Forest Classifiers. *Remote Sens.* **2019**, *11*, 2788. [CrossRef]

52. Congalton, R.G. A review of assessing the accuracy of classifications of remotely sensed data. *Remote Sens. Environ.* **1991**, *37*, 35–46. [CrossRef]

53. Cohen, J. A Coefficient of Agreement for Nominal Scales. *Educ. Psychol. Meas.* **1960**, *20*, 37–46. [CrossRef]

54. Pontius, R.G.; Millones, M. Death to Kappa: Birth of quantity disagreement and allocation disagreement for accuracy assessment. *Int. J. Remote Sens.* **2011**, *32*, 4407–4429. [CrossRef]

55. Gebreslasie, M.T.; Ahmed, F.B.; van Aardt, J.A.N. Extracting structural attributes from IKONOS imagery for Eucalyptus plantation forests in KwaZulu-Natal, South Africa, using image texture analysis and artificial neural networks. *Int. J. Remote Sens.* **2011**, *32*, 7677–7701. [CrossRef]

56. Hall-Beyer, M. Practical guidelines for choosing GLCM textures to use in landscape classification tasks over a range of moderate spatial scales. *Int. J. Remote Sens.* **2017**, *38*, 1312–1338. [CrossRef]

57. Attarchi, S.; Gloaguen, R. Classifying Complex Mountainous Forests with L-Band SAR and Landsat Data Integration: A Comparison among Different Machine Learning Methods in the Hyrcanian Forest. *Remote Sens.* **2014**, *6*, 3624–3647. [CrossRef]

58. Wang, H.; Zhao, Y.; Pu, R.; Zhang, Z. Mapping Robinia Pseudoacacia Forest Health Conditions by Using Combined Spectral, Spatial, and Textural Information Extracted from IKONOS Imagery and Random Forest Classifier. *Remote Sens.* **2015**, *7*, 9020–9044. [CrossRef]

59. Wang, M.; Fei, X.; Zhang, Y.; Chen, Z.; Wang, X.; Tsou, J.Y.; Liu, D.; Lu, X. Assessing Texture Features to Classify Coastal Wetland Vegetation from High Spatial Resolution Imagery Using Completed Local Binary Patterns (CLBP). *Remote Sens.* **2018**, *10*, 778. [CrossRef]

60. Ozdemir, I.; Karnieli, A. Predicting forest structural parameters using the image texture derived from WorldView-2 multispectral imagery in a dryland forest, Israel. *Int. J. Appl. Earth Obs. Geoinf.* **2011**, *13*, 701–710. [CrossRef]

61. Spracklen, B.D.; Spracklen, D.V. Identifying European Old-Growth Forests using Remote Sensing: A Study in the Ukrainian Carpathians. *Forests* **2019**, *10*, 127. [CrossRef]

62. Shaban, M.A.; Dikshit, O. Improvement of classification in urban areas by the use of textural features: The case study of Lucknow city, Uttar Pradesh. *Int. J. Remote Sens.* **2001**, *22*, 565–593. [CrossRef]

63. Yang, G.; Zhao, Y.; Li, B.; Ma, Y.; Li, R.; Jing, J.; Dian, Y. Tree Species Classification by Employing Multiple Features Acquired from Integrated Sensors. *J. Sens.* **2019**, *2019*, 1–12. [CrossRef]

64. Sabat-Tomala, A.; Raczko, E.; Zagajewski, B. Comparison of Support Vector Machine and Random Forest Algorithms for Invasive and Expansive Species Classification Using Airborne Hyperspectral Data. *Remote Sens.* **2020**, *12*, 516. [CrossRef]

65. Dye, M.; Mutanga, O.; Ismail, R. Combining spectral and textural remote sensing variables using the random forest ensemble: Predicting the age of Pinus patula forests in KwaZulu-Natal, South Africa. *Spat. Sci.* **2012**, *57*, 197–215.

66. Zhang, L.; Liu, Z.; Ren, T.; Liu, D.; Ma, Z.; Tong, L.; Zhang, C.; Zhou, T.; Zhang, X.; Li, S. Identification of Seed Maize Fields With High Spatial Resolution and Multiple Spectral Remote Sensing Using Random Forest Classifier. *Remote Sens.* **2020**, *12*, 362. [CrossRef]

67. Kupidura, P. The Comparison of Different Methods of Texture Analysis for Their Efficacy for Land Use Classification in Satellite Imagery. *Remote Sens.* **2019**, *11*, 1233. [CrossRef]

68. Nezami, S.; Khoramshahi, E.; Nevalainen, O.; Pölönen, I.; Honkavaara, E. Tree Species Classification of Drone Hyperspectral and RGB Imagery with Deep Learning Convolutional Neural Networks. *Remote Sens.* **2020**, *12*, 1070. [CrossRef]

69. Miyoshi, G.T.; Arruda, M.S.; Osco, L.P.; Junior, J.M.; Gonçalves, D.N.; Imai, N.N.; Tommaselli, A.M.G.; Honkavaara, E.; Gonçalves, W.N. A Novel Deep Learning Method to Identify Single Tree Species in UAV-Based Hyperspectral Images. *Remote Sens.* **2020**, *12*, 1294. [CrossRef]

70. Schiefer, F.; Kattenborn, T.; Frick, A.; Frey, J.; Schall, P.; Koch, B.; Schmidtlein, S. Mapping forest tree species in high resolution UAV-based RGB-imagery by means of convolutional neural networks. *ISPRS J. Photogramm. Remote Sens.* **2020**, *170*, 205–215. [CrossRef]

71. Grinias, I.; Panagiotakis, C.; Tziritas, G. MRF-based Segmentation and Unsupervised Classification for Building and Road Detection in Peri-urban Areas of High-resolution. *ISPRS J. Photogramm. Remote Sens.* **2016**, *122*, 145–166. [CrossRef]

 *remote sensing* 

Article

Mapping of the Canopy Openings in Mixed Beech–Fir Forest at Sentinel-2 Subpixel Level Using UAV and Machine Learning Approach

Ivan Pilaš [1],*, Mateo Gašparović [2], Alan Novkinić [3] and Damir Klobučar [3]

1   Croatian Forest Research Institute, Division of Ecology, Cvjetno Naselje 41, 10450 Jastrebarsko, Croatia
2   Faculty of Geodesy, University of Zagreb, Kačićeva 26, 10000 Zagreb, Croatia; mateo.gasparovic@geof.unizg.hr
3   Croatian Forests Ltd., Ivana Meštrovića 28, 48000 Koprivnica, Croatia; alan.novkinic@hrsume.hr (A.N.); damir.klobucar@hrsume.hr (D.K.)
*   Correspondence: ivanp@sumins.hr; Tel.: +385-98-326-714

Received: 30 October 2020; Accepted: 27 November 2020; Published: 30 November 2020

Abstract: The presented study demonstrates a bi-sensor approach suitable for rapid and precise up-to-date mapping of forest canopy gaps for the larger spatial extent. The approach makes use of Unmanned Aerial Vehicle (UAV) red, green and blue (RGB) images on smaller areas for highly precise forest canopy mask creation. Sentinel-2 was used as a scaling platform for transferring information from the UAV to a wider spatial extent. Various approaches to an improvement in the predictive performance were examined: (I) the highest R^2 of the single satellite index was 0.57, (II) the highest R^2 using multiple features obtained from the single-date, S-2 image was 0.624, and (III) the highest R^2 on the multitemporal set of S-2 images was 0.697. Satellite indices such as Atmospherically Resistant Vegetation Index (ARVI), Infrared Percentage Vegetation Index (IPVI), Normalized Difference Index (NDI45), Pigment-Specific Simple Ratio Index (PSSRa), Modified Chlorophyll Absorption Ratio Index (MCARI), Color Index (CI), Redness Index (RI), and Normalized Difference Turbidity Index (NDTI) were the dominant predictors in most of the Machine Learning (ML) algorithms. The more complex ML algorithms such as the Support Vector Machines (SVM), Random Forest (RF), Stochastic Gradient Boosting (GBM), Extreme Gradient Boosting (XGBoost), and Catboost that provided the best performance on the training set exhibited weaker generalization capabilities. Therefore, a simpler and more robust Elastic Net (ENET) algorithm was chosen for the final map creation.

Keywords: Sentinel-2; UAV; DJI drone; machine learning; forest canopy; canopy gaps; canopy openings percentage; satellite indices; Elastic Net; beech–fir forests

1. Introduction

Forest canopy gaps, as a consequence of management activities or natural disturbances, are the main drivers that affect forest dynamics in most continuous-cover, close-to-nature silvicultural systems. Gap-based interventions to create a more open-stand structure represent common silvicultural practices for regeneration, as well as forest exploitation in the mixed silver fir (*Abies alba* Mill.) and European beech (*Fagus sylvatica* L.) forests in Central Europe. The size and position of crown openings determine the amount of direct light able to penetrate to the forest floor, thus creating conditions that could favor the shade-tolerant silver fir or the regeneration of the more competitive European beech in conditions of supplementary light availability [1]. The creation of treefall gaps accelerates tree regeneration and increases tree species diversity in mixed beech–fir forests [2]. The creation of canopy openings through natural tree mortality processes, wind-driven gap formation, or manmade activities also has a significant influence on changes in microclimatic conditions and below-ground processes. Canopy openings have an impact on increases in air and soil temperatures and decreases in soil moisture in larger gaps [3]. Humification processes prevail in small gaps, i.e., an increase in soil organic content, while, in large gaps, the mineralization processes related to decreases in organic content are more dominant [4,5]. The questions of size, shape, and topographic position of forest openings have been extensively studied [6–13]. They have been done so as to emulate the natural disturbance processes and to discover the most appropriate silvicultural operations that would imitate natural regeneration, particularly with the aim of conversion of various monocultures toward close-to-nature forests [14].

The estimation of forest canopy gaps or openings is commonly obtained from measures of canopy cover that can be defined as a percentage of forest area occupied by the vertical projection of tree crowns, unlike the canopy closure that has an analogous meaning but is represented by the proportion of sky hemisphere obscured by vegetation when viewed from a single point. There are three different groups of approaches [15] for measuring or estimating the crown canopy density in a forest: (I) ground measurement at the study area, (II) statistical approaches, if the information such as basal area or diameter at breast height and the number of trees per area is available, and (III) remote sensing data such as aerial photographs, satellite data, or laser scanner data. When making ground measurements, devices such as densiometers and hemispherical digital photographs are used or they are based on pure ocular assessment [16,17]. Ground measurement methods, as the most accurate techniques, are labor-intensive and cannot rapidly provide canopy cover estimates [17]. Recently, remote sensing methods, in particular light detection and ranging (LiDAR), have been proven as the most advantageous approaches in terms of accuracy and ease of data acquisition [18–21]. UAV (unmanned aerial vehicle)-based LiDAR, despite observed shortcomings such as aircraft instability, is lower in cost, more convenient in terms of operation, and has a more flexible flight route design, as well as unique advantages in the possibilities of better discernment of the stand canopy details than airborne LiDAR [19]. Satellite remote sensing, medium-resolution data from Landsat are traditionally a very appropriate source of information for crown cover estimation, most often comprising the supervised classification approaches based on ground truth validation sets obtained from field measurements, airborne or UAV aero-photo imagery, and LiDAR or very high-resolution satellite imagery [22–28]. The main drawback of medium-resolution satellite data such as Landsat 8 is that its spatial resolution of 30 m is too coarse for precise forest gap detection. One of the most prominent examples of canopy closure estimation worth mentioning is tree cover density product in 20 m resolution, developed for the Pan-European scale in 2012, 2015, and 2018, as part of the Copernicus Land Monitoring Service [29].

By launching a pair of Sentinel-2 satellites in 2015 and 2017, the European Space Agency (ESA) made a considerable breakthrough in improving the ability of the global monitoring of conditions and changes in forest cover primarily due to cost-free data availability, as well as highly advanced sensor characteristics: very good spectral properties (13 bands from 492.1 to 2185.7 nm, central wavelengths), radiometric resolution (12 bit; a potential range of brightness levels from 0–4095), spatial resolution (10 m for read, green, blue and near infra red bands), and revisit time (5 days under cloud-free conditions) [30]. These characteristics enable improved monitoring of the conditions and changes in the land cover and its dynamics on a global scale. Sentinel 2 Multispectral Instrument (MSI) with a 10 m spatial resolution shows some improvement in the possibility of detection of selective logging gaps in relation to Landsat 8 [31–33]. On the other hand, the possibilities of the employment of Sentinel 2 in operational forestry to its full performance have not been fully recognized and valorized up to the present. Apparently, a major disadvantage of the Sentinel 2 MSI sensor, which limits its greater use in forestry, is its 10 m spatial resolution which prevents exact geolocation and delineation of individual trees, which is important for many silvicultural practices; for this reason, other more precise survey methods such as LiDAR, UAV images, and very high resolution satellite imagery are often more attractive and competitive. Finer spatial resolution is especially important in the management of uneven-aged mixed forests where silvicultural interventions take place at the level of an individual tree or a smaller group of trees, and we rarely find larger interrupted areas (except in the case of natural disasters such as windbreaks or ice breaks) that can be clearly identified on satellite Sentinel 2 or Landsat 8 images [34].

For the successful implementation of remote sensing products in operational forestry, several prerequisites need to be met, related to their compliance on the scale of which the various forest management planning activities are carried out. Given the temporal scale, we can usually distinguish medium-term forest management planning, usually on a decadal time scale, on the basis of the conducted forest inventory. In addition, there is more operational, annual planning that requires more current, timely information from the forests according to which operational silvicultural decisions are implemented in the short term. In the case of continuous-cover mixed forests, it is of great importance to mark locations within the forest stand with an overly dense canopy structure where a selective cut for regeneration purposes or thinning should be carried out. Further information on the location of newly created forest gaps, after a natural disaster, windbreak, ice breakage, or biotic damages, can be used in the planning of the restoration of these areas, construction of the logging trails, calculation of the man-hours of the forest workers and machines, etc. Regarding the spatial scale, the extent of the supporting data layer should comply with the extent of the forest area that is considered in the operational planning, usually on the scale of the forest managerial unit. In Croatia, with the usual size being between 3000 and 6000 hectares, the forest managerial units (FMUs) are the lowest organizational tiers, usually consisting of forest areas with similar characteristics, for which forest management plans (every 10 years) and yearly operational plans that more precisely define required silvicultural, forest protection, timber extraction, and other related activities are conducted. Given the above prerequisites (timeliness, spatial extent, precision), it is unlikely that a single remote sensing source alone is suitable for providing information about the actual state of the forest canopy and the location of the canopy openings, i.e., stand density, to be suitable for cost-efficient, operative applications in forestry. Airborne LiDAR and aerial photos (visible and multispectral) can provide high-precision information about the crown surface; however, they are relatively expensive surveys and are more often used for cyclical inventories (every 10 years). UAV presents a more practical solution to an airborne survey; however, a large spatial extent (3000–6000 hectares) of the FMU often represents a limitation for its operational use where the maximal daily UAV survey limit is up to 400 hectares. Very-high-resolution satellite data fulfill the requirements of spatial extent and timeliness but can be very expensive for everyday operational use. On the other hand, the cost-free Sentinel 2 data have the constraints of insufficient resolution for precise forest canopy gap detection.

This study aims to evaluate a novel, cost-efficient approach of precise mapping of the current, actual state of the forest canopy openings percentage (COP) suitable for the silvicultural decision-making process in forest management on an intra-annual operational scale. It is based on the bi sensor approach; it takes advantage of Sentinel 2 as an auxiliary source of information, very suitable for large-scale mapping, and it derives a high-precision forest mask from the UAV imagery as ground truth, obtained on predefined training/validation locations inside FMUs consisting of the mixed beech–fir forest. This approach of Sentinel-2 and UAV integration has been recently broadly implemented in various domains: in forestry to estimate the aboveground biomass of Mangrove plantations [35], in agriculture to estimate rice crop damages [36], and for water quality monitoring [37]. This study aims to improve the spatial precision of the Sentinel-2, using its enhanced spectral properties that are suitable for capturing and quantifying the very fine shifts in the forest canopy cover at a subpixel level. This is provided by using a standard machine learning framework that consists of the features engineering, training and validation set creation, machine learning algorithm examination and validation by fine-tuning, selection of the final model, and final map creation. In this predictive modeling framework, the specific objectives of the research that are closely related to the improvement of the predictive capabilities and performance of the models emerge: (I) the suitability of a single-source Sentinel-2 image versus the multitemporal set of Sentinel-2 images, closest as possible to the date of UAV image acquisition, (II) the predictive power of the various satellite indices, i.e., spectral features, and (III) the most suitable machine learning algorithm among existing ones, having in mind the predictive performance and overfitting.

2. Materials and Methods

2.1. Study Area

The study area was located in the northeastern, continental part of the Dinaric Alps, a massif in southern and southeastern Europe which stretches for 645 km and separates the continental area of the Balkan Peninsula from the Adriatic Sea. The exact location (Figure 1) was near the town of Vrbovsko, Gorski Kotar region, Croatia (45.4282 latitude and 15.0714 longitude). The research covered the Litorić forest managerial unit (FMU), with an area of 3181.82 hectares, which administratively belongs to the Vrbovsko Forest Office, Forest District Delnice, as an integral part of the state-owned forests managed by the Croatian Forests state company. The FMU Litorić is located at an altitude range of 400 to 900 m and it has 91 compartments. The total wood stock is 965,636 m^3 (313 m^3/ha), and the total annual increment is 202,015 m^3 (6.5 m^3/ha). FMU Litorić consists of 35.45% common beech with 342,297 m^3 of wood stock and 50.11% common fir with stocks of 483,894 m^3. In terms of other species, there are also 8.78% mountain maple, 2.88% spruce, and other hardwoods. According to vegetation typology, the research area is part of a very broad range of European mountain beech forests, a subgroup of Illyrian mountainous beech forests [38] of the Dinaric Alps, where the common beech occurs in combination with the common fir and forms special mixed vegetation forests with high biodiversity and economic value. The lithological base is predominantly composed of limestone and dolomite with associated brown soils.

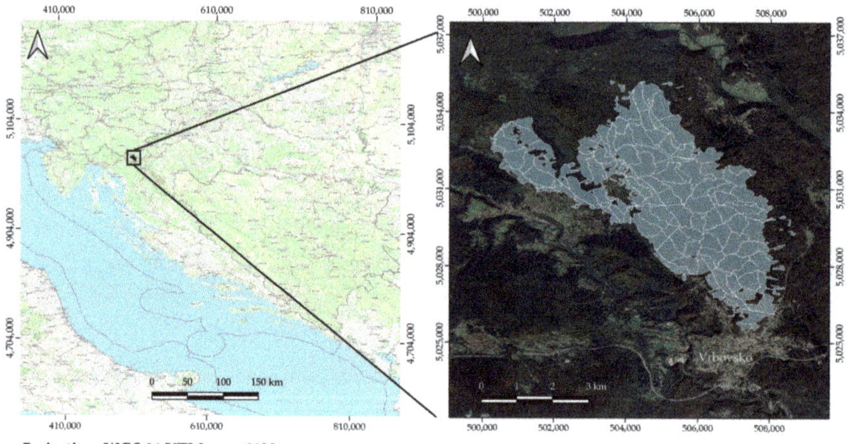

Figure 1. The location of the study area, forest management unit (FMU) "Litorić".

From the perspective of the prevailing end-users in the area, forest managers in the past decade have mostly been concerned with increases in biotic and abiotic damage. Forests in FMU have been influenced by a prolonged outbreak of spruce bark beetle (*Pityokteines curvidens*), which causes a weakening of the vitality of trees and forest stands with the occurrence of occasional individual and group dieback of trees. Additionally, most likely as a consequence of climate change, very fierce storms with strong winds have appeared in the last few years causing windbreaks and windthrows. The storm that took place at the beginning of November 2017, followed by a few strong wind events in 2018, had a particularly strong impact in terms of the spatial extent of forest damage. During these incidents, mostly fir trees were damaged due to the resistance provided by their canopy, while beech trees were damaged to a much lesser extent. Considering that classical terrestrial methods of observation could not accurately determine the spatial effect of wind damage, subsequent log hauling, and surface remediation for the scale of the whole FMU, the COP estimation method was implemented by combined remote sensing means and the methodology presented in this study.

2.2. UAV Survey of the Training and the Testing Area

A detailed survey of the parts of FMU Litorić using UAV was performed at the end of July 2018. The exact time of the survey, the end of July 2018, was selected due to the requirements of the forestry service. Since forest restoration activities at that time were carried out over the entire FMU (logging of remaining fallen trees after windthrows during the winter period and afforestation of the clearings), the interest of the forestry service was focused on obtaining information on the spatial distribution of gaps throughout the FMU to perform the aforementioned silvicultural activities. For this purpose, a UAV survey was first performed at two independent locations, used in this study, where the occurrence of larger windthrows was detected. No particular sampling strategy or design was taken into account when selecting these two sites, with only some preliminary knowledge about forest gap occurrence. After that, Sentinel 2 imagery was preliminarily used, which led to the idea of developing a new methodological approach that would combine both sensing methods, which is presented in this paper. For the survey, a forest service-owned UAV was used with a simple RGB camera, which is otherwise more often used only for visual observation purposes.

One UAV survey area, with a size of 470 hectares, was selected as a training area, and the other, with a size of 310 hectares, was selected as a testing area for selected modeling algorithms (Figure 2). The images in the visible RGB channel were taken using a DJI INSPIRE 2 drone with a ZENMUSE X5S camera with gyroscopic stabilization at a height of 300 m from the ground. All recorded material was transformed in the WGS84 coordinate system. Esri Drone2Map software was used for photo processing and digital orthophoto creation, while cartographic processing was done in the ArcMap 10.6.1. program. The final product presents ortho-rectified high-precision images of the parts of FMU with a spatial grid resolution of 7 cm, through which highly precise canopy boundaries could be delineated and masked out in further processing.

2.3. Satellite Image Processing and Index Extraction

For this study, 4 Sentinel-2 Level 2A (S2_L2A) cloud-free, bottom-of-atmosphere reflectance products were used which consisted of 100 km by 100 km squared ortho-images in UTM/WGS84 projection. Cloud-free imagery, temporarily consistent as much as possible with performed UAV surveys (within an approximately 2 month period), was selected and retrieved from the Copernicus Open Access Hub [38]. Furthermore, all four S2_L2A images had an almost identical time of acquisition and relative orbit position (Table 1). For the purpose of analysis, no additional image enhancement processing of S2_L2A was performed. After the acquisition, due to the differences in pixel resolution, all S2_L2A bands were resampled to a 10 m extent using default resampling processor parameters in the Sentinel Application Platform (SNAP), open-source software provided by the European Space Agency (ESA). Prior to further analysis, the collocation of images was performed, i.e., pixel values of S2_L2A images were resampled into the geographical raster of the master image from 25 July 2018

Table 1. Details of Sentinel 2 Level 2A products used in this study.

Platform ID	Year	Month	Day	Time of Acquisition	Relative Orbit Number	Tile Number Field
S2A	2018	7	25	10:00 a.m.	R122	T33TVL
S2B	2018	8	1	10:00 a.m.	R122	T33TVL
S2B	2018	8	29	10:00 a.m.	R122	T33TVL
S2B	2018	9	28	10:00 a.m.	R122	T33TVL

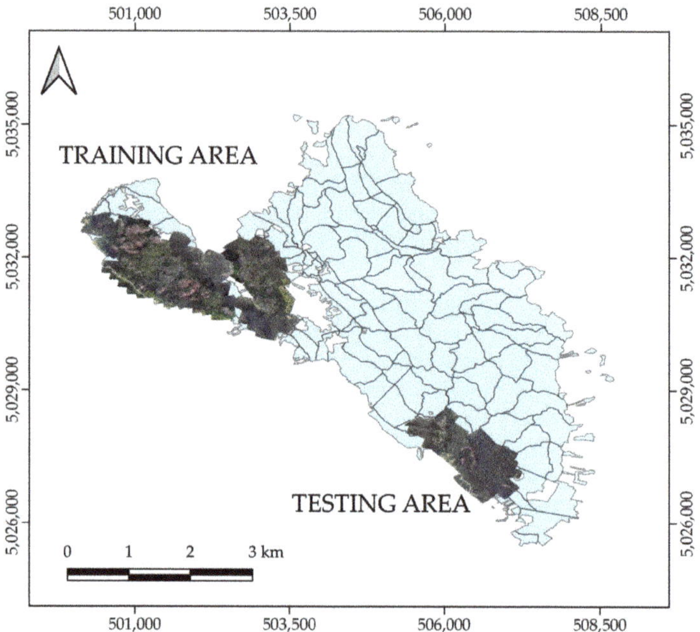

Figure 2. The locations of training and testing areas.

The Sentinel 2 multispectral instrument (MSI) has 13 spectral channels in the visible/near-infrared (VNIR) and short-wave infrared spectral range (SWIR) that facilitate the possibility of the construction of a wide range of spectral indices. The development of satellite platforms and sensors over the past decades has also been accompanied by an intensive contrivance of various satellite indices aimed at better discrimination of land surface categories and objects [39]. The fundamental principle of the differentiation of vegetation cover on multispectral images is based on the ratio between the red part of the spectrum that is absorbed by the chlorophyll in the leaves and the infrared part of the spectrum that is reflected by vegetation. Vegetation indices were developed with the intention of clearer quantification of the intensity of vegetation activity with the best possible reduction of accompanying noise. Noise that exists in satellite images and derived indices is usually a consequence of topography, reflection from bare soil, soil color, atmosphere, spectral characteristics of sensors, view, and solar zenith angels [40].

Depending on the approach of noise separation, especially soil line differentiation, vegetation indices are further divided into slope-based, distance-based, and orthogonal [41]. In contrast to a wide range of vegetation indices, soil radiometric indices were primarily developed with the aim of more complete quantification of soil optical properties. Primarily, this refers to the determination of different degrees of soil brightness as a result of viewing geometry, the surface roughness, the organic matter, water contents, and spectral features occurring in the visible domain, i.e., soil color [42]. The more intrinsic insight into the structure and photosynthetic processes of vegetation provide biophysical variables such as Leaf Area Index (LAI), Fraction of Absorbed Photosynthetically Active Radiation (FAPAR), Fraction of Vegetation Cover (FVC), Chlorophyll Content in the Leaf (CAB), and Canopy Water Content (CWC) that present a proxy for the estimation of canopy cover, photosynthetic dynamics, leaf water content can be effectively derived from Sentinel 2 bands [43,44]. A special group of features, developed from digital image analysis, is based on the detection and quantification of gray textures and tones in the images. This can significantly improve the separation of objects and structures in images and is also very suitable for use in various machine learning methods [45].

Considering very heterogeneous surface conditions in the study area of FMU Litorić with various possible spectral feedbacks (areas of undisturbed mixed species canopy, fragments of bare soil surface after logging, open forest gaps with understory shroud of plants and shrubs, occasional groups of sprawled trees, etc.), it was necessarily to use a broader variety of spectral indices. Furthermore, the statistical and machine learning methods can quite successfully isolate the contribution of individual input features on the final predictive outcome. Therefore, various groups of spectral indices such as vegetation indices, soil reflection indices, biophysical variables, and textural properties (gray-level co-occurrence matrix) were used in this study (Table 2). The selection of the applied indices depended on the predefined catalog of indices that can be automatically retrieved through an in-built S-2 index processor in SNAP software. For each of the considered satellite images, (S2_20180725, S2_20180801, S2_20180829, S2_20180928) a total of 155 predictors were calculated: four soil radiometric indices, 21 vegetation radiometric indices, five water radiometric indices, five biophysical indices, and 10 gray-level co-occurrence matrix (GLCM) parameters for each of the S-2 bands (B1-B12), i.e., 120 GLCM layers per S-2 image.

2.4. Integration of Sentinel 2 and UAV Data

One of the fundamental challenges in this research was the integration of satellite S2_L2A and photogrammetric UAV images on the same spatial scale. The S2_L2A multispectral product consists of bands with a spatial resolution of 10 m (band 2, band 3, band 4, and band 8), 20 m (band 5, band 6, band 7, band 11, and band 12), and 60 m (band 1, band 9 and band 10). Before further processing, all bands were rescaled to a 10 m resolution which is a routine procedure in software such as SNAP. All S2_L2A multitemporal products were also collocated on the same spatial grid. The spatial resolution of the UAV images was approximately 7 cm, i.e., 143 times finer than the S2_L2A such that one Sentinel pixel of 10 × 10m contains 20,449 UAV pixels. Although there are available analytical tools that allow rescaling raster images to finer spatial resolution, such processing methods, due to extremely large differences in spatial resolutions of the imagery, were not suitable because they require high-performance computing capabilities and long-term computer processing time. Moreover, the use of R analytical tools in this research, with exceptional capabilities for statistical or machine learning processing, required rearrangement of input spatial data in the standardized R form (a form of matrix or data frame). In that sense, preparation and aggregation of data followed the principles of "tidy" data standardization described by [46], who defines tidy data as follows: each variable forms a column, each observation forms a row, and each type of observational unit forms a table.

Before data aggregation, canopy boundaries were extracted from RGB UAV bands, and a canopy mask layer was created. Acquired aerial imagery was processed similarly to the methodology presented in the research [47]. A generated high-resolution digital orthophoto was used to create various spectral indices that were tested for fast and accurate canopy mask layer creation. In accordance with [48], the best spectral indices were chosen. Although many indices were tested, such as Brightness Index (BI) [49], Redness Index (RI) [50], Color Index (CI) [51], Green Leaf Index (GLI) [52], Normalized Green Red Difference Index (NGRDI) [53], and Visual Atmospheric Resistance Index (VARI) [54], just three of them were selected. NGRDI, VARI, and GLI. Specifically, the mentioned indices mostly distinguish the healthy forest from bare land, windthrows, or canopy openings. The high-resolution UAV-based canopy mask layer was created using the R program language.

The percentage of canopy density for every S2_L2A pixel overlaying the UAV imagery was made by first producing an S2_L2A fishnet grid (regular polygon grid without attributes) with a cell size of 10 × 10 m. Since all S2_L2A images were collocated and resampled, the derived fishnet grid was used on all multitemporal imagery sets. The fishnet grid was then overlapped with regular points i.e., 10 × 10 points (100 points per S2_L2A grid cell). A sampling of the UAV images was performed according to binary outcomes such that the values of the mask layer (forest canopy) were marked as 0 while the areas outside the canopy (forest gaps) were marked as 1. The percentage (COP) for each of the S2_L2A pixels was obtained from the ratio of the sum of 1 (outside the canopy mask) and the total number of points (100) in each grid cell. Then, for each S2_L2A polygon in the fishnet, a centroid was calculated to which the COP value was assigned as a percentage such that each of the pixels on the satellite image was represented with only one point and the corresponding COP value. Further preparation included a random sampling of 10% points on the S2_L2A polygon grid to minimize the effect of spatial correlation between the sampling points on the images. At each selected random centroid or point, sampling of derived satellite indices was performed on multitemporal imagery. After preparation and sampling, all data were compiled into data frames (a training data frame with 4653 observations and 155 predictor variables for a single S2_L2A image and 620 predictors for the entire multitemporal set). The test set consisted of an equal number of input variables and 3596 observations.

Table 2. Derived satellite indices and textural features.

Satellite Index	Spectral Bands	Sentinel 2 Bands
Soil radiometric indices		
BI—Brightness Index	red, green	B4, B3
BI2—The Second Brightness Index	red, green, NIR(near infrared)	B4, B3, B8
RI—Redness Index	red, green	B4, B3
CI—Color Index	red, green	B4, B3
Vegetation radiometric indices		
SAVI—Soil-Adjusted Vegetation Index	red, NIR	B4, B8
NDVI—Normalized Difference Vegetation Index	red, NIR	B4, B8
TSAVI—Transformed Soil-Adjusted Vegetation Index	red, NIR	B4, B8
MSAVI—Modified Soil-Adjusted Vegetation Index	red, NIR	B4, B8
MSAVI2—The Second Modified Soil-Adjusted Vegetation Index	red, NIR	B4, B8
DVI—Difference Vegetation Index	red, NIR	B4, B8
RVI—Ratio Vegetation Index	red, NIR	B4, B8
PVI—Perpendicular Vegetation Index	red, NIR	B4, B8
IPVI—Infrared Percentage Vegetation Index	red, NIR	B4, B8
WDVI—Weighted Difference Vegetation Index	red, NIR	B4, B8
TNDVI—Transformed Normalized Difference Vegetation Index	red, NIR	B4, B8
GNDVI—Green Normalized Difference Vegetation Index	green, NIR	B3, B7
GEMI—Global Environmental Monitoring Index	red, NIR	B4, B8A
ARVI—Atmospherically Resistant Vegetation Index	red, blue, NIR	B4, B2, B8
NDI45—Normalized Difference Index	red, red edge	B4, B5
MTCI—Meris Terrestrial Chlorophyll Index	red, red edge, NIR	B4, B5, B6
MCARI—Modified Chlorophyll Absorption Ratio Index	red, red edge, green	B4, B5, B3
REIP—Red-Edge Inflection Point Index	red, red edge, red edge, NIR	B4, B5, B6, B7
S2REP—Red-Edge Position Index	red, red edge, red edge, NIR	B4, B5, B6, B7
IRECI—Inverted Red-Edge Chlorophyll Index	red, red edge, red edge, NIR	B4, B5, B6, B7
PSSRa—Pigment-Specific Simple Ratio Index	red, NIR	B4, B7
Water Radiometric Indices		
NDWI—Normalized Difference Water Index	NIR, MIR(mid-infrared)	B8, B12
NDWI2—Second Normalized Difference Water Index	green, NIR	B3, B8
MNDWI—Modified Normalized Difference Water Index	green, MIR	B3, B12
NDPI—Normalized Difference Pond Index	green, MIR	B3, B12
NDTI—Normalized Difference Turbidity Index	red, green	B4, B3
Biophysical indices		
LAI—Leaf Area Index		B3, B4, B5, B6, B7, B8a, B11,
FAPAR—Fraction of Absorbed Photosynthetically Active Radiation		B12, cos(viewing_zenith),
FVC—Fraction of Vegetation Cover		cos(sun_zenith),
CAB—Chlorophyll Content in the Leaf		cos(sun_zenith),
CWC—Canopy Water Content		cos(relative_azimuth_angle)
Texture (Gray-Level Co-occurrence Matrix, GLCM)		
Contrast		
Dissimilarity		
Homogeneity		
Angular Second Moment		
Energy		
Maximum Probability		
Entropy		
GLCM Mean		
GLCM Variance		
GLCM Correlation		

2.5. Model Building and Validation

Various statistical and Machine Learning (ML) algorithms of different levels of complexity were examined for the prediction and mapping of the COP. The construction of the models was performed on the training set of observations, retrieved from the selected UAV training area. The performance of different algorithms was examined on the training set across the range of fine-tuning parameters using the 10-fold cross-validation (CV) resampling method and standard metrics that are used in regression analysis: root-mean-square error (RMSE) and the coefficient of determination (R^2). The 10-fold cross-validation technique that was used concerning other resampling methods gives the best estimate of the actual RMSE and the most favorable trade-off between the bias and variance of the model [55]. It is also one of the fastest resampling techniques, which is important for the optimization of the computational processing time of the training of some more complex algorithms. All models were also validated on the test set from the physically separate UAV survey area to examine the effect of the spatial extrapolation, using the selected model out of the training area. The procedure of the model construction was repeated twice: the first time using only predictors from the single-date S2_L2A image from 25 July 2018 and the second time using a multitemporal set of the four S2_L2A images, with close dates of acquisition (25 July 2018, 1 August 2018, 29 August 2018, and 28 September 2018). As previously mentioned, the response variable is the COP percentage on the interval quantitative scale (full canopy = 0%; open ground = 100%). All computation was performed in the generic R modeling package Caret, which has the possibility of preprocessing input data, training, testing, and precise adjustment of a large number of statistical and machine learning algorithms [56]. In the R Caret modeling interface, a suitable R package was additionally installed for each of the considered algorithms: *stats* package for ordinary least squares, *pls* package for partial least squares, *elasticnet* package for ridge regression and Elastic Net, *nnet* package for neural networks, *e1071* package for support vector machines, *randomForest* package for random forest, *gbm* package for stochastic gradient boosting, *xgboost* for extreme gradient boosting, and *catboost* package for CatBoost algorithm. Prior to starting the modeling process, the relationship between COP and individual S2_L2A indices was assessed. Dependency was firstly visually examined on the scatterplots with locally estimated scatterplot smoothing (LOESS) and afterward on simple regression metrics between the COP and the spectral indices. This was performed only on indices from a single S2_L2A 25 July 2018 image which was closest to the date of acquisition of the UAV aero-photo imagery.

2.5.1. Ordinary Least Squares (OLS) Linear Regression

The OLS linear regression method was used as a benchmark model, the results of which served for comparison with other more complex algorithms. Before OLS construction, training data were preprocessed using centering and scaling of predictor variables (zero mean and standard deviation of one). Improvements in the fit by decorrelation and reduction of predictors set using principal component analysis (PCA) and fitting of the PCA components in the OLS model were also examined. OLS, in short, minimizes the sum of squared errors (SSE) between the observed and predicted response, where the parameter estimates that minimize SSE are those with the least bias. OLS is a very attractive technique with regard to the interpretability of the coefficients but has a drawback in that it is sensitive to the collinearity among predictors and is not suitable when the data have a curvature or nonlinear structure.

2.5.2. Partial Least Squares (PLS)

The PLS method is a very suitable technique in the case of high correlation among predictors where OLS is unstable and not appropriate. The principle behind PLS is that it finds the linear combinations of predictors, called components or latent variables, that maximally summarize the variation in the predictor space. Additionally, the determined components have a maximum correlation with the response. PLS has one tuning parameter, i.e., the number of components to retain.

2.5.3. Ridge Regression (RR)

Ridge regression is a linear regularization method suitable where there are issues with collinearity and in cases of overfitting of the data. It belongs to a group of shrinkage methods that add penalties or shrink the parameter estimates which do not contribute significantly to the reduction of the SSE. In the bias–variance trade-off, ridge regression reduces the mean square error by slightly increasing the bias by adding appropriate penalties. The penalty presents a tuning parameter that was optimized by cross-validation during the model training process.

2.5.4. Elastic Net (ENET)

Elastic Net is a linear regularization technique that also effectively deals with highly correlated predictors. It combines two different penalties of the ridge regression and the lasso model. It is an advancement of the ridge regression that also makes possible variable selection that is a characteristic of the lasso model. In the training framework, it is tuned over the various penalty values via cross-validation.

2.5.5. Feed-Forward Neural Networks (NNET)

Neural networks are nonlinear techniques which use hidden variables or hidden units that present linear combinations of the original predictors. Predictors are transformed by a nonlinear function (logistic, sigmoidal) and each unit is related to the outcome. They are characterized by a large number of estimation parameters which are quite uninterpretable. Due to the large number of regression coefficients, neural networks have a tendency of overfitting. The weight decay is a commonly used penalization method that moderates the overfitting problem. In this study, the simplest neural network architecture, a single-layer feed-forward network was used. The tuning parameters used for cross-validation resampling are the number of hidden units and the amount of weight decay.

2.5.6. Support Vector Machine (SVM)

SVM is a nonlinear regression and classification technique that relies on data points (support vectors) above a certain threshold with high residuals. Observations with residuals within the threshold do not contribute to the regression fit, i.e., samples that the model fits well have no effect on the regression equation. This makes SVM insensitive to the outliers, which is very powerful and robust in nature. SVM uses various kernel functions (linear, polynomial, radial basis) that have the ability to model a nonlinear relationship. SVM tuning parameters are the cost parameter that adjusts the complexity of the model (large cost = more flexible model; small cost = "stiffened" model), the size of the threshold for the selection of the vectors, and the kernel parameter.

2.5.7. Random Forest (RF)

Random forest is a tree-based regression and classification ensemble technique that significantly improves the performance of tree-based methods regarding the reduction in variance and bias. It uses a random subset of the predictors at each tree split, which reduces between-tree correlation. The final prediction comprises an average prediction of the models in the ensemble. Models in the ensemble form independent "strong learners" that yield an improvement in error rates. The RF tuning parameter is the number of randomly selected predictors to choose from at each split.

2.5.8. Boosting (GBM, XGBoost, Catboost)

Boosting presents a group of powerful prediction tools, usually outperforming any individual model. The gradient boosting approach relies on the single tree as a base learner and the addition of the "weak learners" which are gradually included in the residuals. The current model is added to the previous model, and the procedure continues for a user-specified number of iterations. When the regression tree is used as the base learner, simple gradient boosting for regression has two tuning parameters: tree depth and the number of iterations. Boosting techniques are currently very popular and are prone to constant improvements. The classical approach presents a Stochastic Gradient Boosting (GBM) algorithm that uses a random sampling scheme. Recently, improved approaches were developed such as XGBoost (Extreme Gradient Boosting) and CatBoost. The advantage of XGBoost over traditional GBM is in its regularization capabilities and more effective tree pruning. However, the new CatBoost model, with improved capabilities of handling categorical variables and overfitting, also shows a significant advantage in the processing speed concerning other boosting methods.

3. Results

3.1. Relationship of UAV Canopy Openings Percentage (COP) and Individual Sentinel-2 Indices

In Figure 3, which shows scatterplots of selected examples, there is clear evidence of an almost linear negative relationship between COP and the intensity of presented vegetation radiometric indices (NDVI, ARVI, IPVI). Similarly, biophysical indices (FAPAR, LAI) also show a negative relationship in which the nonlinearity between variables is somewhat more pronounced. Soil radiometric indices such as CI and RI show a positive relationship to COP, while the relationship between COP and textural features varies to a certain extent from parameter to parameter.

Figure 4 shows S-2 indices sorted by the coefficient of determination (R^2) from the single regression analysis with COP. The best linear relationship with COP ($R^2 = 0.57$) was determined for the ARVI index (Atmospherically Resistant Vegetation Index), followed by the NDVI (Normalized Difference Vegetation Index; $R^2 = 0.56$), and NDVI-derived indices such as IPVI (Infrared Percentage Vegetation Index; $R^2 = 0.56$), TNDVI (Transformed Normalized Difference Vegetation Index; $R^2 = 0.56$), and NDTI ($R^2 = 0.54$). From the so-called soil group of indices, the most significant was CI (Color Index; $R^2 = 0.54$), and, from the water indices, the most significant was NDTI (Normalized Difference Turbidity Index; $R^2 = 0.54$). Derived biophysical indices such as FAPAR (Fraction of Adsorbed Photosynthetically Active Radiation) and LAI (Leaf Area Index) showed a slightly weaker linear association with the dependent variable (both $R^2 = 0.44$). Of the textural features, the most significant relationship was found in the GLCM (gray-level co-occurrence matrix) variance and mean in the red B4 channel ($R^2 = 0.37$, $R^2 = 0.35$).

3.2. Training and Validation of Predictive Models

The results of the model training process of COP and indices from single S-2 image from 25 July 2018 using 10-fold CV are presented in Table 3. It shows the root-mean-square error (RMSE), coefficient of determination (R^2), mean absolute error (MAE), and final tuning parameters for each selected model obtained by resampling technique. Compared to simple regression results, the inclusion of multiple predictors from a single S-2 image led to an improvement in prediction, but not to a larger extent. The basic OLS model clearly overfitted the data as is noticeable from the decrease in R^2 (from 0.603 to 0.571) when using PCA preprocessing. The addition of other algorithms such as PLS, RR, ENET, NNET, SVM, RF, GBM, XGBoost, and Catboost only slightly improved the prediction results. Complex ensemble models such as XGBoost showed superior results on the training set ($R^2 = 0.624$, RMSE = 15.148, MAE = 11.070). On the other hand, the highly complex NNET algorithm performed below average ($R^2 = 0.591$, RMSE = 15.864, MAE = 11.531).

Model building on the multitemporal S-2 data significantly improved prediction results as is obvious from Table 4. The best results on the training set were provided by the SVM algorithm ($R^2 = 0.697$, RMSE = 13.756, MAE = 9.872) followed by GBM, XGBoost, NNET, RF, CatBoost, and ENET. Simpler algorithms such as PLS and RR performed significantly worse, while the large number of predictors (620) appeared unsuitable for the OLS algorithm (RMSE = 87.413). However, the overall performance of the models of higher complexity showed only small, mostly insignificant improvements.

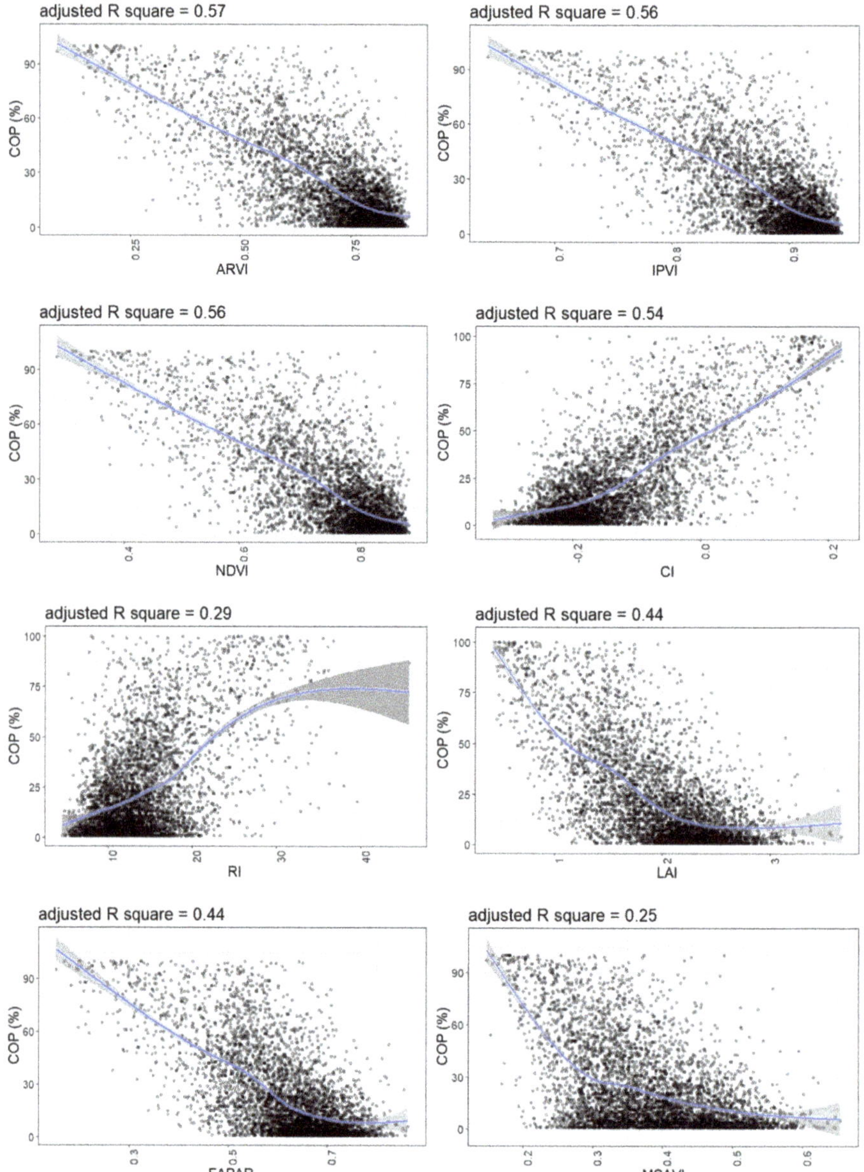

Figure 3. The relationship between individual predictors and canopy openings percentage (COP) with included local regression line (LOESS).

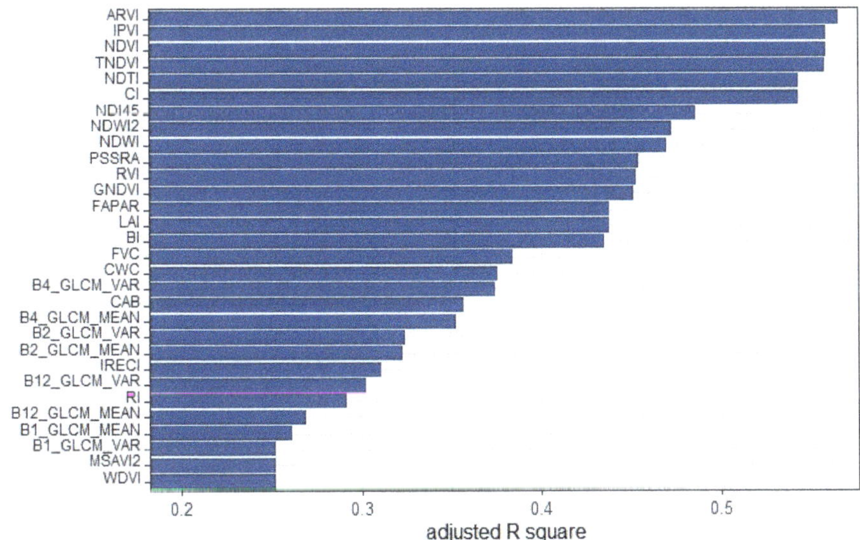

Figure 4. The adjusted coefficient of determination of the 30 most dominant predictors (indices) using simple linear ordinary least squares (OLS) regression.

Table 3. The results of 10-fold cross-validation (CV) on the training set using a single S-2 image. RMSE, root-mean-square error; MAE, mean absolute error.

10-Fold CV Training Set: Single Sentinel-2 Image (S2 25 July 2018)				
Model	**RMSE**	**R^2**	**MAE**	**Tuning Parameters**
Multiple regression (Ordinary Least Squares)—OLS	15.684	0.603	11.366	
Multiple regression (Ordinary Least Squares) with PCA pre-processing—OLS with PCA	16.301	0.571	11.909	
Partial Least Squares—PLS	15.618	0.604	11.397	ncomp = 20
Ridge Regression RR	15.611	0.606	11.305	lambda = 0.007142857
Elastic Net—ENET	15.629	0.605	11.314	fraction = 1 lambda = 0.01
Model Averaged Neural Network—NNET	15.864	0.591	11.531	size = 5 decay = 0.01
Support Vector Machines with Radial Basis Function Kernel—SVM	15.344	0.622	10.784	sigma = 0.007735318 C = 2
Random Forest—RF	15.394	0.616	11.374	mtry = 10
Stochastic Gradient Boosting—GBM	15.376	0.616	11.206	n.trees = 910 interaction.depth = 7 shrinkage = 0.01 n.minobsinnode = 20
Extreme Gradient Boosting—XGBoost	15.148	0.624	11.070	nrounds = 550　　　max_depth = 5 eta = 0.025
Catboost—Cboost	15.308	0.619	11.197	depth = 8　　　learning_rate = 0.1 leaf_reg = 0.001　　　rsm = 0.95

Table 4. The results of 10-fold CV on the training set using multitemporal S-2 images.

10-Fold CV Training Set: Sentinel 2 Multitemporal (S2 25 July 2018; S2 1 August 2018; S2 29 August 2018; S2 28 September 2018)				
Model	RMSE	R^2	MAE	Tuning Parameters
Multiple regression (Ordinary Least Squares)—OLS	87.413	0.593	14.458	
Partial Least Squares—PLS	15.020	0.650	10.770	ncomp = 10
Ridge Regression—RR	41.520	0.611	11.890	lambda = 0.1
Elastic Net—ENET	14.360	0.669	10.543	fraction = 0.2 lambda = 0.01
Model Averaged Neural Network—NNET	14.230	0.680	10.520	size = 11 decay = 0.1
Support Vector Machines with Radial Basis Function Kernel—SVM	13.756	0.697	9.872	sigma = 0.001703831 C = 2
Random Forest—RF	14.265	0.673	10.732	mtry = 213
Stochastic Gradient Boosting—GBM	13.952	0.685	10.369	n.trees = 910 interaction.depth = 7 shrinkage = 0.01 n.minobsinnode = 30
Extreme Gradient Boosting—XGBoost	13.991	0.683	10.348	nrounds = 650 max_depth = 4 eta = 0.05
Catboost—Cboost	14.195	0.676	10.460	depth = 6 learning_rate = 0.1 leaf_reg = 0.001 rsm = 0.95

Validation of the models on the spatially independent test data overall showed much worse results for all the compared models. This is especially evident from the validation results of models based on the single S-2 image, presented in Table 5. The obtained validation metrics, particularly R^2, which ranged from 0.380 for SVM to 0.420 for RR and ENET, were significantly lower than the results obtained in the training set. With the addition of the multitemporal S-2 data, there was only a slight improvement in the model performances, as shown in Table 6. The highest R^2 value, 0.445, was obtained by the ENET model, followed by other algorithms: CatBoost ($R^2 = 0.411$) and GBM, RF, RR ($R^2 = 0.440$). These results emphasize the problem of spatial extrapolation of the model results or the suitability of the model for prediction outside of the training areas, which is a very common problem in spatial analysis. Furthermore, in the case of extrapolation, the suitability of the simpler, robust, linear regularization algorithms such as ENET and RR was equal to or slightly better than the suitability of the complex ensemble algorithms, which showed better realization only on the training data due to probable overfitting.

Table 5. Model validation on the testing set: a single S-2 image (S2 25 July 2018).

Test Set: Single Sentinel-2 Image (S2 25 July 2018)			
Model	RMSE	R^2	MAE
Multiple regression (Ordinary Least Squares)—OLS	15.264	0.406	11.655
Partial Least Squares—PLS	15.220	0.410	11.450
Ridge Regression—RR	15.110	0.420	11.230
Elastic Net—ENET	15.090	0.420	11.190
Model Averaged Neural Network—NNET	15.360	0.410	11.090
Support Vector Machines with Radial Basis Function Kernel—SVM	16.110	0.380	11.390
Random Forest—RF	15.290	0.410	11.240
Stochastic Gradient Boosting—GBM	15.580	0.400	11.150
Extreme Gradient Boosting—XGBoost	15.645	0.387	11.233
Catboost—Cboost	15.488	0.403	11.171

Table 6. Model validation on the testing set: multitemporal S-2 images.

Test Set—Sentinel-2 Multitemporal (S2 25 July 2018; S2 1 August 2018; S2 29 August 2018; S2 28 September 2018)			
Model	RMSE	R^2	MAE
Multiple regression (Ordinary Least Squares)—OLS	15.264	0.406	11.655
Partial Least Squares—PLS	15.303	0.425	11.190
Ridge Regression—RR	15.359	0.440	11.140
Elastic Net—ENET	14.999	0.445	10.916
Model Averaged Neural Network—NNET	15.846	0.402	11.447
Support Vector Machines with Radial Basis Function Kernel—SVM	16.020	0.410	11.240
Random Forest—RF	14.900	0.440	10.935
Stochastic Gradient Boosting—GBM	14.940	0.440	10.730
Extreme Gradient Boosting—XGBoost	15.102	0.428	10.874
Catboost—Cboost	14.951	0.441	10.793

3.3. The Importance of Satellite Indices in Prediction

Inferences about the contribution of particular satellite indices in the COP prediction are usually not the main goal in predictive modeling. However, for the purpose of simplifying the building process, particularly retrieval and preprocessing of the huge number of S-2 variables in eventual future similar studies, it is recommended to focus only on the group of the most valuable indices or features that significantly contribute to models. The common drawback of highly flexible modeling techniques such as neural networks, SVM, random forest, and boosting is the lack of interpretability of the model in terms of inferences about the input variables. The important advantage of the R Caret framework is in its capability of quantification of the variable importance in various models. The variable importance is commonly estimated from the reduction in squared error due to each predictor. In ensembles, improvement values for each predictor are averaged across the entire ensemble to yield an overall importance value. The scores of the variable importance estimation for the best-performing models on the multitemporal set in this study are provided in Table 7.

The importance profile of provided scores in Table 7 provides some clues about how models incorporate the data. GBM and XGBoost had a very steep slope of declining scores from top to bottom which is the usual for boosting algorithms. This is due to the fact that trees from boosting are dependent on each other; they have correlated structures as the method follows by the gradient. As a result, many of the same predictors were selected across the trees, increasing their contribution to the importance metric. The RF algorithm relies on the decorrelation of trees and the selection of a random number of variables at each tree node; therefore, it has slower decline in scores. What could be distinguished among the models was the order of incorporating indices from the multitemporal S-2 set across the importance profile. ENET (as well as RR) gradually incorporated the most dominant predictors following the chronology of the multi-temporal S-2 set: first incorporating variables from the first S-2 image, then the second and third (with the exception of the RVI from image four). Incorporation of the most dominant variables from the images in chronological order was probably the learning principle of the regularization algorithm. It slid from image to image, repeatedly incorporated the most significant predictors that reduced SSE, and shrunk or penalized the others. What is also evident is that the ENET model relied only on the first three images in the multitemporal set, and the predictive information from the last set was probably redundant for the model construction. The RF algorithm behaved quite oppositely to the ENET algorithm in that it alternately incorporated variables from the first, then fourth and second S-2 images without some discernible pattern.

Table 7. Variable importance scores for the top 20 satellite indices and related models. The S-2 columns refer to the positioning of the index variable in the multitemporal set (I = 25 July 2018, II = 1 August 2018, III = 29 August 2018, IV = 28 September 2018).

CBoost	S-2	Score	GBM	S-2	Score	XGBoost	S-2	Score
IPVI	I	100.00	CI	I	100.00	CI	I	100.00
NDTI	I	92.57	ARVI	I	36.53	ARVI	I	33.58
TNDVI	I	75.02	NDI45	III	32.07	IPVI	I	33.48
NDI45	IV	71.79	PSSRA	I	29.16	NDI45	III	19.74
NDI45	II	54.44	NDI45	II	26.5	PSSRA	II	13.47
NDTI	II	49.79	IPVI	I	22.65	NDI45	I	11.85
NDVI	IV	41.68	NDI45	IV	16.43	NDI45	IV	11.73
MCARI	I	40.58	CI	II	15.6	CI	II	10.79
ARVI	I	37.70	NDI45	I	15.02	IPVI	IV	5.63
RI	II	36.04	CI	IV	10.75	CI	IV	5.61
TNDVI	IV	34.42	IPVI	IV	7.27	NDI45	I	5.50
NDWI2	IV	32.80	ARVI	II	7.22	RI	II	4.62
CI	II	32.24	RI	II	7.08	MCARI	IV	4.18
PSSRA	I	31.53	PSSRA	II	6.04	MCARI	I	3.8
NDTI	IV	30.57	CI	III	5.99	ARVI	II	3.48
NDTI	III	29.08	MCARI	IV	5.65	CI	III	3.24
GLCM_VAR_B1	IV	28.21	MCARI	I	4.66	GLCM_VAR_B1	IV	2.54
ARVI	IV	27.97	PSSRA	II	4.61	ARVI	IV	2.46
CI	III	27.32	NDWI2	IV	4.08	NDWI2	IV	2.46
NDVI	I	26.11	ARVI	IV	3.64	PSSRA	II	1.78
ENET	**S-2**	**Score**	**RF**	**S-2**	**Score**	**SVM**	**S-2**	**Score**
ARVI	I	100	NDI45	IV	100	ARVI	I	100
TNDVI	I	98.85	ARVI	I	80.22	TNDVI	I	98.85
IPVI	I	98.85	MCARI	IV	79.45	IPVI	I	98.85
NDVI	I	98.85	RI	II	75.48	NDVI	I	98.85
RVI	IV	98.45	IPVI	IV	73	RVI	IV	98.45
RVI	I	98.45	PSSRA	I	71.2	RVI	I	98.45
ARVI	II	97.85	MCARI	I	69.72	ARVI	II	97.85
CI	I	97.73	CI	IV	69.61	NDTI	I	97.73
NDTI	I	97.73	CI	II	65.47	CI	I	97.73
PSSRA	I	97.46	CI	I	65.11	PSSRA	I	97.46
CI	II	96.73	NDVI	IV	63.8	NDTI	II	96.73
NDTI	II	96.73	NDTI	I	63.28	CI	II	96.73
IPVI	II	96.4	TNDVI	IV	62.23	IPVI	II	96.4
NDVI	II	96.4	NDTI	IV	59.98	NDVI	II	96.4
TNDVI	II	96.4	RVI	IV	59.08	TNDVI	II	96.4
PSSRA	II	96.28	NDTI	II	57.78	PSSRA	II	96.28
RVI	II	95.78	RVI	I	56.28	RVI	II	95.78
PSSRA	III	93.33	ARVI	IV	55.73	PSSRA	III	93.33
ARVI	III	92.24	NDVI	I	55.57	ARVI	IV	92.24
NDVI	III	91.45	TNDVI	I	54.88	NDVI	IV	91.45

Given the spectral indices used, there were also certain peculiarities amongst the models. In most cases, particular vegetation and soil radiometric indices played a dominant role. From the vegetation indices, the most significant were ARVI, IPVI, NDI45, PSSRA, and MCARI, those of soil radiometric indices were CI and RI, and that of water radiometric indices was NDTI. Of the textural features, only GLCM_variance appeared on channel B1 on the last S-2 set. It is also important to note that boosting techniques (GBM, XGBoost) first discriminated sites according to reflectance from the bare soil (CI index) and then from vegetation, unlike other models that built first on reflectance from the vegetation surface.

3.4. Final Model Building and COP Map Production

On the basis of the results of training and validation, the ENET algorithm was chosen to create the final model. The main reasons for choosing the ENET model were (I) simplicity of the algorithm, (II) robustness, with best validation results on the test set, (III) one of the shortest computing times amongst all models, and (IV) requirement of the smallest number of S-2 images (three) in the multitemporal set compared to other methods. The final model was produced using all 8222 samples from the training and validation sets, as well as all predictors from the S-2 multitemporal set, 620 in total. The validation metrics of the final ENET model were as follows: RMSE = 14.427, R^2 = 0.603, and MAE = 10.821. The model was then used for the prediction of COP over the whole FMU Litorić. Due to the use of regression settings for the prediction of percentages in the interval between 0% and 100%, the final model also contained protruding predictions that were beyond this interval. In such cases, negative values were simply constrained to 0 and values above were constrained to the upper limit of 100. The final result of the whole modeling process, a map of COP percentage for FMU Litorić, is presented in Figure 5. The precision of the produced map can be best perceived by the visual examination of individual details, as presented in Figure 6, which shows a mixed segment with alternating forest cover and clearings.

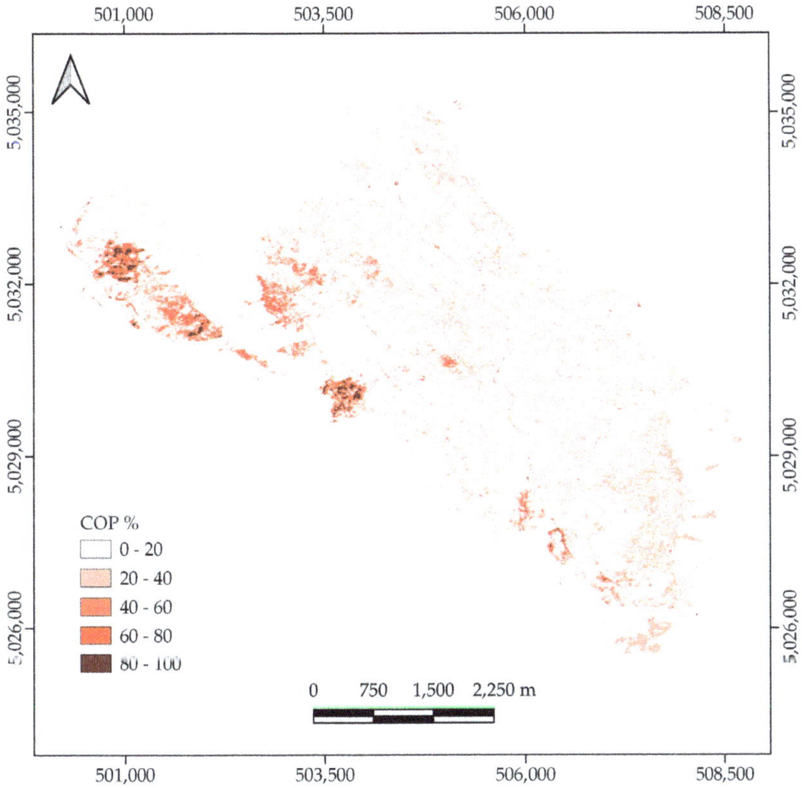

Figure 5. The final map of canopy openings percentage for FMU Litorić.

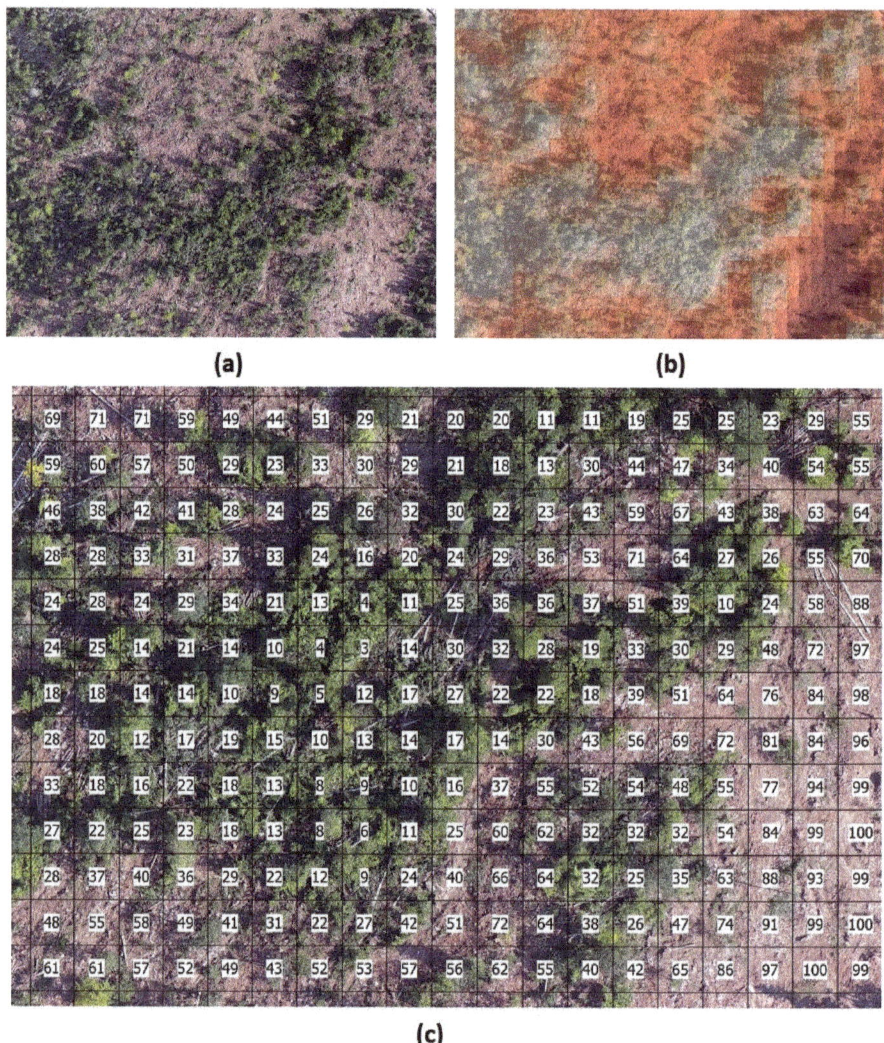

Figure 6. Representation of map precision. (a) Detail of the area obtained using UAV image; (b) the same detail with overlaid COP map including marked areas in red that indicate disturbances in the canopy; (c) precision of the map can be also perceived from the labels that display COP values in percentages across the S-2 polygon grid.

3.5. Analysis of Residuals and Observed Shortcomings of the Model

The relationship of observed vs. predicted values from the final ENET model and the distribution of the residuals are presented in Figure 7.

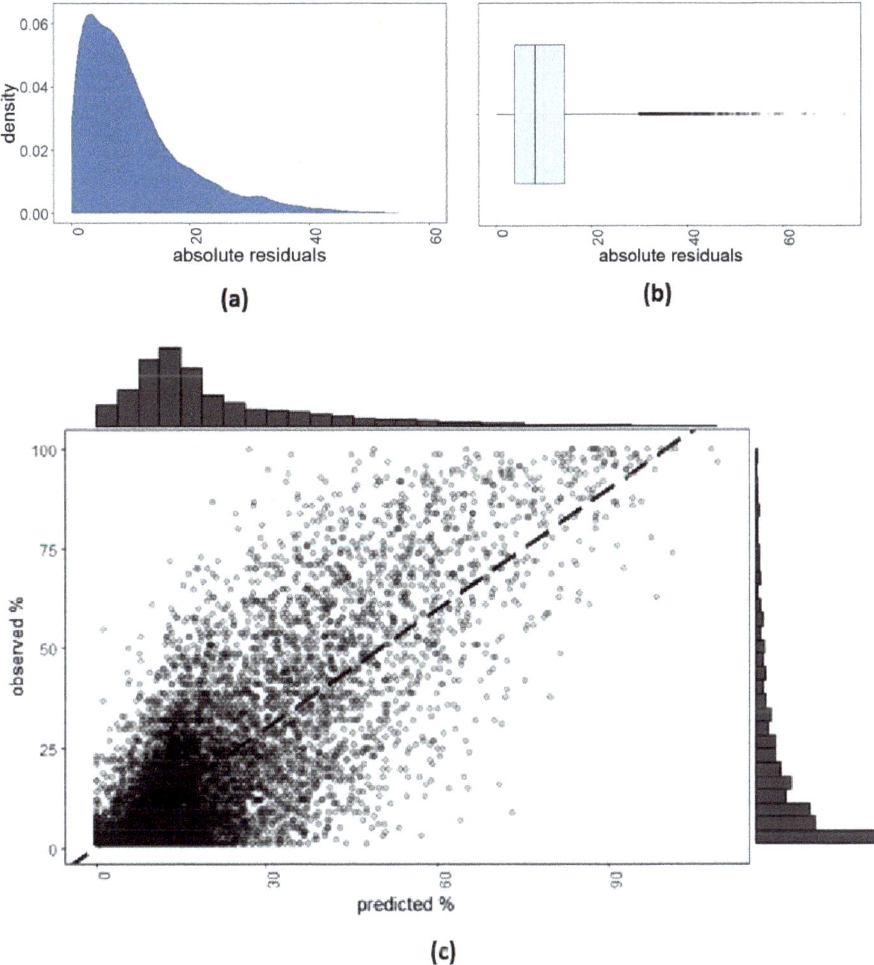

(a)

(b)

(c)

Figure 7. Assessment of the final model residuals. (a) Density plot of absolute residuals of the ENET model, (b) boxplot of absolute residuals (median value = 8.2%); (c) scatterplot of the predicted vs observed values with distribution histograms.

The model residuals were mainly spatially randomly distributed. In addition, we tried to explain some observed shortcomings of the model with respect to the limited degree of explanation of the variance of response (max R^2 of 0.697 in the training set and 0.603 in the final set). This was performed first by gaining insight into spatial patterns of observations with extreme positive and negative residual values and then with further visual examination of the potential reasons for these discrepancies in UAV images. One such case is presented on Figure 8, which shows two larger clearings within the forest on which trees were completely absent. The clearing on the left side of the image contains a group of observations with positive residuals (model over estimation), while the clearing on the right contains a group with negative residuals (model under estimation) Deeper insight is shown in Figure 9.

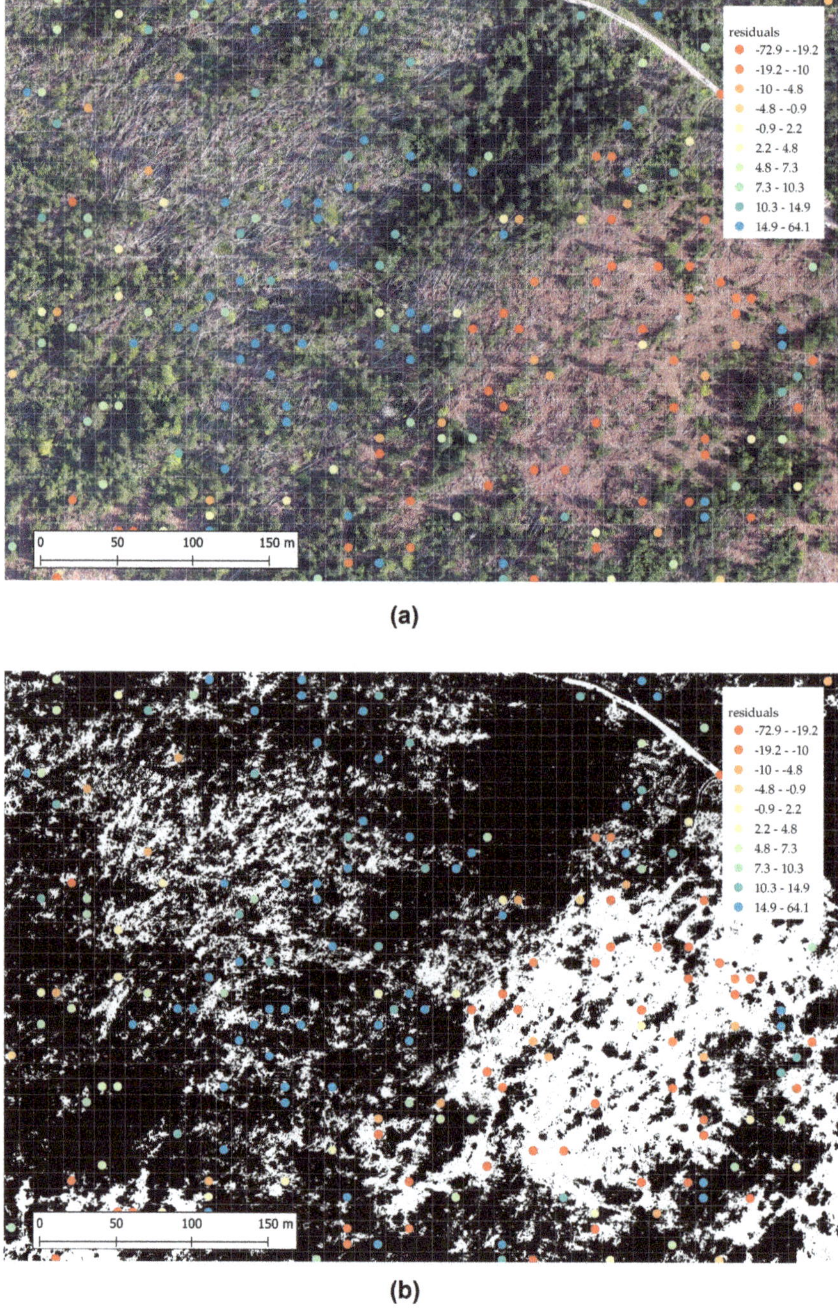

Figure 8. Representation of two cases of model discrepancies. (**a**) Contrasting areas of COP overestimation (blue points) and underestimation (red points); (**b**) the same forest segments with the UAV mask overlay with visible dissimilarity in abundance of open areas marked in white.

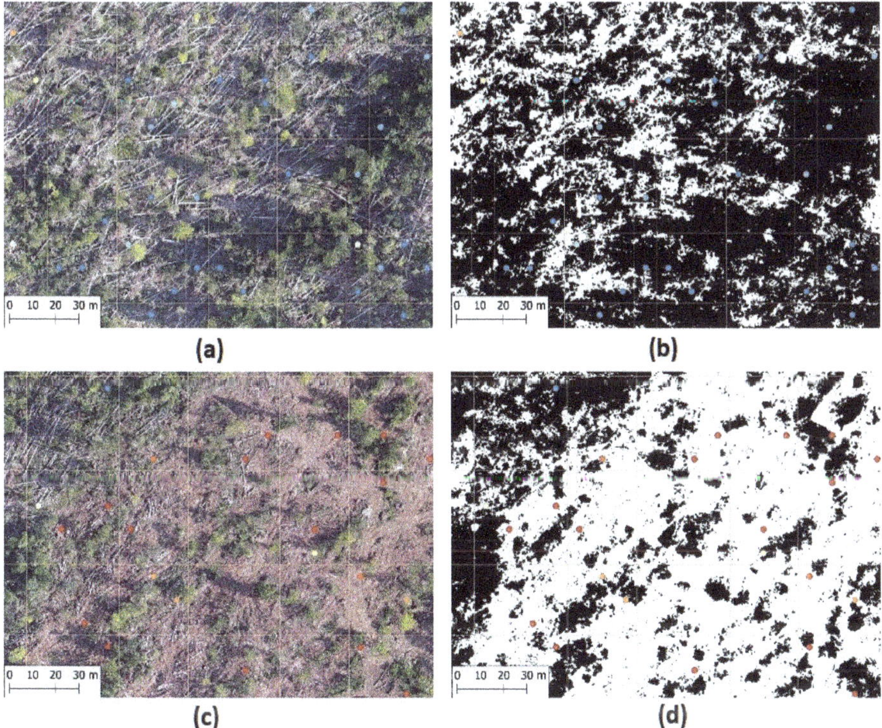

Figure 9. Deeper insight into two contrasting areas. (**a**) An area with fallen trees after a windstorm; (**b**) comparative UAV mask that overestimates real forest cover; (**c**) forest clearing with removed logs and noticeable bare ground with distinguishable soil color; (**d**) comparative UAV mask of the forest clearing with scarce canopy cover.

A deeper insight into the above locations revealed the difference in the surface cover. Specifically, in the image on the left, the fallen trees, as a result of the recent windbreak in the area of FMU Litorić, were noticeable, while, in the image on the right, the trees were completely cleared and the bare soil surface came to expression. Fallen trees, somewhere still with a green canopy, as well as the shadows of the forest edge, were affected by the process of automatic production of the UAV forest mask, which in such cases showed a much lower COP percentage than in reality. The ENET model in such cases provided somewhat realistic COP values that were much higher than the UAV mask. In the figure on the right, the opposite occurred; on bare clearings, the ENET model estimated significantly lower values than in reality, causing negative residuals.

Another case of discrepancy was observed on locations where damaged trees with a higher degree of crown discoloration occurred. These trees that still formed a complete canopy cover were falsely recognized as open areas in the constructed UAV RGB forest mask (Figure 10). This could have caused an unrealistic overestimation of the COP prediction.

Figure 10. Example of false canopy gap creation from UAV RGB images over the damaged trees with crown discoloration. (**a**) UAV image segment of trees with the damaged canopy; (**b**) related UAV mask with false open spaces presented in white.

4. Discussion

From the presented results of this research, the suitability of the combined method of forest gaps (COP) estimation using UAV and Sentinel 2 images, intended for precise rapid mapping of COP on a large forest management unit scale, was demonstrated. The results of the research show that, with the synergistic effect of two, in many respects, very different Earth observation (EO) systems, we can significantly improve the current practice of monitoring the state and disturbances of forest cover. This approach enables relatively fast and very precise insight into forest canopy over relatively larger areas. There are two particularly important aspects where the significant benefits of using the aforementioned approach could be identified. The first element is related to the improvement of forestry planning and operational activities that could take advantage of the availability of timely and accurate information on the locations and magnitude of forest gaps throughout the FMU. The second element is predominantly contribution in the remote sensing domain, which refers to a proxy way of improvement of the capabilities of Sentinel-2 for very fine gradation of canopy cover at subpixel precision. The effectiveness of the combined use of these two sensors stems from the blending of their comparative advantages and the elimination of their weaknesses. Sentinel-2 is currently one of the most advanced satellite EO systems whose most important features are global coverage, high spectral resolution, and short revisit time, although its spatial resolution is too coarse for the discrimination of canopy openings smaller than 10 m. UAV systems, with ultra-high spatial resolutions and flexibility of manipulation, are currently unsurpassed in capabilities of precise detection of stand coverage, although they are limited in the operational area they can cover. By using the UAV system as ground truth on preselected smaller training and validation areas and Sentinel-2 for wall-to-wall prediction with ML techniques, we extended, in a surrogate fashion, information from the UAV over the whole FMU. Using a simple UAV system with only an RGB camera, which has some limitations in canopy segmentation, has advantages in wide operational applicability. Currently, systems such as DJI drones with RGB cameras are by far the most widespread of all UAV systems, while the processing of such images is far simpler than more complex multispectral, hyperspectral, thermal, and LiDAR images, which greatly facilitates the acquisition of information. As the main purpose of the presented approach is applicability in operational forest management or silvicultural decision-making processes, the method has certain trade-offs that resulted in a somewhat weaker overall prediction accuracy at the expense of the promptness of the required data acquisition and the short time required for the final product manufacturing, which is usually one of the main requirements of end-users in operational forestry for this kind of EO product (e.g., when assessing large-scale calamity caused by windthrows, ice breaks, biotic damages, etc.). The applied approach relies entirely on information from remote sensors in different image resolutions, while it minimizes field work which is in line with current trends of advanced smart forestry technologies.

For the integration of different sensors, UAV and Sentinel-2, the predictive modeling approach was applied, which consisted of feature engineering, the creation of training and validation sets of data, training of various algorithms using resampling techniques, validation of models on the test set, model selection, construction of the final model, and prediction of the new samples. The model building process brought about the improvement in predictive performance on the training set in successive steps; (I) the highest R^2 of the single feature (derived satellite index) using simple linear regression was 0.57, (II) the highest R^2 using multiple features obtained from the single-date, S-2 image and the best ML algorithm was 0.624, and (III) the highest R^2 on the multi-temporal set of four consecutive S-2 images, using the best ML algorithm was 0.697.

From the wider range of satellite indices or features examined in this study, several exhibited superior properties in differentiating the COP gradient and were, therefore, incorporated as the main foundation in the majority of ML models. NDVI (Normalized Difference Vegetation Index), which most often represents the core index for determining variations in the structure of vegetation cover, did not prove to be the primary choice of ML algorithms examined in this research. Instead, as a primary choice, they selected indices closely based on NDVI but with enhanced properties in certain segments. ARVI (Atmospherically Resistant Vegetation Index), as with NDVI, is based on the normalized difference between the near-infrared and red channel, but it corrects the atmospheric effect on the red channel by using the difference in the radiance between the blue and red band. It has a similar dynamic range to NDVI but, on average, is four times less sensitive to atmospheric effects. IPVI (Infrared Percentage Vegetation Index) is functionally equivalent to NDVI with the only difference that it includes infrared and red factors in the NDVI equation which results in simplification of the calculation process. The NDI45 (Normalized Difference Index) algorithm uses the same relations and bands but is more linear and usually less saturated at higher values than the NDVI. In addition to the above indices based on the improved NDVI principle, simple ratio indices that are sensitive to variations in chlorophyll concentrations in canopies were also shown to be the dominant choice for ML algorithms. PSSRa (Pigment-Specific Simple Ratio, or chlorophyll index) consists of a simple ratio of the near-infrared and red bands. It is a narrow-band pigment index that has the strongest and most linear relationship with canopy concentration per unit area of chlorophyll a, chlorophyll b, and carotenoids. MCARI (Modified Chlorophyll Absorption Ratio Index) is the ratio of differences in bands in the visible spectra (red, green, blue), responsive to chlorophyll variation, and ground reflectance. In addition to the mentioned vegetation radiometric indices, a few ML algorithms (GBM and XGBoost), as a first source, used a CI (Color Index) which belongs to the group of soil radiometric indices. CI was primarily developed to differentiate soils in the field, and, in this study, it probably had the best properties to differentiate areas with bare soil within significantly disturbed canopy cover caused by recent windthrows. In most cases, it gave complementary information to the NDVI. RI (Redness Index), as with CI, enables identifying areas with bare soil, which, in the FMU Litorić, have a commonly reddish hue, characteristic for soils developed on hard limestone. Of the other used indices, of greater importance was the NDTI (Normalized Difference Turbidity Index) which belongs to the water radiometric group of indices commonly used for the measurement of water turbidity or water clarity. Since the NDTI algorithm is color-sensitive, with a structure very similar to the CI algorithm, its representation in the models was complementary to the CI and RI indices.

The use of a multitemporal set of S-2 data, compared to information from the single S-2 image, contributed most significantly to an increase in the prediction results compared to all other examined means of prediction improvement. The highest achieved R^2 using 10-fold CV on the training set with data from a single S-2 image from 25 July 2018, which was closest to the UAV image acquisition, was 0.624. However, by including variables from the multitemporal set, a significant increase in the explained COP variance from the multitemporal predictors was achieved (R^2 of 0.697). On the other hand, the use of a multitemporal set led to an only slight improvement in RMSE (13.756) compared to single-date variables (15.148), which ultimately made a difference of only 1.3% of COP. The advantages of using a multitemporal set of Sentinel 2 images vs. a single image for the discrimination of forest cover are in accordance with recently published results [57,58]. The most likely reason for the enhanced results of the multitemporal set is that, although the S-2 images were obtained within a short time interval (approximately 2 months), they were able to discern certain variations in phenology that are very significant for discriminating small changes in the forest cover structure. According to the results [58], classifications made using multitemporal Sentinel 2 imagery for two contrasting seasons (spring and autumn) were the most accurate, while classifications based on the single-season spring images were the second most accurate. Classifications made using other combinations of seasons and feature types varied but yielded significantly lower accuracy. Incorporation of the Sentinel-2 image from the early spring aspect could most probably improve the COP prediction, but at the risk of loss of up-to-date insight into the current situation of the forest cover, which is often crucial for operational silvicultural decisions.

The addition of ML models with varying degrees of complexity was one of the options in which an improvement in COP prediction was sought. The included ML algorithms on the single temporal S-2 training set with 10-fold CV only partially contributed to improved prediction. Compared to the basic OLS algorithm (RMSE = 15.684; R^2 = 0.603), the best XGBoost model (RMSE = 15.148; R^2 = 0.624) only slightly improved the prediction results (a decrease in RMSE of 0.536 and an increase in R^2 of 0.021). Although they showed a very small improvement of model metrics, a few ML algorithms such as SVM, RF, GBM, XGBoost, and CatBoost showed somewhat better results, whereas some algorithms such as NNET underperformed and others such as PLS, RR, and ENET had similar capabilities to OLS. A significant difference between the algorithms was noticeable on the multitemporal set between the basic OLS model (RMSE = 87.413; R^2 = 0.593) and the best SVM model (RMSE = 13.756; R^2 = 0.669). It should be noted that the basic OLS algorithm, due to a large number of predictors, exhibited a problem with convergence and instability (high RMSE) and, therefore, was unsuitable for the multitemporal set.

One of the observed shortcomings of the method is that it was less suitable for extrapolation of COP outside the training area, i.e., lower prediction capabilities for COP mapping of the entire FMU area. This was obvious from the obtained metric from the spatially remote validation dataset. The accuracy achieved on the test set was significantly lower than the accuracy obtained using 10-fold CV resampling on the training set. The best obtained RMSE and R^2 on a single S-2 image were 15.09 and 0.42, while those on the multitemporal set were 14.900 and 0.445, respectively, significantly worse than the 10-fold CV from the training data. Furthermore, more complex algorithms that were somewhat superior on the training set did not validate their advantages on the independent set. The benefit of using more complex ML algorithms is that they allow better adjustment to local conditions, while this advantage is lost in cases of extrapolation to new conditions out of the training area. In other words, complex ML algorithms generally overfitted the local condition, while, outside the training area, they were comparable to simpler algorithms with greater robustness and generalization ability. These results indicate the inadequate sampling design of the ground-truth UAV data acquisition used for model creation and mapping of the whole FMU area in this study (the sampling design of UAV locations was not considered preliminary). Poorer generalization capabilities of the model could be remedied by the acquisition of UAV ground-truth images from a larger number of smaller training areas across the FMU to incorporate the variety of conditions (stratified random sampling). The ML model constructed in this way, in addition to encompassing a wider range of diversity, would be used primarily for interpolation between the UAV survey locations rather than extrapolation as was the case in this study. Moreover, the improved sampling design will enable much better performance of more complex ML models (RF and boosting) which are superb for the representation of conditions that are not beyond the range of data used for learning.

Precise and reliable delineation of the crown edge on UAV images presented one of the most important factors affecting the reliability of the model. The problems observed in this study stemmed from the limited capabilities of UAV images only in the visible RGB channel to accurately delineate the crown projection, which was then used as reference values for COP modeling. According to our knowledge, research on the combined application of Sentinel-2 and UAV in similar, highly heterogeneous forest types such as mixed beech–fir forests has not been conducted so far; thus, a direct comparison of the obtained results is not possible. On the other hand, there are results in which a somewhat similar approach was used but in somewhat different forest conditions. In a similar study evaluating tree canopy cover [44], using combined Sentinel-2, Google Earth imagery, and ML methods, significantly better prediction was achieved (R^2 = 82.8%, RMSE = 8.68%) than in our study. However, higher accuracy of the assessment was obtained in the semiarid Mediterranean silvopastoral system (*Quercus ilex*, *Quercus suber*) with sparse tree density and very contrasting bare soil with a distinctive color. In such conditions, it was possible to identify more clearly the crown boundary and produce a more accurate canopy mask layer, which were the most likely reasons for better assessment. In conditions of extremely complex canopy structure, as is the case with mixed beech–fir forests, special attention should be paid when creating a canopy mask layer. Using images only from the visible part of the spectrum, relying solely on differences in color between the canopy and ground, is often very unreliable due to noise from shadows at gap edges, groups of fallen trees after windthrows, etc. Discoloration and defoliation of individual, partially damaged trees, which are falsely rejected from the canopy mask using an automatic classification based only on color indices from the visible spectra, also represent an observed problem that probably affected the accuracy of the method.

5. Conclusions

The development of new remote sensing methods such as UAV, as well as new satellite sensors and the increasing amount and availability of high-quality and free-of-charge satellite data, opens up new opportunities for improving decision-making support in operational forest management. Satellite data have so far, through a large number of different examples, shown their distinct advantages for the detection of changes in forests, but most often in large areas and in coarser spatial resolution, where their robust nature comes to the fore. The use of satellite Earth observations to obtain more accurate information of forest structure is most often hampered by the need for field collection of ground-truth information required for better representation of forest characteristics, which is often very slow and costly. The so-called change detection methods that have been used most often to quantify differences caused by anthropogenic or natural changes in the forest structure provide only proxy, indirect representations inherited from the properties of the used satellite index. On the other hand, the development of unmanned aerial vehicles and their applications in forest photogrammetry have made it possible to obtain very precise, near-real-time information on the forest structure in a very cost-efficient way. The high precision of photogrammetric UAV methods, due to the immense amount of information, and the limited flying range simultaneously represent constraints in the use of this method over larger areas.

Sentinel-2 MSI, due to its advanced spectral characteristics, frequent revisit time, and free-of-charge data availability, represents a highly usable system for various forestry applications. A prerequisite for more efficient use of Sentinel-2 in operational forestry, as well as other optical sensors, is the establishment of stochastic relationships between obtained reflectance in the various spectral bands on multispectral images as surrogate information from forests and common measurable stand parameters used in forest management.

In this study, we demonstrated the effectiveness but also emphasized detected weaknesses of the combined approach to assessment of one of the basic silvicultural parameters such as canopy openings percentage (inverse of the canopy density percentage) using a series of Sentinel-2 imagery and simple UAV RGB images. With the combined approach, the existing limitations of each sensor were annulled: the coarser spatial resolution of Sentinel 2 and the limited flying range of UAV. UAV imagery was used for precise representation of the canopy cover, and Sentinel-2 was used as a scaling platform for transferring information from UAV to a wider spatial extent, while the connection was established using an adequate ML algorithm. A fairly reliable estimate was obtained using a series of four consecutive Sentinel-2 images collected over a period of just over 2 months, which overlapped in time with the UAV recording. Using only one Sentinel-2 image to compare with UAV gave much poorer results. Additionally, for extrapolation of information over the whole FMU, slightly better final results were obtained by using a simpler and more robust linear regularization algorithm such as Elastic Net rather than other more complex and robust ML methods (random forest, boosting, support vector machines, neural networks).

Given its applicability in forestry, the method can be used for operational applications in support of silvicultural planning and decision-making on an annual basis or any time when there is a need from users for up-to-date information of the state of the forest canopy, such as in case of identified disturbances (windthrows, ice breakage), as well as for control of silvicultural or other activities (forest clearings, illegal loggings, etc.). Although the approach was demonstrated in very heterogeneous conditions such as those prevalent in the uneven-aged mixed beech–fir forest, it is possible to extend it to various other forest types and environmental conditions. Furthermore, this approach is suitable for combining Sentinel-2 with other structural and functional parameters of stands that can be obtained directly from UAV sensors (LiDAR, multispectral, hyperspectral) in which there is a stochastic relationship with Sentinel-2 images.

Author Contributions: Conceptualization, I.P. and D.K.; methodology, I.P. and D.K.; validation, I.P. and D.K.; formal analysis, I.P., M.G., and A.N.; writing—original draft preparation, I.P., M.G., and A.N.; writing—review and editing, I.P., M.G., and D.K. All authors have read and agreed to the published version of the manuscript.

Funding: This research received no external funding.

Acknowledgments: This research was undertaken as an additional activity within the Horizon 2020 project "My Sustainable Forest—Earth Observation services for silviculture".

Conflicts of Interest: The authors declare no conflict of interest.

References

1. Čater, M.; Diaci, J.; Roženbergar, D. Gap size and position influence variable response of *Fagus sylvatica* L. and *Abies alba* Mill. *For. Ecol. Manag.* **2014**, *325*, 128–135. [CrossRef]
2. Dobrowolska, D.; Veblen, T.T. Treefall-gap structure and regeneration in mixed *Abies alba* stands in central Poland. *For. Ecol. Manag.* **2008**, *255*, 3469–3476. [CrossRef]
3. Ugarković, D.; Tikvić, I.; Popić, K.; Malnar, J.; Stankić, I. Microclimate and natural regeneration of forest gaps as a consequence of silver fir (*Abies alba* Mill.) dieback. *J. For.* **2018**, *5–6*, 235–245.
4. Muscolo, A.; Sidari, M.; Mercurio, R. Variations in soil chemical properties and microbial biomass in artificial gaps in silver fir stands. *Eur. J. For. Res.* **2007**, *126*, 59–65. [CrossRef]
5. Muscolo, A.; Sidari, M.; Bagnato, S.; Mallamaci, C.; Mercurio, R. Gap size effects on above and below-ground processes in a silver fir stand. *Eur. J. For. Res.* **2010**, *129*, 355–365. [CrossRef]
6. Albanesi, E.; Gugliotta, O.I.; Mercurio, I.; Mercurio, R. Effects of gap size and within-gap position on seedlings establishment in silver fir stands. *Forest* **2005**, *2*, 358–366. [CrossRef]
7. Nagel, T.A.; Svoboda, M. Gap disturbance regime in an old-growth Fagus-Abies forest in the Dinaric Mountains, Bosnia-Hercegovina. *Can. J. For. Res.* **2008**, *38*, 2728–2737. [CrossRef]
8. Nagel, T.A.; Svoboda, M.; Diaci, J. Regeneration patterns after intermediate wind disturbance in an old-growth Fagus-Abies forest in Southeastern Slovenia. *For. Ecol. Manag.* **2006**, *226*, 268–278. [CrossRef]
9. Diaci, J.; Pisek, R.; Boncina, A. Regeneration in experimental gaps of subalpine *Picea abies* forest in Slovenian Alps. *Eur. J. For. Res.* **2005**, *124*, 29–36. [CrossRef]
10. Malcolm, D.C.; Mason, W.I.; Clarke, G.C. The transformation of conifer forests in Britain—Regeneration, gap size and silvicultural systems. *For. Ecol. Manag.* **2001**, *151*, 7–23. [CrossRef]
11. Navarro-Cerrillo, R.M.; Camarero, J.J.; Menzanedo, R.D.; Sánchez-Cuesta, R.; Quintanilla, J.L.; Sánchez, S. Regeneration of Abies pinsapo within gaps created by Heterobasidion annosuminduced tree mortality in southern Spain. *iForest* **2014**, *7*, 209–215. [CrossRef]

12. Vilhar, U.; Roženbergar, D.; Simončić, P.; Diaci, J. Variation in irradiance, soil features and regeneration patterns in experimental forest canopy gaps. *Ann. For. Sci.* **2015**, *72*, 253–266. [CrossRef]

13. Streit, K.; Wunder, K.; Brang, P. Slit-shaped gaps are a successful silvicultural technique to promote *Picea abies* regeneration in mountain forests of the Swiss Alps. *For. Ecol. Manag.* **2009**, *257*, 1902–1909. [CrossRef]

14. Thurnher, C.; Klopf, M.; Hasenauer, H. Forest in Transition: A Harvesting Model for Uneven-aged Mixed Species Forest in Austria. *Forestry* **2011**, *84*, 517–526. [CrossRef]

15. Abdollahnejad, A.; Panagiotidis, D.; Surový, P. Forest canopy density assessment using different approaches—Review. *J. For. Sci.* **2017**, *63*, 106–115. [CrossRef]

16. Paletto, A.; Tosi, V. Forest canopy cover and canopy closure: Comparison of assessment techniques. *Eur. J. For. Res.* **2009**, *128*, 265–272. [CrossRef]

17. Karhonen, L.; Karhonen, K.T.; Rautiainen, M.; Stenberg, P. Estimation of forest canopy cover: A Comparison of Field Measurement Techniques. *Silva Fenn.* **2006**, *40*, 577–588. [CrossRef]

18. Arumäe, T.; Lang, M. Estimation of canopy cover in dense mixed-species forests using airborne lidar data. *Eur. J. Remote Sens.* **2017**, *51*, 132–141. [CrossRef]

19. Wu, X.; Shen, X.; Cao, L.; Wang, G.; Cao, F. Assessment of Individual Tree Detection and Canopy Cover Estimation using Unmanned Aerial Vehicle based Light Detection and Ranging (UAV-LiDAR) Data in Planted Forests. *Remote Sens.* **2019**, *11*, 908. [CrossRef]

20. Deutcher, J.; Granica, K.; Steinegger, M.; Hirschmugl, M.; Perko, R.; Schardt, M. Updating Lidar-derived Crown Cover Density Products with Sentinel-2. *IGARSS* **2017**, 2867–2870. [CrossRef]

21. Karna, K.Y.; Penman, T.D.; Aponte, C.; Bennett, L.T. Assessing Legacy Effects of Wildfires on the Crown Structure of Fire-Tolerant Eucalypt Trees Using Airborne LiDAR Data. *Remote Sens.* **2019**, *11*, 2433. [CrossRef]

22. Huang, C.; Yang, L.; Wylie, B.; Homer, C. A strategy for estimating tree canopy density using Landsat 7 ETM+ and high resolution images over large areas. In Proceedings of the Third International Conference on Geospatial Information in Agriculture and Forestry, Denver, CO, USA, 5–7 November 2001.

23. Mon, M.S.; Mizoue, N.; Htun, N.Z.; Kajisa, T.; Yoshida, S. Estimating forest canopy density of tropical mixed deciduous vegetation using Landsat data; a comparison of three classification approaches. *Int. J. Remote Sens.* **2012**, *33*, 1042–1057. [CrossRef]

24. Baynes, J. Assessing forest canopy density in a highly variable landscape using Landsat data and FCD Mapper software. *Aust. For.* **2004**, *67*, 247–253. [CrossRef]

25. Joshi, C.; De Leeuw, J.; Skidmore, A.K.; van Duren, I.C.; van Oosten, H. Remote sensed estimation of forest canopy density: A comparison of the performance of four methods. *Int. J. Appl. Earth Obs. Geoinf.* **2006**, *8*, 84–95. [CrossRef]

26. Korhonen, L.; Hadi Packalen, P.; Rautiainen, M. Comparison of Sentinel-2 and Landsat 8 in the estimation of boreal forest canopy cover and leaf area index. *Remote Sens. Environ.* **2017**, *195*, 259–274. [CrossRef]

27. Mulatu, K.A.; Decuyper, M.; Brede, B.; Kooistra, L.; Reiche, J.; Mora, B.; Herold, M. Linking terrestrial LiDAR scanner and conventional forest structure measurements with multi-modal satellite data. *Forests* **2019**, *10*, 291. [CrossRef]

28. Revill, A.; Florence, A.; MacArthur, A.; Hoad, S.; Rees, R.; Williams, M. Quantifying uncertainity and bridging the scaling ga pin the retrieval of leaf area indeks by coupling Sentinel-2 and UAV observations. *Remote Sens.* **2020**, *12*, 1843. [CrossRef]

29. Copernicus Land Monitoring Service. Available online: https://land.copernicus.eu/ (accessed on 15 August 2020).

30. *Sentinel 2 User Handbook*; Rev 2; European Space Agency: Paris, France, 2015; p. 64. Available online: https://sentinel.esa.int/documents/247904/685211/Sentinel-2_User_Handbook (accessed on 18 August 2020).

31. Lima, T.A.; Beuchle, R.; Langner, A.; Grecchi, R.C.; Griess, V.C.; Achard, F. Comparing Sentinel-2 MSI and Landsat 8 OLI imagery for monitoring selective logging in the Brazilian Amazon. *Remote Sens.* **2019**, *11*, 961. [CrossRef]

32. Masiliūnas, D. Evaluating the potential of Sentinel-2 and Landsat image time series for detecting selective logging in the Amazon. In *Geo-Information Science and Remote Sensing*; Thesis Report GIRS-2017-27; Wageningen University and Research Centre: Wageningen, The Netherlands, 2017; p. 54.

33. Barton, I.; Király, G.; Czimber, K.; Hollaus, M.; Pfeifer, N. Treefall gap mapping using Sentinel-2 images. *Forests* **2017**, *8*, 426. [CrossRef]

34. Šimić Milas, A.; Rupasinghe, P.; Balenović, I.; Grosevski, P. Assessment of Forest Damage in Croatia using Landsat-8 OLI Images. *SEEFOR* **2015**, *6*, 159–169. [CrossRef]
35. Navarro, J.A.; Algeet, N.; Fernández-Landa, A.; Esteban, J.; Rodríguez-Noriega, P.; Guillén-Climent, M.L. Integration of UAV, Sentinel-1, and Sentinel-2 Data for Mangrove Plantation Aboveground Biomass Monitoring in Senegal. *Remote Sens.* **2019**, *11*, 77. [CrossRef]
36. Pla, M.; Bota, G.; Duane, A.; Balagué, J.; Curcó, A.; Gutiérrez, R.; Brotons, L. Calibrating Sentinel-2 imagery with multispectral UAV derived information to quantify damages in Mediterranean rice crops caused by Western Swamphen (*Porphyrio porphyrio*). *Drones* **2019**, *3*, 45. [CrossRef]
37. Castro, C.C.; Gómez, J.A.D.; Martin, J.D.; Sánchez, B.A.H.; Arango, J.L.C.; Tuya, F.A.C.; Diaz-Varela, R. An UAV and satellite multispectral data approach to monitor water quality in small reservoirs. *Remote Sens.* **2020**, *12*, 1514. [CrossRef]
38. Copernicus Open Access Hub. Available online: https://scihub.copernicus.eu/ (accessed on 12 February 2020).
39. Xue, J.; Su, B. Significant Remote Sensing Vegetation Indices: A Review of Developments and Applications. *J. Sens.* **2017**, *17*. [CrossRef]
40. Bannari, A.; Morin, D.; Bonn, F. A Review of vegetation indices. *Remote Sens. Rev.* **1995**, *13*, 95–120. [CrossRef]
41. Silleos, N.G.; Alexandridis, T.K.; Gitas, I.Z.; Perakis, K. Vegetation Indices: Advances Made in Biomass Estimation and Vegetation Monitoring in the Last 30 Years. *Geocarto Int.* **2006**, *21*, 21–28. [CrossRef]
42. Escadafal, R.; Girard, M.C.; Courault, D. Munsell soil color and soil reflectance in the visible spectral bands of landsat MSS and TM data. *Remote Sens. Environ.* **1989**, *27*, 37–46. [CrossRef]
43. Djamai, N.; Zhong, D.; Fernandes, R.; Zhou, F. Evaluation of Vegetation Biophysical variables Time Series Derived from Synthetic Sentinel-2 Images. *Remote Sens.* **2019**, *11*, 1547. [CrossRef]
44. Godinho, S.; Guiomar, N.; Gil, A. Estimating tree canopy cover percentage in a mediterranean silvopastoral systems using Sentinel-2A imagery and the stochastic gradient boosting algorithm. *Int. J. Remote Sens.* **2018**, *39*, 4640–4662. [CrossRef]
45. Haralick, R.; Shanmugam, K.; Dinstein, I. Textural Features for Image Classification. *IEEE Trans. Syst. Man Cybern.* **1973**, *3*, 610–621. [CrossRef]
46. Wickham, H. Tidy data. *J. Stat. Softw.* **2014**, *59*, 1–23. [CrossRef]
47. Gašparović, M.; Seletković, A.; Berta, A.; Balenović, I. The evaluation of photogrammetry-based DSM from low-cost UAV by LiDAR-based DSM. *SEEFOR* **2017**, *8*, 117–125. [CrossRef]
48. Gašparović, M.; Zrinjski, M.; Barković, Đ.; Radočaj, D. An automatic method for weed mapping in oat fields based on UAV imagery. *Comput Electron. Agric.* **2020**, *173*, 105385. [CrossRef]
49. Mathieu, R.; Pouget, M.; Cervelle, B.; Escadafal, R. Relationships between satellite-based radiometric indices simulated using laboratory reflectance data and typic soil color of an arid environment. *Remote Sens. Environ.* **1998**, *66*, 17–28. [CrossRef]
50. Madeira, J.; Bedidi, A.; Cervelle, B.; Pouget, M.; Flay, N. Visible spectrometric indices of hematite (Hm) and goethite (Gt) content in lateritic soils: The application of a Thematic Mapper (TM) image for soil-mapping in Brasilia, Brazil. *Int. J. Remote Sens.* **1997**, *18*, 2835–2852. [CrossRef]
51. Escadafal, R.; Huete, A. Etude des propriétés spectrales des sols arides appliquée à l'amélioration des indices de végétation obtenus par télédétection. In *Comptes Rendus de l'Académie des Sciences. Série 2, Mécanique, Physique, Chimie, Sciences de l'univers, Sciences de la Terre*; Gauthier-Villars: Paris, France, 1991; Volume 312, pp. 1385–1391.
52. Louhaichi, M.; Borman, M.M.; Johnson, D.E. Spatially located platform and aerial photography for documentation of grazing impacts on wheat. *Geocarto Int.* **2001**, *16*, 65–70. [CrossRef]
53. Tucker, C.J. Red and photographic infrared linear combinations for monitoring vegetation. *Remote Sens. Environ.* **1979**, *8*, 127–150. [CrossRef]
54. Gitelson, A.A.; Kaufman, Y.J.; Stark, R.; Rundquist, D. Novel algorithms for remote estimation of vegetation fraction. *Remote Sens. Environ.* **2002**, *80*, 76–87. [CrossRef]
55. Kuhn, M.; Johnson, K. *Applied Predictive Modeling*; Springer: New York, NY, USA, 2013; p. 615.
56. The Caret Package. Available online: http://topepo.github.io/caret/index.html (accessed on 18 May 2020).
57. Wittke, S.; Yu, X.; Karjalainen, M.; Hyyppä, J.; Puttonen, E. Comparison of two-dimensional multitemporal Sentinel-2 data with three dimensional remote sensing data sources for forest inventory parameter estimation over a boreal forest. *Int. J. Appl. Earth Obs. Geoinf.* **2019**, *76*, 167–178. [CrossRef]

Remote Sens. **2020**, *12*, 3925

58. Macintyre, P.; van Niekerk, A.; Mucina, L. Efficacy of multi-season Sentinel-2 imagery for compositional vegetation classification. *Int. J. Appl. Earth. Obs. Geoinf.* **2020**, *85*, 101980. [CrossRef]

Publisher's Note: MDPI stays neutral with regard to jurisdictional claims in published maps and institutional affiliations.

 *remote sensing*

Article

Determining Optimal Location for Mangrove Planting Using Remote Sensing and Climate Model Projection in Southeast Asia

Luri Nurlaila Syahid [1,2], Anjar Dimara Sakti [1,2,*], Riantini Virtriana [1,2], Ketut Wikantika [1,2], Wiwin Windupranata [3], Satoshi Tsuyuki [4], Rezzy Eko Caraka [5] and Rudhi Pribadi [6]

[1] Remote Sensing and Geographic Information Science Research Group, Faculty of Earth Science and Technology, Institut Teknologi Bandung, Bandung 40132, Indonesia; luri.nurlaila@students.itb.ac.id (L.N.S.); riantini@gd.itb.ac.id (R.V.); ketut@gd.itb.ac.id (K.W.)

[2] Center for Remote Sensing, Institut Teknologi Bandung, Bandung 40132, Indonesia

[3] Hydrography Division, Faculty of Earth Science and Technology, Institut Teknologi Bandung, Bandung 40132, Indonesia; w.windupranata@itb.ac.id

[4] Global Forest Environmental Studies, Graduate School of Agricultural and Life Sciences, The University of Tokyo, Tokyo 113-8654, Japan; tsuyuki@fr.a.u-tokyo.ac.jp

[5] Departement of Statistics, Seoul National University, Gwanak-gu, Seoul 151-742, Korea; rezzy94@snu.ac.kr

[6] Marine Science Department, Faculty of Fisheries and Marine Science, Diponegoro University, Semarang 50275, Indonesia; rudhi_pribadi@yahoo.co.uk

* Correspondence: anjards@gd.itb.ac.id

Received: 31 August 2020; Accepted: 11 November 2020; Published: 13 November 2020

Abstract: The decreasing area of mangroves is an ongoing problem since, between 1980 and 2005, one-third of the world's mangroves were lost. Rehabilitation and restoration strategies are required to address this situation. However, mangroves do not always respond well to these strategies and have high mortality due to several growth limiting parameters. This study developed a land suitability map for new mangrove plantations in different Southeast Asian countries for both current and future climates at a 250-m resolution. Hydrodynamic, geomorphological, climatic, and socio-economic parameters and three representative concentration pathway (RCP) scenarios (RCP 2.6, 4.5, and 8.5) for 2050 and 2070 with two global climate model datasets (the Centre National de Recherches Météorologiques Climate model version 5 [CNRM-CM5.1] and the Model for Interdisciplinary Research on Climate [MIROC5]) were used to predict suitable areas for mangrove planting. An analytical hierarchy process (AHP) was used to determine the level of importance for each parameter. To test the accuracy of the results, the mangrove land suitability analysis were further compared using different weights in every parameter. The sensitivity test using the Wilcoxon test was also carried out to test which variables had changed with the first weight and the AHP weight. The land suitability products from this study were compared with those from previous studies. The differences in land suitability for each country in Southeast Asia in 2050 and 2070 to analyze the differences in each RCP scenario and their effects on the mangrove land suitability were also assessed. Currently, there is 398,000 ha of potentially suitable land for mangrove planting in Southeast Asia, and this study shows that it will increase between now and 2070. Indonesia account for 67.34% of the total land area in the "very suitable" and "suitable" class categories. The RCP 8.5 scenario in 2070, with both the MIROC5 and CNRM-CM5.1 models, resulted in the largest area of a "very suitable" class category for mangrove planting. This study provides information for the migration of mangrove forests to the land, alleviating many drawbacks, especially for ecosystems.

Keywords: mangrove; Southeast Asia; replanting; restoration; analytic hierarchy process

1. Introduction

Mangroves are defined as the vegetation that grows along the intertidal zones in both tropical and subtropical countries [1–4], and they are beneficial for humans and their surrounding ecosystems. For example, they can protect people from natural hazards, such as storm surges and tsunamis [5–7], act as water purifiers, and prevent coastal erosion and abrasion [5]. Furthermore, mangrove ecosystems have an important functional role in determining the balance of biological and nutrient cycling [8–10] since they are often used as nursery habitats [9,10]. Moreover, they play a significant role in preventing climate change, as they sequester 1.023 Mg of carbon per hectare [5,8] and have five times the carbon of tropical, boreal, and temperate forests [11].

In the 20th century, mangroves covered approximately 181,000 km^2 globally [12], and almost half of the world's current mangroves are in Asia [12,13]. However, according to the Food and Agriculture Organization (FAO; 14), approximately one-third of the global mangrove area was lost from 1980 to 2005, and, from this, the largest loss was in Asia [14]. In Southeast Asia, more than 110,000 ha of mangroves have been deforested of which approximately 97,000 ha were lost from 2000 to 2012 [15]. Over half the known mangrove species (36 and 46 out of 70) are found in the Indo-Malay Philippine Archipelago [16]. In these regions, less than 15% of the mangrove species are threatened [16].

Anthropogenic activities are one of the main factors leading to the decline of mangrove areas [8,15]. According to Richards and Friess (2016, [15]), the conversion of mangroves into aquaculture and rice fields is the biggest cause of mangrove deforestation in Southeast Asia. Additionally, climate change is also causing loss in some areas because of drought, sea level rise, and a drastic increase in temperature and salinity, to which mangroves find it difficult to adapt [17–19]. For instance, during the transition from El Niño to La Niña in 2016, the mortality rate of seedlings increased dramatically, owing to an increase in the salinity levels and the tidal inundation on the Pacific coast of Columbia [20]. The loss of mangroves has many negative consequences, and, for some countries, results in economic decline [5,21]. For example, the decrease of mangrove areas in the Gulf of California is predicted to reduce the productivity of fish and crabs by approximately 37,500 USD per hectare [21]. Furthermore, mangrove loss can also disrupt coastal ecosystems [5].

Rehabilitation and restoration strategies are required to address the global decreases in mangrove areas [5,6,22]. However, some mangroves do not respond well to these strategies and have high mortality rates, owing to several parameters that hinder their growth [23–30]. For example, in the Philippines, during mangrove rehabilitation practices, seedlings experienced a mortality rate of 10%–20% because of unsuitability of the location selected [25]. This highlights one of the main reasons for mangrove restoration failures, as unsuitable target areas often do not have the required tidal inundation for mangrove growth [24,26]. The presence of high waves erodes the coastal area and can result in the death of new mangrove plants [23,28]. According to previous investigations [31], the beach slope and elevation affect the frequency of tidal inundation and the impact strength of the waves. This indicates that the elevation and slope parameters are important when selecting suitable areas for mangrove restoration. In addition, in Fujian and Zhejiang, the failure of mangrove rehabilitation was partly due to their unsuitable climates [32].

Two of the important climatic parameters for mangrove growth are precipitation and air temperature. Future climate models project that there will be extreme variations in precipitation [33,34]. Schewe and Levermann (2012, [35]) have predicted that an increase in temperature at the end of the 21st century and early 22nd, will cause changes in the distribution of rainy seasons to 70% below normal. This will cause the rainy seasons in Southeast Asia to be delayed [36,37]. Changes in precipitation can affect the growth, distribution, and area of mangroves [17]. According to Fischer and Knutti [38], an increase in precipitation will result in the death of plants. Conversely, Eslami-Andargoli et al. [39] found that an increase in precipitation could increase the number of mangrove areas due to the expansion of mangrove forests inland. Increases in precipitation could also lead to decreases in pore water salinity and sulfate concentrations, which could increase mangrove productivity [40]. In addition, future climate models also project that there will be an increase in air temperature of 10%

°C^{-1} in Southeast Asia [41,42]. According to the Intergovernmental Panel on Climate Change (IPCC) (2013, [34]), with scenario representative concentration pathway 8.5 (RCP 8.5), global temperatures are expected to increase by 4.8 °C from 2081–2100. An increase in temperature will affect the rate of evaporation and transpiration, which adversely affects plants [43]. Furthermore, a decrease in precipitation and an increase in evaporation will cause an increase in soil salinity. This will result in decreased seedling survival and increase mangrove loss, as their current areas will become hypersaline mud plains [19,44,45]. The occurrence of increased salinity and drought may also affect the species diversity, size, and productivity of mangrove forests [46,47]. Research is, thus, required to identify suitable areas for mangrove restoration in future climatic conditions using various RCP scenarios in Southeast Asia. Moreover, social and economic factors also greatly influence the selection of land for mangrove restoration. This is because coastal urban areas have grown at a faster rate than non-urban coastal areas [48,49]. The coastal population growth and rate of urbanization, which have outpaced demographic development in inland areas, have been driven by rapid economic growth and migration to the coast [50,51]. Hence, the existence of rapid population growth and a coastal economy are important parameters to consider when identifying suitable land for mangrove restoration.

Remote sensing techniques, such as mapping and monitoring, have been used extensively in mangrove research. For example, Vo et al. and Zhen et al. [52,53] used remote sensing to map mangrove forests, whereas Liu et al. and Fauzi et al. [54,55] used remote sensing to monitor the changes in mangrove forests and analyze the cause of their deforestation. Once potential areas have been identified using methods like remote sensing, their suitability needs to be ranked so the optimal areas can be selected for rehabilitation and restoration [56,57]. There are several weighting techniques for land suitability analysis, such as artificial neural networks [58,59], multivariate applications [60], multi-criteria analysis outperforming competitors [61], boolean classification methods [62], root quadratic and multilevel methods [63,64], productivity index [65], a pairwise comparison matrix [66], and an analytical hierarchy process (AHP) [67–78]. This study utilized AHP as its weighting method. AHP is a model that is widely used in decision-making processes, such as when selecting potential land areas and calculating their risk and vulnerability [56–78]. The large number of recent investigations into mangroves makes the AHP model the most suitable for this investigation. This is because this method uses the criteria suggested in previous studies and expert opinions to decide the weight of each of its own criterion. In addition, AHP can consider the relative priority of alternatives as well as represent the best alternative, as it determines the effects of certain weights based on the comparison of paired parameters, according to relative importance [79]. AHP analysis also calculates the consistency value of the index, so that the weights generated for each alternative are consistent with one another [79].

The objective of this study was to assess the amount of land suitable for mangrove restoration today and in the future, according to the different climate change scenarios for 2050 and 2070, using the AHP method with remote sensing, model, and statistical data. The study areas were in Southeast Asia and included Myanmar, Thailand, Cambodia, Vietnam, the Philippines, Malaysia, Singapore, Brunei Darussalam, and Indonesia. To the best of our knowledge, this is the first study to combine climate, hydrodynamic, geomorphological, and socio-economic data to determine the suitability of mangrove land for mangroves in the regional areas using remote sensing, model, and statistical data.

2. Materials and Methods

To determine if the land now, and in the future, is suitable for planting mangroves, this study used hydrodynamic, geomorphological, and climatic parameters. The geomorphological parameters included elevation and slope sub-parameters. The hydrodynamic parameters included tidal inundation sub-parameters, and the climatic parameters included air temperature and precipitation sub-parameters. AHP was used as a method of weighting and determining the importance of each parameter and sub-parameter. After identifying the land suitable for planting mangroves, this study analyzed the socio-economic parameters of the area and their influences on the land's suitability. The socio-economic parameters represented the human pressures that hinder mangrove restoration and included parameters for land cover, population, gross domestic product (GDP), and night light. To understand the effects of using the AHP method to weigh the selections for mangrove land suitability, the results were compared with the results of the land selection using equal weights (i.e., without AHP). They were also compared with the land suitability products of other studies to analyze the uncertainties in this study. The suitable sites for mangrove growth in 2050 and 2070 were assessed using climate model data (precipitation and average air temperature parameters) from the Centre National de Recherches Météorologiques Climate model version 5 (CNRM-CM5.1) and the Model for Interdisciplinary Research on Climate (MIROC). Three RCP scenarios (RCP 2.6, 4.5, and 8.5) were used for each climate parameter. This study compared the results of the current land suitability with that in 2050 and 2070 for each RCP scenario with the two climate models to determine the differences in each scenario for mangrove planting.

The methodology used in this study was divided into stages. The first stage was to determine the parameters and build their hierarchy. The second stage was to create the base map. The third stage was to classify both the parameters and the sub-parameters into several classes. The fourth stage was to determine the weight of each parameter and sub-parameter using the AHP method. Finally, the fifth stage was to create scenarios to determine the land's suitability for mangroves. Figure 1 shows the data-processing schemes for the land suitability analysis for mangrove planting.

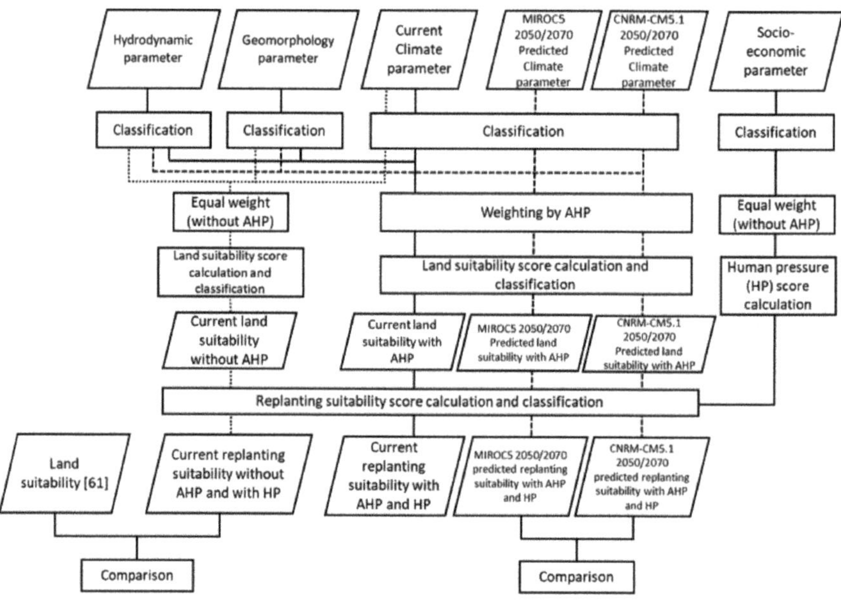

Figure 1. Data-processing scheme of the land suitability analysis for mangrove planting. AHP: analytic hierarchy process. MIROC5: Model for Interdisciplinary Research on Climate 5. CNRM-CM5.1: Centre National de Recherches Météorologiques Climate model version 5.

2.1. Determination of Parameters and Hierarchy Building

The determination of the appropriate parameters was the most important step for identifying land suitability. The parameters were first selected from a literature review and then the parameters were divided into three categories. The first category included the hydrodynamic, geomorphological, and climatic parameters that made the land suitable for planting mangroves. Moreover, the sub-parameters used in this study were those that had the highest weight when compared with the other sub-parameters, when weighted using the AHP method, and those with very low weights were not used in this study. The second category included parameters that make a land unsuitable for planting mangroves, owing to intensive social and economic anthropogenic activities. In this study, these parameters were classified as human pressure parameters. Finally, the third category included climate prediction model parameters. These parameters were used to predict the land suitability for the future growth of mangroves, while assuming that there would be no changes in the hydrodynamic, geomorphological, and human pressure parameters.

2.1.1. Hydrodynamic

The seawater parameter had one sub-parameter, which was tidal inundation. The tidal inundation data used in this study were obtained from tidal data models. The tidal data model was created from TPXO 9.0 data. TPXO is a tidal model based on bathymetry, which assimilates various data sources and is processed using OTIS software [80,81]. The tidal model was verified using tidal station data across Indonesia, which are available from the Indonesian Geospatial Information Agency. Subsequently, the tidal inundation was calculated by multiplying the amplitude value by two, which results in a tidal inundation map of 2018 that had a spatial resolution of 1°.

2.1.2. Geomorphology

The geomorphology parameter had two sub-parameters, elevation, and slope. The elevation and slope sub-parameter data used in this study were obtained from the Multi Error Removed Improved Terrain Digital Elevation Model (MERIT DEM). MERIT DEM is the result of the development of DEM data from SRTM3 v2.1 and AW3D-30m v.1, which eliminated several error components, including absolute bias, noise stripe, noise speckle, and height of the tree bias. These data had a spatial resolution of 90 m [82].

2.1.3. Climate

The climatic parameters had two sub-parameters, namely, average air temperature and precipitation. The average air temperature data used in this study were the 2 m air temperature data obtained from the European Center for Medium-Range Weather Forecasts (ECMWF). These air temperature data pertained to 2 m above the land, sea, or water surface. The 2-m air temperature was calculated by interpolating between the lowest level model and the surface of the earth, while taking atmospheric conditions into account. The unit of measurement for this parameter was kelvin (K). The data used were monthly data derived from daily data accumulation with a spatial resolution of 0.125°.

The precipitation data used in this study were obtained from the Climate Hazards Group InfraRed Precipitation with Stations version 2.0 (CHIRPS v.2.0). The CHIRPS data were a combination of pentadal rainfall data, global geostationary TIR satellite observations from Climate Prediction Center (CPC) and the National Climatic Data Center (NCDC), rainfall atmospheric models from the NOAA Climate Forecast System version 2 (CFSv2), and rainfall data collected from in situ observations at each station [83,84]. This CHIRPS product had a 0.05° spatial resolution with pentadal, decadal, and monthly temporal resolutions [85].

2.1.4. Human Pressure

Human pressure was the parameter that made an area unsuitable for planting mangroves owing to over-population and development of urban areas. The human pressure parameter had four sub-parameters, land cover, population, gross domestic product (GDP), and night light. The land cover data used in this study were climate change land cover initiative (CCI-LC) data from the European Space Agency (ESA). These data describe the surface of the earth in 37 classes of the original land cover based on the United Nations Land Cover Classification System (UN-LCCS) with a spatial resolution of 300 m [86]. The population data used in this study were gridded populations sourced from the world version 4 (GPWv4) data. These data provide information on the population density (number of people per km^2) with a 30″ × 30″ grid size [87].

The GDP data used in this study were obtained from Kummu et al. [88]. They included USD GDP data, obtained from the GDP per capita multiplied by the History Database of the Global Environment (HYDE) population data 3.2, which had a grid size of 5′ × 5′ [88]. The night light data used in this study were the black marble nighttime light (VNP46) data of the National Aeronautics and Space Administration (NASA). These data had a spatial resolution of 500 m. NASA developed black marble as a series of daily products that were calibrated, correlated, and validated, so that the night light data could be used effectively for scientific observations [89].

2.1.5. Climate Prediction Model

The climate prediction data model had two sub-parameters, namely, average air temperature and precipitation. These sub-parameters were obtained from the CNRM-CM5.1 and the MIROC5 models. CNRM-CM5.1 and MIROC5 are the earth system models designed to facilitate simulations that contribute to phase five of the Coupled Model Intercomparison Project (CMIP5). CNRM-CM5.1 consists of several models, namely, the atmospheric model from ARPEGE-Climat version 5.2 and the marine model from The Nucleus for European Modelling of the Ocean version 3.2 (NEMO v3.2) and includes the Interaction Sol-Biosphère-Atmosphère (ISBA) land surface and the sea ice model from Global Experimental Leads and sea ice for Atmosphere and Ocean version 5 (GELATO v5). These models were designed independently and combined through OASIS v.3 software [90]. The CNRM-CM5.1 data had a spatial resolution of 1.4° × 1.4° and it is one of the best global climate models for temperature and rainfall parameters data in Southeast Asia [91]. MIROC5 is a climate prediction model created by the Japanese research community. It has 40 standard vertical resolution levels of up to 3 hPa with a spatial resolution of 1.4° × 1.4°. MIROC5 has better climatological features (precipitation, average zonal atmospheric fields, equatorial subsurface fields, and El Niño-Southern Oscillation simulations) than the previous version [92]. Both CNRM-CM5.1 and MIROC5 included three scenarios, namely, RCP 2.6, 4.5, and 8.5. In RCP 2.6, the greenhouse gas concentration was very low, while RCP 4.5 meant that the total radiative forcing was stabilized shortly. In RCP 8.5, the greenhouse gas emissions increase over time, resulting in higher concentrations. In this study, the results of the mangrove suitability site analyses were compared in the context of these RCPs to determine the scenarios that included the most suitable sites for planting mangroves.

2.2. Creation of the Basemap

Before initiating the data processing, a base map was constructed from coastline data and mangrove distributions and buffered 10 km inland. The mangrove distribution data used in this study were obtained from Giri et al. [13]. The primary data for the mangrove distribution resulted from a combination of the global land survey (GLS) data from 1997 to 2000, with a 30-m resolution, and the Landsat imagery available from the United States Geological Survey (USGS). The secondary data used were global mangrove data [14] and data from the national and local mangrove databases. The mangrove distribution data were buffered 10 km inland. This base map was then used for land selection and to analyze land suitability.

2.3. Classification of the Parameter

To select suitable land for planting mangroves, each parameter needed to be classified and scored. The classification and scoring in this study involved two stages. The first stage was the division of the classes of each land suitability parameter. Each parameter could be classified into four classes (very suitable, suitable, moderate, and unsuitable). The classes were devised based on the results of a literature review (Table 1). The second stage involved assigning a score to each class. Very suitable and suitable classes were assigned scores of four and three, respectively, whereas moderate and unsuitable classes were assigned scores of two and one, respectively. In addition, class divisions and scores were applied to each human pressure parameter. The human pressure parameters could be categorized into either of four classes (low, medium, high, and very high) except for the land cover parameter. The land cover parameter included two classes (urban and non-urban). Low, medium, high, and very high classes were assigned scores of one, two, three, and four, respectively.

Table 1. Classification of the land suitability and human pressure parameters.

	Parameter	Sub-Parameter	Class Value	Class	Reference
Land suitability parameters	Climate	Air temperature (°C)	28–30	Very suitable	[93]
			26–28, 30–32	Suitable	
			8–26, 32–42	Moderate	
			<8, >42	Unsuitable	
		Precipitation (mm)	1400–3750	Very suitable	[94]
			1200–1400, 3750–4500	Suitable	
			0–1200, 4500–7500	Moderate	
			>7500	Unsuitable	
	Geomorphology	Elevation (m)	(−0.25)–1.5	Very suitable	[95]
			(−0.20)–(−0.25), 1.5–1.8	Suitable	
			(−0.20)–(−0.4), 1.8–2.8	Moderate	
			<(−0.4), >2.8	Unsuitable	
		Slope (%)	0–2	Very suitable	[96]
			2–2.15	Suitable	
			2.15–2.5	Moderate	
			>2.5	Unsuitable	
	Hydrodynamic	Tidal inundation (m)	≤0.4	Very suitable	[97,98]
			0.4–0.3	Suitable	
			0.3–1.27	Moderate	
			>1.27	Unsuitable	
Human pressure parameters	Land cover	-	189–191	Very high	-
			10–189, 191–210	Low	
	Population	-	> 150	Very high	-
			50–150	High	
			10–50.0	Medium	
			<10	Low	
	GDP per capita (PPP)	-	>12.375	Very high	[99]
			3.996–12.375	High	
			1.026–3.996	Medium	
			<1.026	Low	
	Nightlight	-	14–255	Very high	-
			10–14.0	High	
			7.0–10	Medium	
			2.0–7.0	Low	

2.4. Determination of the Parameters Weight

The AHP method was used to determine the parameter weights in this study. AHP is a decision-making technique developed by Saaty [79] and was carried out in three stages. First, a pairwise comparison matrix was created by creating a scale with values from one to nine for the hydrodynamic, geomorphological, and climate parameters (Table 2). The scale of importance between one parameter and another was determined by expert judgment. Second, normalize the results of the pairwise comparison matrix were normalized and a column vector created that had n units with n components, so that the weight value and the total weight vector for each parameter could be obtained. Third, the consistency ratios were estimated. To calculate the consistency ratio, the lambda (λ) and consistency index (CI) parameters were required. The λ value was the average value of the consistency vector values of all parameters. The consistency vector value was obtained by dividing the number vector by the weight of each parameter with the following formula.

$$\text{Vector of consistency}\left(VC_{n^{th}\ parameter}\right) = \frac{Number\ of\ vectors_{n^{th}\ parameter}}{Weight_{n^{th}\ parameter}} \tag{1}$$

$$\text{Average consistency}\ (\lambda) = \frac{VC_{1^{st}\ parameter} + VC_{2^{nd}\ parameter} + VC_{n^{th}\ parameter}}{Number\ of\ parameters} \tag{2}$$

Table 2. Pairwise comparison matrix to determine the weights of the main parameters.

Parameter	Hydrodynamic	Geomorphology	Climate
Hydrodynamic	1.000	1/3	2.000
Geomorphology	1.000	1.000	2.000
Climate	1.000	1/3	1.000
n = 3, λ = 2.91, CI = −0.046, RI = 0.58, CR = −0.01			

The λ value was used to calculate the *CI* value, as described in Equation (3). The *CI* calculation was based on the observation that λ is always greater than or equal to the number of the criteria considered (*n*). If the paired comparison matrix is a consistent matrix, then it has a positive value, a reciprocal matrix, and λ = *n*. Therefore, if λ-*n*, then it can be considered as a measure of the inconsistency level. For more details, the CI formula can be written as follows.

$$CI = \frac{\lambda - n}{n - 1} \tag{3}$$

The CI value in Equation (3) can be used to calculate the consistency ratio (*CR*), as described in Equation (4) below.

$$CR = \frac{CI}{RI} \tag{4}$$

where *RI* (random index) is the consistency index of a paired comparison matrix that is randomly generated. If a *CR* value of less than 0.10 was obtained, it indicated a level of rational/reasonable consistency in pairwise comparisons. However, if the *CR* value obtained was greater than 0.10, the ratio values indicated inconsistent ratings. If the assessment was inconsistent, a correction needed to be made to the scoring when the comparison matrix was paired. Table 2 shows the pairwise comparison matrix of the main parameters and the value of the *CR* from this study. Moreover, pairwise comparison matrixes for the sub-parameters are shown in Tables A1–A7. The results of the weighting method using AHP for each parameter and sub-parameter are shown in Table 3.

Table 3. The weight of each parameter and sub-parameter.

Parameter		Sub-Parameter		
Parameter	Weight of Parameter	Sub-Parameter	Weight of Sub-Parameter	Class Value
Climate	0.25	Air temperature (°C)	0.21	28–30 8–28, 30–42 6–8, 42–44 <6, >44
		Precipitation (cm)	0.17	140–375 0–140, 375–750 750–850 >8500
Geomorphology	0.44	Elevation (m)	0.20	(−0.25)–1.5 (−0.4)–(−0.25), 1.5–2.8 (−1.5)–(−0.4), 2.9–3.5 <(−1.5), >3.5
		Slope (%)	0.16	0–2 2–2.5 2.5–3 3.0–4.0
Hydrodynamic	0.31	Tidal inundation (m)	0.25	≤0.4 0.4–1.27 1.27–2 2.0–3.0

2.5. Scenario Generation of Land Suitability

In this study, five scenarios were created to analyze the suitability of the land for mangrove planting in Southeast Asia. In the first scenario, a land suitability map for mangrove planting was created with the use of the AHP method. In the second scenario, the AHP method was utilized and the influences of the human pressure parameters were also considered. In the third scenario, all parameters were assumed to have the same level of importance, and the AHP method was not used. The influence of the human pressure parameters was not considered, as described in Equation (5). In the fourth scenario, all parameters were assumed to have the same level of importance, and the influence of the human pressure parameters was considered. Finally, in the fifth scenario, a land suitability map for mangrove planting in 2050 and 2070, using the data from the two models (CNRM-CM5.1 and MIROC5) was constructed with the use of the AHP method and influence of the human pressure parameters, as represented in Equations (6) and (7). The scenarios in which the human pressure parameters were considered aimed to assess their influence and how essential it was to consider them when selecting suitable areas for mangroves. The scenarios that did not utilize AHP were used to determine the accuracy of the weight results obtained from the AHP method based on the subjectivity of the experts' judgment. To assess the differences between the AHP effects, the Wilcoxon test was carried out as described previously by Chakraborty et al. [57]. The Wilcoxon test can also be used to compare the mean values of a variable from two paired sample data [100,101], whereas the Wilcoxon signed rank test is used only for interval or ratio type data that does not follow a normal distribution. In addition, Equations (5)–(7), respectively, represent the calculations for the replanting suitability of the mangroves' sites.

$$Replanting\ suitability\ score = Land\ suitability\ score - Human\ pressure\ score \tag{5}$$

$$Total\ score\ with\ AHP = \sum_{i=1}^{m} \left(wpi \cdot \sum_{j=1}^{n} (wspij \cdot wrij) \right) \tag{6}$$

$$Total\ score\ without\ AHP = \sum_{i=1}^{m} \left(\frac{1}{m} \cdot \sum_{j=1}^{n} \left(\frac{1}{n} \cdot wspij \cdot wrij \right) \right) \tag{7}$$

where:

m = number of sub-parameters,
n = number of sub-parameters,
wpi = weight of parameter I,
$wspij$ = weight of sub-parameter j in parameter I,
$wrij$ = ranking weight of the pixel's sub parameter j class in parameter i.

3. Results

3.1. Land Suitability for Mangrove Planting as Determined with the AHP Method

This study produced a land suitability map for mangrove planting in Southeast Asia, using the AHP technique, with a 250-m spatial resolution. The potential land suitability area were divided into three classes (very suitable, suitable, and moderate). The detailed results are shown for a selected region of Southeast Asia in Figure 2.

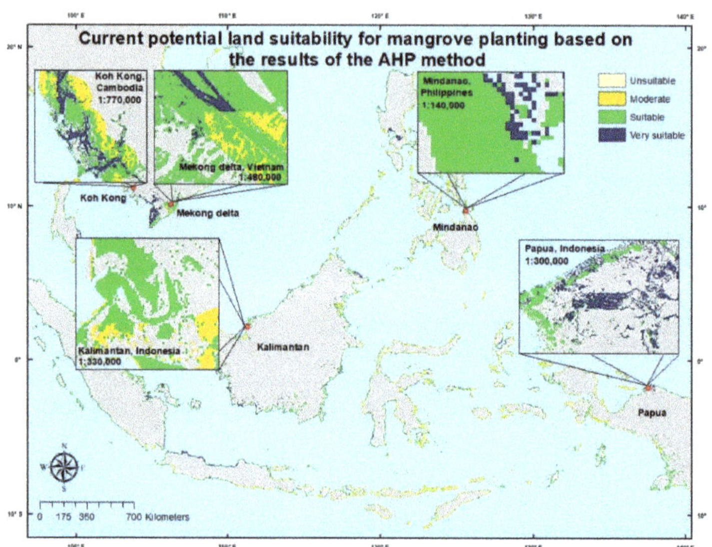

Figure 2. Map of land suitability for mangrove planting based on the results of the analytic hierarchy process (AHP) method.

According to the land suitability map for mangrove planting that was generated using the AHP technique (Figure 2), approximately 3,960,000 ha of land have the potential to be very suitable for mangrove planting in Southeast Asia. The land that was classified as suitable and moderate for mangrove planting extended over 27,791,000 ha and 16,357,000 ha, respectively. The country that had the greatest potential was Indonesia, as it accounted for 57.38% of the total land area, including the very suitable and suitable categories (approximately 18,220,000 ha). Indonesia was followed by the Philippines, Malaysia, Vietnam, Thailand, and Myanmar with areas of 4,483,000 ha (14.12%), 2,725,000 ha (8.58%), 2,637,000 ha (8.31%), 1,679,000 ha (5.29%), and 1,530,000 ha (4.82%), respectively. The countries with the least land potential for mangrove planting were Cambodia (1.06%), Brunei Darussalam (0.29%), and Singapore (0.15%).

The inclusion of the human pressure parameters resulted in a significant change in the potential land area that could be suitable for mangrove planting in Southeast Asia. Based on the land suitability map for mangrove planting that was generated using the AHP technique combined with the human pressure parameters (Figure 3), there were approximately 398,000 ha of land suitable for mangrove planting in Southeast Asia. The land that was classified as suitable and moderate for mangrove planting was 4,771,000 ha and 20,123,000 ha, respectively.

Additionally, the order of the countries with the highest land potential for mangrove planting changed. The country with the largest potential land area was still Indonesia, which accounted for 67.34% of the total land area, including land in the very suitable and suitable categories, which was approximately 2,926,000 ha. Indonesia was followed by Vietnam, the Philippines, Thailand, Malaysia, and Cambodia with areas of 457,000 ha (10.52%), 287,000 ha (6.62%), 234,000 ha (5.38%), 182,000 ha (4.18%), and 143,000 ha (3.29%), respectively. The countries with the smallest area of land for mangroves planting were Myanmar (2.57%), Brunei Darussalam (0.09%), and Singapore (0.02%).

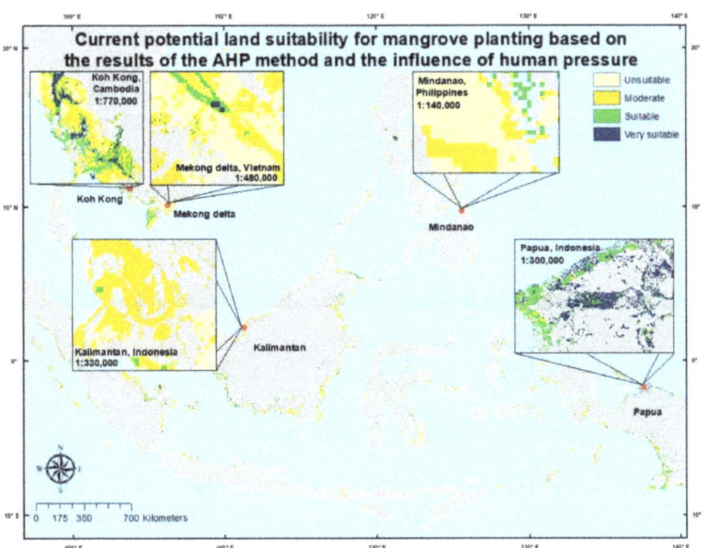

Figure 3. Map of land suitability for mangrove planting based on the results of the analytic hierarchy process (AHP) method and the application of the human pressure parameters.

3.2. Current Potential Land Suitability for Mangrove Planting without the AHP Method

The differences in the potentially suitable land for mangrove planting with and without the use of the AHP technique (Figures 4 and 5), which was used to determine the weight of each parameter, were compared. This comparative analysis aimed to determine the amount of influence the weighting produced by the AHP technique exerted on the potential suitability of the land for mangrove planting.

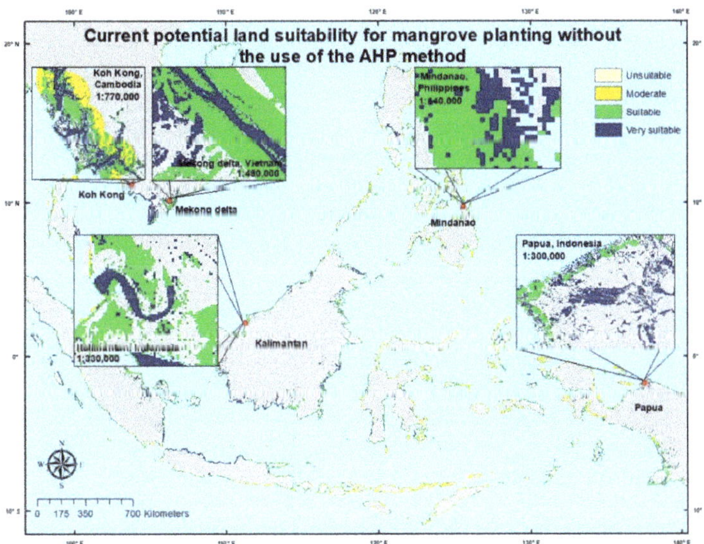

Figure 4. Map of the current potential land suitability for mangrove planting without the use of the analytic hierarchy process (AHP) method.

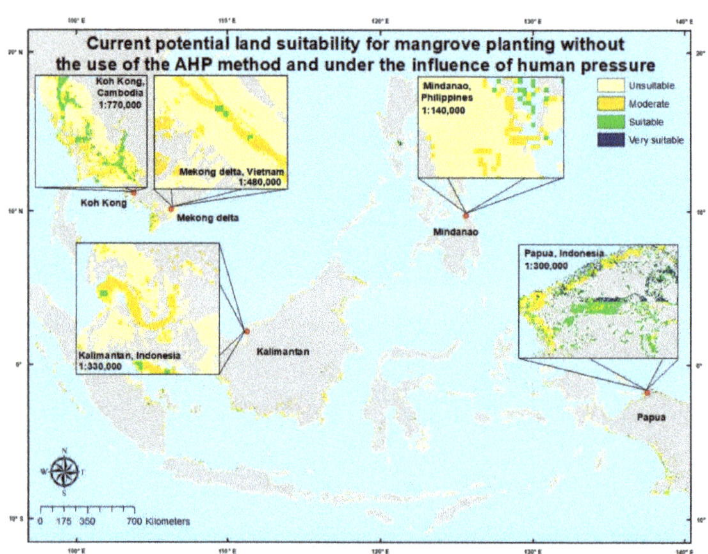

Figure 5. Map of the current potential land suitability for mangrove planting without the use of the analytic hierarchy process (AHP) method, but with the influence of human pressure.

Based on the map of the land suitability potential without the use of the AHP technique (Figure 4), approximately 1,149,781 ha of land had the potential to be very suitable for planting mangroves in Southeast Asia. The land that was very suitable and suitable extended over 9,648,900 ha and 23,320,444 ha, respectively. The country with the greatest land potential for mangrove planting was Indonesia, as it accounted for 44.819% of the total land area, including very suitable and suitable categories of approximately 4,840,000 ha. Indonesia was followed by Myanmar, Vietnam, the Philippines, Thailand, and Malaysia with areas of 2,045,000 ha (18.940%), 1,331,000 ha (12.327%), 809,000 ha (7.494%), 781,000 ha (7.231%), and 753,000 ha (6.972%), respectively. Meanwhile, the countries with the least land potential for mangrove planting were Cambodia (1.956%), Brunei Darussalam (0.248%), and Singapore (0.014%).

In addition, this study mapped the potential land suitability for mangrove planting without using AHP, but considering the influence of the human pressure parameters (Figure 5). The resultant map of the land suitability potential showed that there appeared to be approximately 131,756 ha of potentially very suitable land in Southeast Asia. The very suitable and suitable land areas for mangrove planting were 939,231 ha and 6,988,350 ha, respectively. The country with the greatest land potential for mangrove planting was Indonesia, as it accounted for 59.326% of the total land area, including approximately 635,375 ha of land in the very suitable and suitable categories. The country with the second greatest land area suitable for mangrove planting was the Philippines with 118,806 ha (11.093%). The Philippines was followed by Vietnam, Thailand, Cambodia, and Malaysia with 104,063 ha (9.716%), 92,600 ha (8.646%), 55,025 ha (5.138%), and 36,488 ha (3.407%), respectively. The countries with the least land potential for mangrove planting were Myanmar (2.652%), Singapore (0.015%), and Brunei Darussalam (0.007%).

It is evident from the results that the analysis without the use of the AHP technique resulted in smaller areas of suitable and very suitable land (Table 4). This was because each parameter had a different weighed value, and these weight differences caused the higher-weighted parameters to be more important than the lower-weighted parameters. The results of the AHP technique indicated that the geomorphological parameters had the highest weight, accounting for approximately 38% of the total weight, which were followed by the hydrodynamic parameters and accounted for 32%

of the total weight. The parameter with the smallest weight was the climate, which accounted for 30% of the total weight. These results showed that the geomorphological parameters were more important than the hydrodynamic and climate parameters. However, the weights of each of these parameters added up to the weights of each sub-parameter, which means that an area that has a large geomorphological parameter value may be no more suitable for mangrove planting than areas with high hydrodynamic values. This proves that the weight of each sub-parameter can be influential on the selection of potentially suitable lands for mangrove planting.

Table 4. Areas of land classified as very suitable, suitable, and moderate in terms of planting mangroves in Southeast Asia both with and without the use of the analytic hierarchy process (AHP).

Weighting Technique	Country	Very Suitable (ha)	Suitable (ha)	Moderate (ha)
With AHP	Brunei Darussalam	0	4.050	26.388
	Indonesia	163.738	2762.350	12,741.319
	Cambodia	20.106	122.694	180.519
	Myanmar	1.006	110.638	1445.525
	Malaysia	769	180.975	1332.781
	Philippines	44.194	243.256	1752.450
	Singapore	100	694	11.138
	Thailand	60.538	173.075	662.563
	Vietnam	1.750	455.238	872.931
Without AHP	Brunei Darussalam	0	75	11.388
	Indonesia	52.688	582.688	4826.194
	Cambodia	4.488	50.538	152.625
	Myanmar	0	28.400	378.800
	Malaysia	13	36.475	428.294
	Philippines	18.206	100.600	339.219
	Singapore	0	156	1.100
	Thailand	56.263	36.338	270.825
	Vietnam	100	103.963	579.906

The results of the AHP technique analysis showed that the tidal inundation sub-parameter had the greatest weight when compared with the other sub-parameters. This was approximately four times the value of the smallest weight. The greater weight of this sub-parameter suggests that an area with tidal inundation classified in the very suitable, suitable, or moderate categories for mangrove planting will be preferred over areas in which other sub-parameters may have greater weights, even if they are classified as very suitable. In addition to the tidal inundation sub-parameter, the air temperature sub-parameter played an important role in the selection of potentially suitable land for mangrove planting. The air temperature sub-parameter ranked second in terms of importance after the tidal inundation sub-parameter and its weight was approximately three times that of the smallest weight. This suggests that the air temperature sub-parameter should be prioritized over the other three sub-parameters when selecting land that is potentially suitable for mangrove planting. The elevation and precipitation sub-parameters had the same weight and were considered equally important. However, when compared with the tidal inundation and air temperature, they were less important. The sub-parameter with the smallest weight, according to the results of the AHP technique, was slope.

The analysis of the potential land suitability for mangrove planting without the use of the AHP technique considered all parameters and sub-parameters as equally important in the selection of potentially suitable land for mangrove planting. The equalization of the parameter and sub-parameter weights gave a less accurate selection of the potentially suitable land than when the AHP method was used. Inaccurate land selection may result in the use of unsuitable land for mangrove planting or growth.

3.3. Land Suitability for Mangrove Planting Using the AHP Method and Climate Models

Two climate models were used to predict the potential land suitability for mangrove planting in Southeast Asia in 2050 and 2070. These were the CNRM-CM5.1 and the MIROC5.

3.3.1. Land Suitability for Mangrove Planting in 2050 and 2070 Using the CNRM-CM5.1 Model and the AHP Method

In the land suitability map for mangrove planting in 2050 constructed using the CNRM-CM5.1 model, the 2.6, 4.5, and 8.5 RCP scenarios were examined (Figure 6). There was a slight difference in the areas of the suitable land for mangrove planting among these three scenarios. RCP 8.5 resulted in the largest potential land area (324,125 ha), which was followed by RCP 4.5 (307,313 ha). RCP 2.6 resulted in the smallest suitable land area (264,906 ha). The difference in the land area of the very suitable category between RCP 8.5 and 2.6 was 59,219 ha.

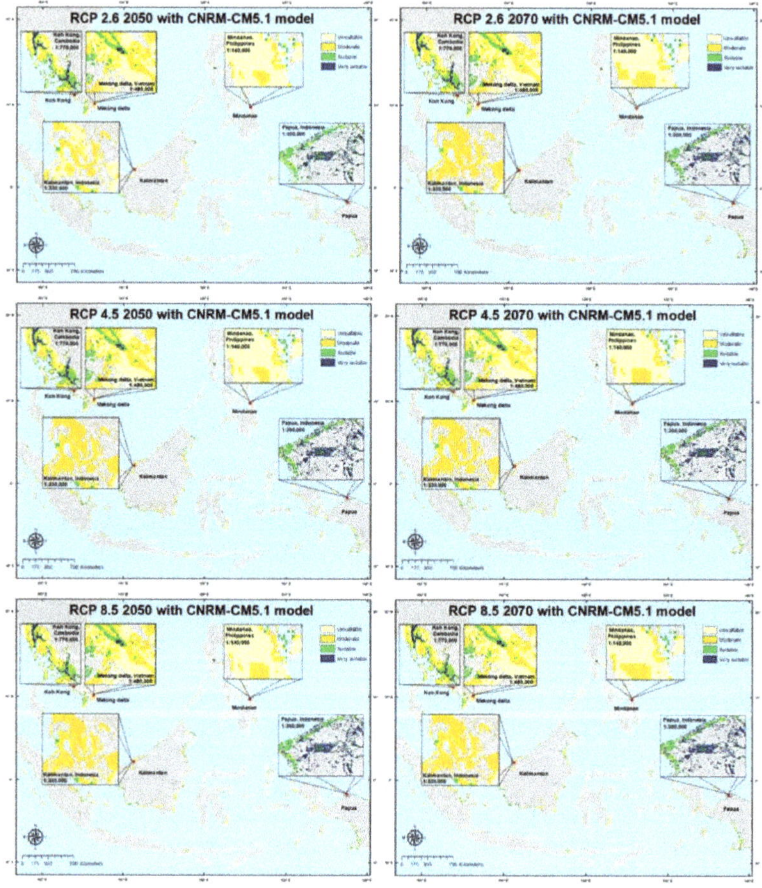

Figure 6. Maps of the land suitability for mangrove planting in 2050 and 2070 using the Centre National de Recherches Météorologiques Climate model version 5 (CNRM-CM5.1) and the analytic hierarchy process (AHP) method with the application of human pressure parameters. RCP: representative concentration pathways.

The results of the three RCPs were also different in terms of the area of the potential land area that was very suitable for mangrove planting in each country (Figure 7). Once again, Indonesia had the highest land suitability in all three RCPs. Additionally, Indonesia showed an increasing area of very suitable land from RCP 2.6 to 8.5 (52,181 ha). This increase was also observed for several other Southeast Asian countries, namely, the Philippines, Myanmar, and Malaysia, with increases of 6650 ha, 1331 ha, and 244 ha, respectively. However, Thailand, Vietnam, and Brunei Darussalam maintained the same area of very suitable potential land with all three RCPs. Singapore had the same potential land area with RCP 2.6 and 4.5, and it increased with RCP 8.5 by 13 ha. The suitable potential land area in Cambodia increased by 156 ha from RCP 2.6 to RCP 4.5. However, there was a drastic reduction of 1200 ha with RCP 8.5.

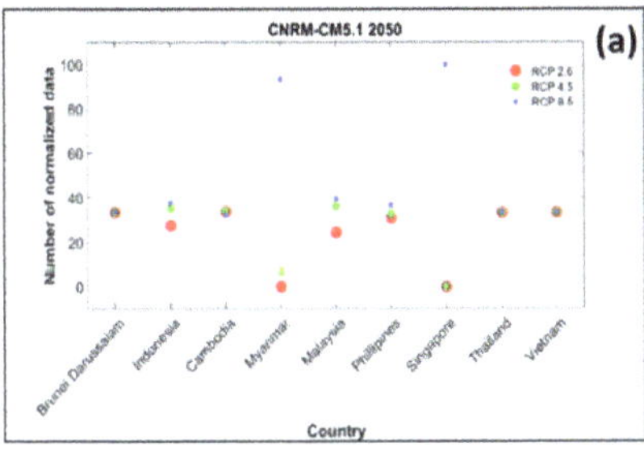
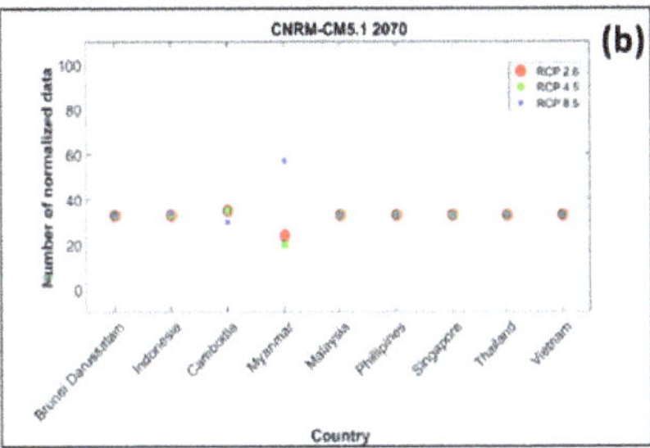

Figure 7. Comparison of the suitable land area for mangrove plating with the RCP 2.6, 4.5, and 8.5 scenarios in (**a**) 2050 and; (**b**) 2070. The data for each country were normalized using the Centre National de Recherches Météorologiques Climate model version 5 (CNRM-CM5.1) and by applying human pressure parameters with the analytic hierarchy process (AHP) method. RCP: representative concentration pathways.

In contrast to 2050, for the three RCPs in 2070 (Figures 8 and 9), the land area that was potentially very suitable for mangrove planting remained relatively similar for each RCP in most countries including Thailand, the Philippines, Vietnam, Malaysia, Brunei Darussalam, and Singapore. Indonesia experienced a decrease of 1219 ha in the potential land area from RCP 2.6 to RCP 4.5. However, the land area that was very suitable for mangrove planting in Indonesia increased by 3356 ha with RCP 8.5. Like Indonesia, the potential land area that was very suitable in Myanmar decreased by 281 ha from RCP 2.6 to RCP 4.5, and increased by 2519 ha with RCP 8.5. Conversely, in Cambodia, the very suitable potential land area decreased by 3813 ha from RCP 2.6 to RCP 8.5.

3.3.2. Potential Land Suitability for Mangrove Planting in 2050 and 2070 Using the MIROC5 Model and the AHP Method

In the land suitability map constructed using the MIROC5 model 2050, the 2.6, 4.5, and 8.5 RCP scenarios were also assessed (Figure 8). There was a difference in the potential land area that was very suitable for mangrove planting among the three scenarios. The largest potential land area for mangrove planting in RCP 8.5 was 324,125 ha, whereas RCP 4.5 had a very suitable potential land area of 307,313 ha. RCP 2.6 had the least very suitable potential land area (264,906 ha). Therefore, the difference in the very suitable land area from RCP 8.5 to 2.6 was 82,838 ha.

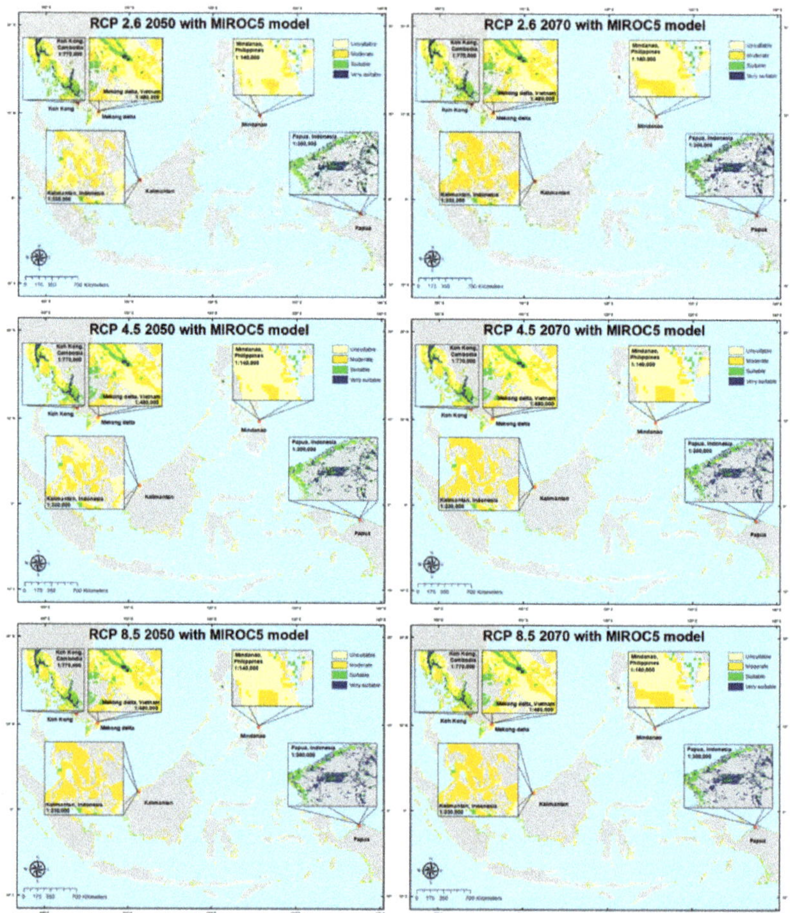

Figure 8. Map of land suitability for mangrove planting in 2050 and 2070 using the Model for Interdisciplinary Research on the Climate (MIROC5) and the analytical hierarchy process (AHP) method with the application of human pressure parameters. RCP: representative concentration pathways.

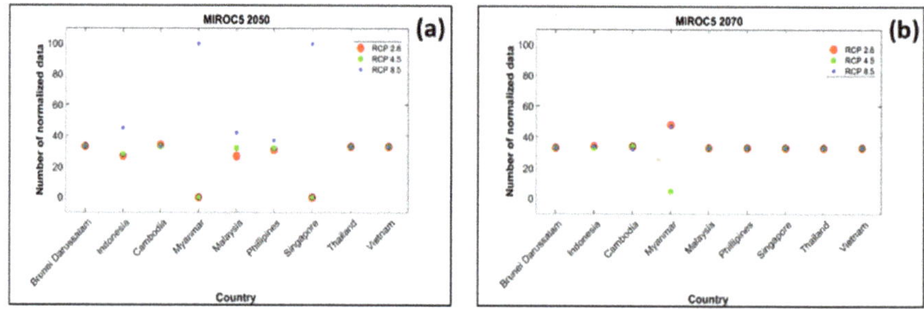

Figure 9. Comparison of the suitable land areas with RCP 2.6, 4.5, and 8.5 in (**a**) 2050 and; (**b**) 2070. The data for each country were normalized using the Model for Interdisciplinary Research on the Climate (MIROC5) and by applying human pressure parameters with the analytical hierarchy process (AHP) method.

The three RCPs also differed in terms of the potential land area that was very suitable in each country (Figure 9). Indonesia had the largest potential land area suitable in all three RCPs, and it increased by 75,800 ha from RCP 2.6 to 8.5. An increase also occurred in several other Southeast Asian countries, namely, the Philippines, Myanmar, Malaysia, and Singapore. These experienced increases of 6413 ha, 1138 ha, 231 ha, and 13 ha, respectively. The area of the very suitable potential land in Vietnam and Brunei Darussalam remained the same in the three RCPs. The very suitable potential land area in Thailand decreased by 69 ha from RCP 2.6 to 4.5 and then increased by 3 ha with RCP 8.5. The suitable potential land area in Cambodia decreased by 719 ha from RCP 2.6 to RCP 8.5.

There were differences in the potentially suitable land area for mangrove planting (Figure 8). The largest area of potential very suitable land was found with RCP 2.6, which was 330,919 ha, whereas RCP 8.5 had the largest area of potential land area in the suitable category (329,381 ha). RCP 4.5 had the smallest land area in the very suitable category (321,506 ha) when compared with the other two RCPs. Therefore, there was a difference of 9413 ha in the very suitable land area from RCP 2.6 to RCP 4.5. Apart from the differences in the area of the very suitable potential land, the results of the three RCPs were relatively stable (Figure 8). The countries where the land area remained the same were Thailand, the Philippines, Vietnam, Malaysia, Brunei Darussalam, and Singapore. In contrast to the other countries, Indonesia experienced a drastic decrease of 5888 ha in a very suitable land area from RCP 2.6 to RCP 4.5. However, it experienced an increase of 5813 ha in the very suitable land area in RCP 8.5. Therefore, the decrease from RCP 2.6 to RCP 8.5 was 75 ha.

4. Discussion

4.1. Comparison of Our Results with Those of Other Studies

An additional feature of this study was the comparison of the results of the potential land suitability for mangroves that were obtained (with or without the use of the AHP method) with the results of Worthington and Spalding (2019, [102]) (Figure 10). The results of Worthington and Spalding (2019, [102]) showed that the total area with restoration potential was 303,708 ha. According to our results, the total area with the greatest potential (very suitable) to plant mangroves (with the use of the AHP technique) was 398,000 ha. Without the use of the AHP technique, this area was reduced to 131,756 ha.

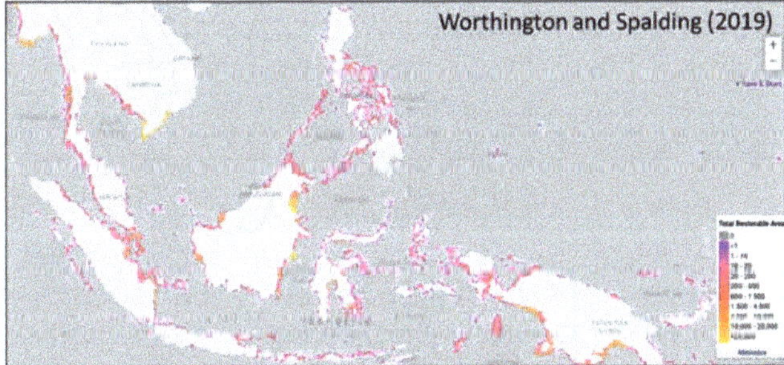

Figure 10. Total restorable mangrove area [102].

In addition, there were differences in the area of land that was determined to be very suitable for mangrove planting in each country (Figure 11). The sequence of countries obtained in this investigation that had the potential for mangrove planting is also different from the previously reported results of Worthington and Spalding (2019, [102]). In this investigation, the sequence of countries, in order from most suitable to least suitable, was Indonesia, Thailand, the Philippines, Cambodia, Vietnam, Myanmar, Malaysia, Singapore, and Brunei Darussalam. The corresponding sequence of the countries obtained without the use of the AHP technique was Thailand, Indonesia, the Philippines, Cambodia, Vietnam,

Malaysia, Myanmar, Brunei Darussalam, and Singapore. According to the results of Worthington and Spalding (2019, [102]), the countries with the highest potentials for mangrove restoration were Indonesia, Myanmar, Thailand, Vietnam, Malaysia, the Philippines, Cambodia, Brunei Darussalam, and Singapore in order of potential. The differences were caused by the variations in the use of environmental parameters during data processing.

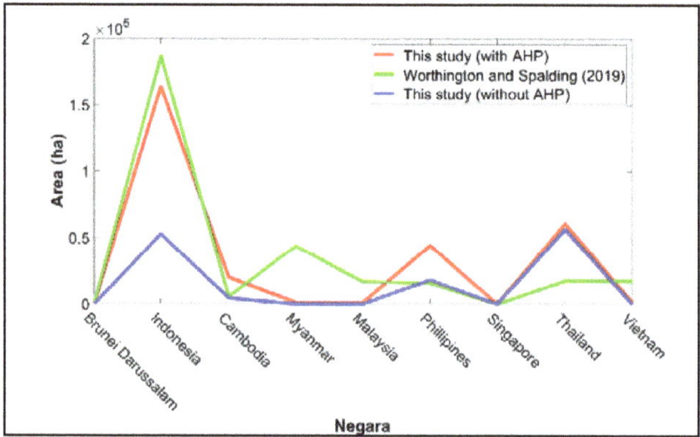

Figure 11. Comparison of the results of the land suitability for mangrove planting in Southeast Asia obtained from this study (with and without the use of the analytic hierarchy process (AHP) method) and the potential restoration results obtained from Worthington and Spalding (2019, [102]).

In addition, in this study, human pressure parameters were included. Therefore, an area that was classified as very suitable was selected not only because of the suitability of the environmental parameters but also because of the low human pressure parameter values. A high human pressure parameter value in a country was caused by high population growth and the GDP. In such a scenario, a larger land area will be required for the construction of socio-economic facilities. The results of this study showed that there were some countries with a smaller land area that were very suitable for mangrove planting when compared with other countries, which may have had larger mangrove areas.

In this study, all parameters were derived from remote sensing data and included environmental (hydrodynamic and geomorphology), climate, and human pressure parameters. However, the parameters used by Worthington and Spalding (2019, [102]) were tidal range, sea level rise, projected future sea level rise, sediment change, average size of mangrove loss, and proximity of the lost area to the remaining mangrove forest.

4.2. Comparison RCPs Scenario in Terms of the Potential Land Suitability for Mangrove Planting in the Future

In this study, two global climate models were used (namely, the CNRM-CM5.1 and the MIROC5) to predict the potential land suitability for mangrove planting in 2050 and 2070 with the RCP 2.6, 4.5, and 8.5. The different uses of the models created a number of different models of potential land suitability for mangrove planting. As shown in Figure 12, RCP 8.5 had the most suitable potential land area. This was due to the fact that its temperature increases were greater than those for the other RCPs (from 18.5 °C to 30.5 °C). In RCP 2.6 and RCP 4.5, the temperature increase was slightly lesser than that in RCP 8.5, from 18 °C to 29.9 °C, and 18.2 °C to 30.1 °C, respectively. Mangroves live and grow optimally at temperatures of 28 °C to 30 °C. This means that most Southeast Asian countries will have very suitable land area for the mangrove in RCP 8.5 in 2050. In contrast, mangrove land suitability predictions for 2070 showed that the majority of Southeast Asian countries maintained the same area across the three RCPs. This occurred because, in 2070, the lowest temperature was predicted to occur in RCP 4.5 (18.5–30.4 °C). RCP 8.5 was predicted to have the highest temperature in 2070, which ranged from 19.2 °C to 31.3 °C. On the other hand, RCP 2.6 had an intermediate temperature between RCP 4.5

and RCP 8.5 (18.5–30.5 °C). The temperature increase from RCP 2.6 to RCP 8.5 both in 2050 and 2070 means that the majority of Southeast Asian countries would have the same pattern in terms of very suitable land area for mangroves planting (Figure 13). Furthermore, the increased air temperature in 2050 and 2070 were not predicted across the temperature limits required for mangrove survival.

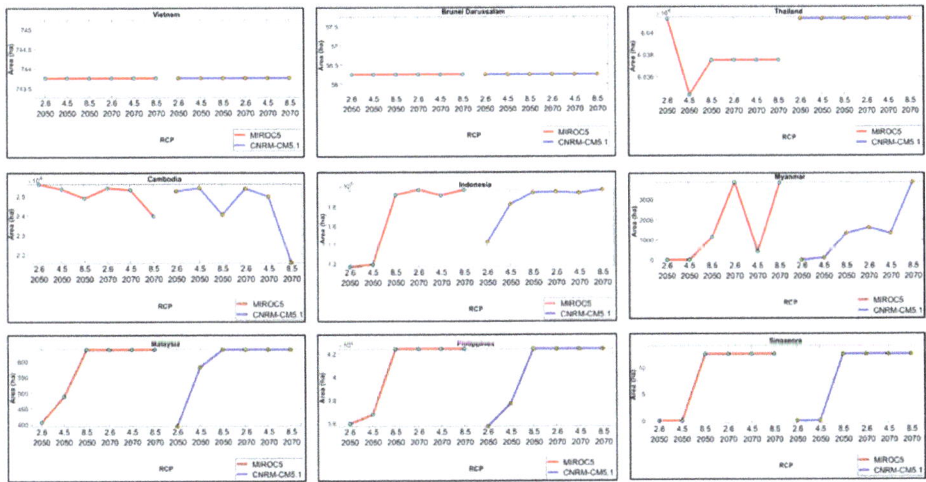

Figure 12. The total area included in the very suitable category for nine Southeast Asian countries in 2050 and 2070 predictions for three representative concentration pathways (RCP), according to the Centre National de Recherches Météorologiques Climate model version 5 (CNRM-CM5.1) and the Model for Interdisciplinary Research on Climate (MIROC5).

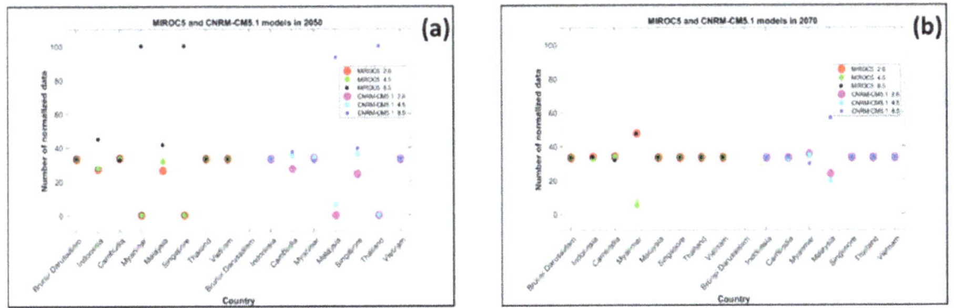

Figure 13. Comparison of the suitable land areas of RCP 2.6, 4.5, and 8.5 in 2050 and 2070. The data of each country were normalized based on the analytic hierarchy process (AHP) method, and the application of human pressure parameters with (**a**) the Centre National de Recherches Météorologiques Climate model version 5 (CNRM-CM5.1), (**b**) the Model for Interdisciplinary Research on Climate (MIROC5). RCP: representative concentration pathways.

According to Duke et al. [45], mangroves will thrive with precipitation levels of 1400 to 3750 mm per year. According to the climate model data of MIROC5 and CNRM-CM5.1 in 2050 and 2070, the amount of precipitation per year in Southeast Asia will range from 0–1257 mm. For RCP 8.5 in 2070, the MIROC5 and CNRM-CM5.1 models had the highest precipitation values (12.574 and 988 mm per year, respectively). The increase in precipitation will also increase the number of areas suitable for planting mangroves in 2070, especially for RCP 8.5. This is consistent with the findings of Eslami-Andargoli et al. (2009, [39]), which increased precipitation, would result in the migration of mangrove forests in land and, consequently, increase the number of areas suitable for planting mangroves. However, the increasing number of inland mangrove areas with RCP 8.5 in 2070 can result in changes to the zonation of mangrove species [17,44,45]. An increase in precipitation could also reduce salinity [44,103,104]. Furthermore, each mangrove species has different hydrodynamic, geomorphological, and climatic criteria. Thus, if there is an increase in precipitation and temperature, some mangrove species would not survive and would eventually become extinct. At the same time, several surviving mangrove species would increase peat production due to decreased salinity and increased freshwater retention [105,106].

4.3. Uncertainties in Selecting Land Suitable for Mangrove Replanting

In this study, hydrodynamic, geomorphological, climatic, and socio-economic remote sensing data were used to produce maps of mangrove land suitability. The use of remote sensing data creates uncertainties related to the selection of the most suitable sites for mangrove planting and restoration. This is due to several reasons. First, the data used had different spatial resolutions. This can affect the interpretation of the suitability of the land for mangrove planting. Second, there were many parameters that affected the suitability of the land in terms of mangrove planting. However, most of these parameters have no available data. Some data were available, but their spatial resolutions were so coarse that they could not be used in this study. Owing to the limitations in the availability of these data, several other land suitability parameters were selected for this study. The parameters chosen had a higher level of importance in terms of the suitability of the land for mangrove planting than that of the other parameters. The AHP method was used to determine the importance level of each parameter. The determination of the level of importance for each parameter in the context of the AHP method resulted from expert judgment. However, some experts had conflicting opinions even though the results of the experts' consistency assessments were consistent among the parameters. For example, expert A said that the tidal inundation parameter was the most influential parameter for selecting land suitable for mangrove planting, whereas expert B said that the most important parameter was slope. Both experts assigned values to each parameter consistently using the same rationale. This situation can also affect the land suitability results. Finally, the maps that predict the mangrove land suitability for 2050 and 2070 were created based on the assumption that hydrodynamic, geomorphological, and socio-economic parameters had fixed values based on the current situation even though this may not be the case. Furthermore, the sea-level rise parameters was not considered in this study. However, the sea-level rise is the most pertinent from global warming because it can change the duration of the swamps, their frequency, and salinity [46,107]. Therefore, the predicted results for 2050 and 2070 may be less accurate.

Table A3 explains that the *CI* value of 0.034 meant that the value of the relationship for each different variable had a fairly different similarity index and was also represented as having been sampled from our domain. Then, the sensitivity test was used to test which variables had changed with the first weight and the AHP weight. After the reduction, the differences in the tidal inundation (0.29), slope (0.19), elevation (0.23), air temperature (0.24), and precipitation (0.21) were obtained. In line with this, the slope parameter had the smallest difference in sensitivity with the total difference in sensitivity $\sum_{i=1}^{5}(S_i) = 1.16$. It can then be interpreted that the use and employment of AHP here was appropriate, as proven by the Wilcoxon Test, which is a non-parametric version of the Student's t-test with value ($W = 25$) and *p*-value = 0.007937 less than the significance level $\alpha = 5\%$. The true location shift is not equal to 0. The weight is the most important aspect in determining model performance, after which it is then necessary to pay attention to the parameters that are significant for model construction. If the weight increases, it takes longer time for a model to reach the maximum solution or convergence. Meanwhile, the Wilcoxon value also proved that the weight obtained was correct.

In addition, the results of this study were compared with several similar studies, namely, by Chakraborty et al. (2019, [56]) and Worthington and Spalding (2019, [102]) (Figure 14). The land suitability of the mangrove in the study by Worthington and Spalding [102] focused on areas lost since 1996. Factors such as tidal ranges, recent sea-level rises, projections of future sea-level rises, and recent sediment changes were added. The time since loss, mean size of patch lost, and proximity of the area lost to the remaining mangrove forests were analyzed using the Delphi method. Meanwhile, the land suitability of mangrove in another study [56] was made based on the climatic, geomorphic, edaphic, and floral conditions and the human interface parameters that were analyzed using the AHP method in the Northern and Central Andaman regions. Although there are some differences in the parameters and the methods used, the results of the mangrove land suitability maps produced by the three studies are very similar. For example, the results of the three studies in the Mayabunder area indicated that this location was relatively unsuitable for mangrove planting, whereas, in the Shyamkund Rangat area, all three studies indicated that most of the area was suitable for mangrove planting. Another example is that all three studies indicated that the Sahari Rangat area was very suitable for planting mangroves (Figure 14). This proves that some of the same parameters used in these three studies have a significant role in selecting suitable locations for mangroves. In addition, there are similarities in several locations suitable for mangrove planting.

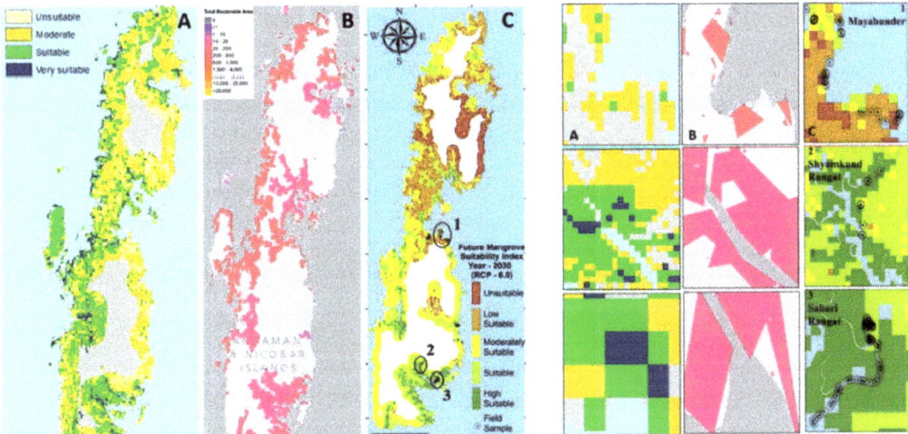

Figure 14. Comparison of land suitability for mangrove restoration in North and Middle Andaman obtained from: (**A**) this study, (**B**) [102], and (**C**) [56].

4.4. Possible Future Directions

The results of this study can be applied in future research both in the environmental and the socio-economic sectors. One of the applications of this study could be to identify suitable land for planting mangroves in Southeast Asia. The reference map of the mangrove land suitability can help select areas that will be subjected to restoration before field surveys are conducted. However, to restore these areas, consideration should be paid to other mangrove land suitability parameters that were not included in this analysis. This will ultimately help minimize failures when planting mangroves. Part of this application includes an effort to implement sustainable development goals (SDGs) 13 and 14 regarding climate change and protecting the marine ecosystem [108].

A range of other important parameters should be included in future studies that are related to the suitability of land for mangrove planting. For example, determining mangrove species that are suitable to plant in areas subjected to either restoration or rehabilitation is of importance. Sea-level rises are also a restoration consideration in many areas. In addition, parameters for the future population growth models can also be added as barriers for the suitability of the land for future mangrove planting. Another research direction that could be followed is the examination of how much carbon will be absorbed if a number of areas are subjected to restoration. The findings of these studies will help elucidate how much greenhouse gas emissions will be reduced if mangrove restoration is carried out in a number of areas. This can support the implementation of greenhouse gas emission reduction targets in accordance with the Paris agreement [109]. The prediction of future mangrove suitability areas in 2050 and 2070 could also be improved by considering long-term changes in the anthropogenic and naturogenic impact [110], such as in agriculture areas [111,112], urban areas [113], and aquaculture practices [114]. These products could be integrated as one input data to make future analyses more realistic. Improving pixel resolution should also be considered to improve future analyses. Several techniques such as a spectral mixture model [115,116] and climate model interpolation [117,118] are key for producing high resolution images of mangrove replanting sites.

5. Conclusions

A map of the land suitable for mangrove planting in Southeast Asia was produced with a spatial resolution of 250 m, and was divided into four scenarios. The first scenario included the land suitability map for mangrove planting with the use of the AHP method. The second scenario included the AHP method and the influence of human pressure parameters. In the third scenario, all parameters were assumed to have the same level of importance and did not use the AHP model. Finally, in the fourth scenario, it was assumed that all parameters had the same level of importance, and the influence of the human pressure parameters was considered, but the AHP method was not used. The first and the second scenarios showed that approximately 3,960,000 ha and 398,000 ha of land, respectively, appeared to have the potential to be very suitable for mangrove planting. In scenarios three and four, approximately 1,149,781 ha and 131,756 ha of land, respectively, appeared to have the potential to be suitable for mangrove planting. All four scenarios showed that the country with the largest land area suitable for mangrove planting was Indonesia, which accounted for approximately 50%–60% of the total land included in the very suitable category in Southeast Asia. Moreover, this study presented a potential land suitability map for mangrove planting for 2050 and 2070 using two climate models (CNRM-CM5.1 and MIROC5) for each year. Each climate model had three RCP scenarios (2.6, 4.5, and 8.5). The results from both models showed that RCP 8.5 resulted in the largest land area suitable for mangrove planting in 2050 in the majority of the Southeast Asian countries. In 2070, almost all Southeast Asian countries had the same land area suitable for mangrove planting with all three RCPs. Future research could use feature selection in machine learning to overcome the shortcomings of the AHP method.

Author Contributions: L.N.S. and A.D.S. were responsible for the overall design of the study. L.N.S. and R.V. were responsible for the Geographic Information System (GIS) method. W.W. was responsible for processing the hydrography data. R.E.C. validated the design of the study and ran statistical analysis. K.W., S.T. and R.P. supported the model design of the study. L.N.S. wrote the paper. All authors have read and agreed to the published version of the manuscript.

Funding: This project was funded by the Pendidikan Magister Menuju Doktor untuk Sarjana Unggul (PMDSU) scholarship from the Ministry of Research, Technology, and Higher Education Indonesia (RisetDikti), and the Indonesian Collaborative Research-World Class University Program, Kurita Asia Research Grant (19Pid017).

Acknowledgments: The authors are grateful to acknowledge the support from the PMDSU scholarship, the Ministry of Research, Technology, and Higher Education Indonesia. The authors thank the experts as respondents to fill the questionnaire. The authors also thank the anonymous reviewers whose valuable comments greatly helped us to prepare an improved and clearer version of this paper. All persons and institutes who kindly made their data available for this analysis are acknowledged.

Conflicts of Interest: The authors declare no conflict of interest. The funders had no role in the design of the study, in the collection, analysis, or interpretation of data, in the writing of the manuscript, or in the decision to publish the results.

Appendix A

Table A1. Pairwise comparison matrix for determination of the weight of hydrodynamic sub-parameters.

Sub-Parameter	Sea Water Temperature	Energy from Sea Wave and Sea Tide	Tidal Inundation
Sea water temperature	1	1/3	1/3
Energy from sea wave and sea tide	3	1	1/3
Tidal inundation	3	2	1
$n = 3$; $\lambda = 2.965$; CI = -0.0177; RI = 0.58; CR = -0.031			

Table A2. The weight of hydrodynamic sub-parameters.

Sub-Parameter	Weight
Sea water temperature	0.15
Energy from sea wave and sea tide	0.31
Tidal inundation	0.54

Table A3. Pairwise comparison matrix for determination of the weight of geomorphology sub-parameters.

Sub-Parameter	Slope	Bathymetry	Elevation
Slope	1.000	3.000	1/3
Bathymetry	1.000	1.000	1/4
Elevation	1.000	2.000	1.000
$n = 3$; $\lambda = 3.067$; CI = 0.034; RI = 0.58; CR = 0.058			

Table A4. The weight of geomorphology sub-parameters.

Sub-Parameter	Weight
Slope	0.35
Bathymetry	0.22
Elevation	0.43

Table A5. Pairwise comparison matrix for determination of the weight of climate sub-parameters.

Sub-Parameter	Air Temperature	Precipitation	Evaporation
Air temperature	1.000	1.000	3.000
Precipitation	1/3	1.000	4.000
Evaporation	1/2	1/4	1.000
$n = 3$; $\lambda = 3.016$; CI = 0.008; RI = 0.58; CR = 0.014			

Table A6. The weight of climate sub-parameters.

Sub-Parameter	Weight
Air temperature	0.45
Precipitation	0.38
Evaporation	0.17

Table A7. The weight of sub-parameters after normalized.

Sub-Parameter	Weight	New weight
Tidal inundation	0.54	0.25
Slope	0.35	0.16
Elevation	0.43	0.20
Air temperature	0.45	0.21
Precipitation	0.38	0.17

References

1. Baran, E. A review of quantified relationships between mangroves and coastal resources. *Phuket Mar. Biol. Cent. Res. Bull.* **1999**, *62*, 57–64.
2. Barbier, E.B.; Edward, B. Valuing the environmental as input: Review of applications to mangrove-fishery linkages. *Ecol. Econ.* **2000**, *35*, 47–61. [CrossRef]
3. Nagelkerken, L.; Blaber, S.J.M.; Bouillon, S.; Green, P.; Haywood, M.; Kirton, L.G.; Meynecke, J.-O.; Pawlik, J.; Penrose, H.M. The habitat function of mangroves for terrestrial and marine fauna: A review. *Aqua Bot.* **2008**, *89*, 155–185. [CrossRef]
4. Cannicci, S.; Burrows, D.; Fratini, S.; Smith III, T.J.; Offenberg, J.; Dahdouh-Guebas, F. Faunal impact on vegetation structure and ecosystem function in mangrove forests: A review. *Aqua Bot.* **2008**, *89*, 186–200. [CrossRef]
5. Barbier, E.B.; Hacker, S.D.; Kebbedy, C.; Loch, E.W.; Stier, A.C.; Silliman, B.R. The value of estuarine and coastal ecosystem services. *Ecol. Soc. Am.* **2011**, *81*, 169–193. [CrossRef]
6. Kathiresan, K.; Rajendran, N. Coastal mangrove forests mitigated tsunami. *Estuar. Coast. Shelf Sci.* **2005**, *65*, 601–606. [CrossRef]
7. Dahdouh-Guebas, F.; Jayatissa, L.P.; Nitto, D.D.; Bosire, J.O.; Seen, D.L.; Koedam, N. How effective were mangroves as a defence against the recent tsunami? *Curr. Biol.* **2005**, *15*, 12. [CrossRef]
8. Donato, D.C.; Kauffman, J.B.; Murdiyarso, D.; Kurnianto, S.; Stidham, M.; Kanninen, M. Mangroves among the most carbon-rich forests in the tropics. *Nat. Geosci.* **2011**, *4*, 293–297. [CrossRef]
9. Robertson, A.I.; Duke, N.C. Mangrove fish-communities in tropical Queensland, Australia: Spatial and temporal patterns in densities, biomass and community structure. *Mar. Biol.* **1990**, *104*, 369–379. [CrossRef]
10. Paillon, C.; Wantiez, L.; Kulbicki, M.; Labonne, M.; Vigliola, L. Extent of mangrove nursery habitats determines the geographic distribution of coral reef fish in a South-Pacific Archipelago. *PLoS ONE* **2014**, *9*, e105158. [CrossRef]
11. Mcleod, E.; Chmura, G.L.; Bouillon, S.; Salm, R.; Björk, M.; Duarte, C.M.; Lovelock, C.E.; Schlesinger, W.H.; Silliman, B.R. A blueprint for blue carbon: Toward an improved understanding of the role of vegetated coastal habitats in sequestering CO_2. *Front. Ecol. Environ.* **2011**, *9*, 552–560. [CrossRef]
12. Spalding, M.D.; Blasco, F.; Field, C.D. *World Mangrove Atlas*, 3rd ed.; The International Society for Mangrove Ecosystems: Okinawa, Japan, 1997; 178p.
13. Giri, C.; Ochieng, E.; Tieszen, L.L.; Zhu, Z.; Singh, A.; Loveland, T.; Masek, J.; Duke, N. Status and distribution of mangrove forests of the world using earth observation satellite data. *Glob. Ecol. Biogeogr.* **2011**, *20*, 154–159. [CrossRef]
14. Food and Agriculture Organization (FAO) of the United Nations. The world's mangroves 1980–2005. *FAO For. Pap.* **2007**, *153*, 18–77.
15. Richards, D.R.; Friess, D.A. Rates and drives of mangrove deforestation in Southeast Asia, 2000–2012. *Proc. Natl. Acad. Sci. USA* **2016**, *113*, 344–349. [CrossRef] [PubMed]
16. Polidoro, B.A.; Carpenter, K.E.; Collins, L.; Duke, N.C.; Ellison, A.M.; Ellison, J.C.; Ellison, J.C.; Farnsworth, E.J.; Fernando, E.S.; Kathiresan, K.; et al. The loss of species: Mangrove extinction risk and geographic areas of global concern. *PLoS ONE* **2010**, *5*, e10095. [CrossRef]
17. Gilman, E.L.; Ellison, J.; Duke, N.C.; Field, C. Threats to mangroves from climate change and adaptation options: A review. *Aqua Bot.* **2008**, *89*, 237–250. [CrossRef]
18. Alongi, D.M. The impact of climate change on mangrove forests. *Curr. Clim. Chang. Rep.* **2015**, *1*, 30–39. [CrossRef]

19. Ward, R.D.; Friess, D.A.; Day, R.H.; Mackenzie, R.A. Impacts of climate change on mangrove ecosystems: A region by region overview. *Ecosyst. Health Sustain.* **2016**, *2*, 1–25. [CrossRef]

20. Riascos, J.M.; Cantera, J.R.; Blanco-Libreros, J.F. Growth and mortality of mangrove seedlings in the wettest neotropical mangrove forests during ENSO: Implications for vulnerability to climate change. *Aquat. Bot.* **2018**, *147*, 34–42. [CrossRef]

21. Aburto-oropeza, O.; Ezcurra, E.; Danemann, G.; Valdez, V.; Murray, J.; Sala, E. Mangroves in the Gulf of California increase fishery yields. *Proc. Natl. Acad. Sci. USA* **2007**, *105*, 10456–10459. [CrossRef]

22. Andradi-Brown, D.A.; Howe, C.; Mace, G.M.; Knight, A.T. Do mangrove forest restoration or rehabilitation activities return biodiversity to pre-impact levels? *Environ. Evid.* **2013**, *2*, 1–8. [CrossRef]

23. Elster, C. Reasons for reforestation success and failure with three mangrove species in Columbia. *For. Ecol. Manag.* **2000**, *131*, 201–214. [CrossRef]

24. Lewis, R.R., III. Ecological engineering for successful management and restoration of mangrove forests. *Ecol. Eng.* **2005**, *24*, 403–418. [CrossRef]

25. Primavera, J.H.; Esteban, J.M.A. A review of mangrove rehabilitation in the Philippines: Successes, failures and future prospects. *Wetl. Ecol. Manag.* **2008**, *16*, 345–358. [CrossRef]

26. Samson, M.S.; Rollon, R.N. Growth performance of planted mangrove in the Philippines: Revisiting forest management strategies. *AMBIO* **2008**, *37*, 234–240. [CrossRef]

27. Zaldívar-Jiménez, M.A.; Herrera-Silveira, J.A.; Teutli-Hernández, C.; Comín, F.A.; Andrade, J.L.; Molina, C.C.; Ceballos, R.P. Conceptual framework for mangrove restoration in the Yucatán Peninsula. *Ecol. Restor.* **2010**, *28*, 333–342. [CrossRef]

28. Winterwerp, J.C.; Erftemeijer, P.L.A.; Suryadiputra, N.; van Eijk, P.; Zhang, L. Defining eco-morphodynamic requirements for rehabilitating eroding mangrove-mud coasts. *Wetlands* **2013**, *33*, 515–526. [CrossRef]

29. Brown, B.; Fadillah, R.; Nurdin, Y.; Soulsby, I.; Ahmad, R. Community based ecological mangrove rehabilitation (CBEMR) in Indonesia. *Surv. Perspect. Integr. Environ. Soc.* **2014**, *7*, 1–11.

30. Kodikara, K.A.S.; Mukherjee, N.; Jayatissa, L.P.; Dahdouh-Guebas, F.; Koedam, N. Have mangrove restoration projects worked? An in-depth study in Sri Lanka. *Restor. Ecol.* **2017**, *25*, 705–716. [CrossRef]

31. Earlie, C.; Masselink, G.; Russell, P. The role of beach morphology on coastal cliff erosion under extreme waves. *Earth Surf. Process Landf.* **2018**, *43*, 1213–1228. [CrossRef]

32. Peng, Y.S.; Zhou, Y.W.; Hou, Y.W.; Chen, G.Z. The restoration of mangrove wetlands: A review. *Acta Ecol. Sin.* **2008**, *28*, 786–797.

33. Maslin, M.; Austin, P. Climate models at their limit? *Nature* **2012**, *486*, 183–184. [CrossRef] [PubMed]

34. Intergovermental Panel on Climate Change (IPCC). Climate change 2013: The physical science basis. In *Contribution of Working Group I to the Fifth Assessment Report of the Intergovermental Panel on Climate Change*; Cambridge University Press: Cambridge, UK; New York, NY, USA, 2013; 1535p.

35. Schewe, J.; Levermann, A. A statistically predictive model for future monsoon failure in India. *Environ. Res. Lett.* **2012**, *7*, 1–9. [CrossRef]

36. Ashfaq, M.; Shi, Y.; Tung, W.; Trapp, R.J.; Gao, X.; Pal, J.S.; Diffenbaugh, N.S. Suppression of South Asian summer monsoon precipitation in the 21st century. *Geophys. Res. Lett.* **2009**, *26*, L01704. [CrossRef]

37. Loo, Y.Y.; Billa, L.; Singh, A. Effect of climate change on seasonal monsoon in Asia and its impact on the variability of monsoon rainfall in Southeast Asia. *Geosci. Front.* **2015**, *6*, 817–823. [CrossRef]

38. Fischer, E.M.; Knutti, R. Detection of spatially aggregated changes in temperature and precipitation extremes. *Geophys. Res. Lett.* **2014**, *41*, 547–554. [CrossRef]

39. Eslami-Andargoli, L.; Dale, P.; Sipe, N.; Chaseling, J. Mangrove expansion and rainfall patterns in Moreton Bay, Southeast Queensland, Australia. *Estuar. Coast. Shelf Sci.* **2009**, *85*, 292–298. [CrossRef]

40. Gilman, E.; Ellison, J. Efficacy of alternative low-cost approach to mangrove restoration, American Samoa. *Estuaries Coasts* **2007**, *30*, 641–651. [CrossRef]

41. Bathiany, S.; Dakos, V.; Scheffer, M.; Lenton, T.M. Climate models predict increasing temperature variability in poor countries. *Sci. Adv.* **2018**, *4*, 1–10. [CrossRef]

42. Coumou, D.; Rahmstorf, S. A decade of seather extremes. *Nat. Clim. Chang.* **2012**, *2*, 491–496. [CrossRef]

43. Arnell, N.W. Climate change and global water resources: SRES emissions and socio-economic scenarios. *Glob. Environ. Chang.* **2004**, *14*, 31–51. [CrossRef]

44. Field, C. Impacts of expected climate change on mangroves. *Hydrobiologia* **1995**, *295*, 75–81. [CrossRef]

45. Duke, N.C.; Ball, M.C.; Ellison, J.C. Factors influencing biodiversity and distributional gradients in mangrove. *Glob. Ecol. Biogeogr.* **1998**, *7*, 27–47. [CrossRef]

46. Smith, T.J., III; Duke, N.C. Physical determinants of inter-estuary variation in mangrove species richness around the tropical coastline of Australia. *J. Biogeogr.* **1987**, *14*, 9–19. [CrossRef]

47. Ball, M.C.; Sobrado, M.A. Ecophysiology of mangroves: Challenges in linking physiological processes with patterns in forest structure. In *Advances in Plant Physiological Ecology*; Press, M.C., Scholes, J.D., Barker, M.G., Eds.; Blackwell Science: Oxford, UK, 2002; pp. 331–346.

48. Small, C.; Nicholls, R.J. A global analysis of human settlement in coastal zones. *J. Coast. Res.* **2003**, *19*, 584–599.

49. Balk, D.; Montgomery, M.R.; McGranahan, G.; Kim, D.; Mara, V.; Todd, M.; Buettner, T.; Dorelien, A. Mapping urban settlements and the risks of climate change in Africa, Asia and South America. *Popul. Dyn. Clim. Chang.* **2009**, *80*, 103.

50. McGranahan, G.; Balk, D.; Anderson, B. The rising tide: Assessing the risks of climate change and human settlements in low elevation coastal zones. *Environ. Urban.* **2007**, *19*, 17–37. [CrossRef]

51. Smith, K. We are sevel billion. *Nat. Clim. Chang.* **2011**, *1*, 331–335. [CrossRef]

52. Vo, Q.T.; Oppelt, N.; Leinenkugel, P.; Kuenzer, C. Remote sensing in mapping mangrove ecosystems-an object-based approach. *Remote Sens.* **2013**, *5*, 183–201. [CrossRef]

53. Zhen, J.; Liao, J.; Shen, G. Mapping mangrove forests of Dongzhaigang nature reserve in China using Landsat 8 and Radarsat-2 polimetric SAR data. *Sensors* **2018**, *18*, 4012. [CrossRef]

54. Liu, K.; Li, X.; Shi, X.; Wang, S. Monitoring mangrove forest changes using remote sensing and GIS data with decision-tree learning. *Wetlands* **2008**, *28*, 336–346. [CrossRef]

55. Fauzi, A.; Sakti, A.; Yayusman, L.; Harto, A.; Prasetyo, L.; Irawan, B.; Kamal, M.; Wikantika, K. Contextualizing mangrove forest deforestation in Southeast Asia using environmental and socio-economic data product. *Forests* **2019**, *10*, 952. [CrossRef]

56. Chakraborty, S.; Sahoo, S.; Majumdar, D.; Saha, S.; Roy, S. Future mangrove suitability assessment of Andaman to strengthen sustainable development. *J. Clean. Prod.* **2019**, *234*, 597–614. [CrossRef]

57. Monsef, H.A.; Hassan, M.A.A.; Shata, S. Using spatial data analysis for delineating existing mangroves stands and siting suitable locations for mangroves plantation. *Comput. Electron. Agric.* **2017**, *141*, 310–326. [CrossRef]

58. Wang, F. The use of artificial neural networks im a geographical information system for agricultural land-suitability assessment. *Environ. Plan.* **1994**, *26*, 265–284. [CrossRef]

59. Bagherzadeh, A.; Ghadiri, E.; Darban, A.R.S.; Gholizadeh, A. Land suitability modeling by parametric-based neural networks and fuzzy methods for soybean production in a semi-arid region. *Modeling Earth Syst. Environ.* **2016**, *2*, 1–11. [CrossRef]

60. Bojórquez-Tapia, L.A.; Diaz-Mondragón, S.; Ezcurra, E. GIS-based approach for participatory decision making and land suitability assessment. *Int. J. Geogr. Inf. Sci.* **2001**, *15*, 129–151. [CrossRef]

61. Joerin, F.; Thériault, M.; Musy, A. Using GIS and outranking multicriteria analysis for land-use suitability assessment. *Int. J. Geogr. Inf. Sci.* **2001**, *15*, 153–174. [CrossRef]

62. Kalogirou, S. Expert systems and GIS: An application of land suitability evaluation. *Comput. Environ. Urban Syst.* **2002**, *26*, 89–112. [CrossRef]

63. Shalaby, A.; Ouma, Y.O.; Tateishi, R. Land suitability assessment for perennial crops using remote sensing and geographic information system: A case study in northwestern Egypt. *Arch. Agron. Soil Sci.* **2006**, *52*, 243–261. [CrossRef]

64. Foshtomi, M.D.; Norouzi, M.; Rezaei, M.; Akef, M.; Akbarzadeh, A. Qualitative and economic land suitability evaluation for tea (*Camellia sinesis* L.). *J. Biol. Environ. Sci.* **2011**, *5*, 135–146.

65. Olayeye, A.O.; Akinbola, G.E.; Marake, V.M.; Molete, S.F.; Mapheshoane, B. Soil in suitability evaluation for irrigated lowland rice culture in Southwestern Nigeria: Management implication for sustainability. *Commun. Soil Sci. Plant Anal.* **2008**, *39*, 2920–2938. [CrossRef]

66. Chandio, I.A.; Matori, A.N.; Yusof, K.; Talpur, M.A.H.; Aminu, M. GIS-basedland suitability analysis of sustainable hillside development. *Procedia Eng.* **2014**, *77*, 87–94. [CrossRef]

67. Bandyopadhyay, S.; Jaiswal, R.K.; Hegde, V.S.; Jayaraman, V. Assessment of land suitability potentials for agriculture using a remote sensing and GIS based approach. *Int. J. Remote Sens.* **2009**, *30*, 879–895. [CrossRef]

68. Cengiz, T.; Akbulak, C. Application of analytical hierarchy process and geographic information systems in land-use suitability evaluation: A case study of Dùmrek village (Canakkale, Turkey). *Int. J. Sustain. Dev. World Ecol.* **2009**, *16*, 286–294. [CrossRef]

69. Jafari, S.; Zaredar, N. Land suitability analysis using multi attributr decision making approach. *Int. J. Environ. Sci. Dev.* **2010**, *1*, 441–445. [CrossRef]

70. Chandio, I.A.; Matori, A.N.; Lawal, D.U.; Sabri, S. GIS- based land suitability analysis using AHP for public parks planning in Larkana City. *Mod. Appl. Sci.* **2011**, *5*, 177–189. [CrossRef]

71. Feizizadeh, B.; Blaschke, T. Land suitability analysis for Tabriz Country, Iran: A multi-criteria evaluation approach using GIS. *J. Environ. Plan. Manag.* **2012**, *56*, 1–23. [CrossRef]

72. Akinci, H.; Özalp, A.Y.; Turgut, B. Agricultural land use suitability analysis using GIS and AHP technique. *Comput. Electron. Agric.* **2013**, *97*, 71–82. [CrossRef]

73. García, J.L.; Alvarado, A.; Blanco, J.; Jiménez, E.; Maldonado, A.A.; Cortés, G. Multi-attribute evaluation and selection of sites for agricultural product warehouses based on an analytic hierarchy process. *Comput. Electron. Agric.* **2014**, *100*, 60–69. [CrossRef]

74. Yi, X.; Wang, L. Land suitability assessment on a watershed of Loess Plateau using the analytic hierarchy process. *PLoS ONE* **2013**, *8*, e69498. [CrossRef]

75. Parry, J.A.; Ganaie, S.A.; Bhat, M.S. GIS based land suitability analysis using AHP model for urban services planning in Srinagar and Jammu urban centers of J&K, India. *J. Urban Manag.* **2018**, *7*, 46–56.

76. Arulbalaji, P.; Padmalal, D.; Sreelash, K. GIS and AHP techniques based delineation of groundwater potential zones: A case study from Southern Western Ghats, India. *Sci. Data* **2019**, *9*, 1–17. [CrossRef] [PubMed]

77. Chen, Z.; Chen, T.; Qu, Z.; Yang, Z.; Ji, X.; Zhou, Y.; Zhang, H. Use of evidential reasoning and AHP to assess regional industrial safety. *PLoS ONE* **2018**, *13*, 1–21. [CrossRef] [PubMed]

78. Papathoma-Köhle, M.; Schlögl, M.; Fuchs, S. Vulnerability indicators for natural hazards: An innovative selection and weighting approach. *Sci. Rep.* **2019**, *9*, 1–14. [CrossRef] [PubMed]

79. Saaty, R.M. The analytic hierarchy process-what it is and how it is used. *Math. Model.* **1987**, *9*, 161–176. [CrossRef]

80. Egbert, G.D.; Bennett, A.F. TOPEX/POSEIDON tides estimated using a global inverse model. *J. Geophys. Res.* **1994**, *99*, 24821–24852. [CrossRef]

81. Egbert, G.D.; Erofeeva, S.Y. Efficient inverse modeling of barotropic ocean tides. *Am. Meteorol. Soc.* **2002**, *19*, 183–203. [CrossRef]

82. Yamazaki, D.; Ikeshima, D.; Tawatari, R.; Yamaguchi, T.; O'Loughlin, F.; Nael, C.J.; Sampson, C.C.; Kanae, S.; Bates, P.D. A high-accuracy map of global terrain elevations. *Geophys. Res. Lett.* **2017**, *44*, 5844–5853. [CrossRef]

83. Janowiak, J.E.; Joyce, R.J.; Yarosh, Y. A real-time global half-hourly pixel-resolution infrared dataset and its applications. *Bull. Am. Meteorol. Soc.* **2001**, *82*, 205–217. [CrossRef]

84. Saha, S.; Moorthi, S.; Pan, H.L.; Wu, X.; Wang, J.; Nadiga, S.; Tripp, P.; Kistler, R.; Woollen, J.; Behringer, D.; et al. The NCEP climate forecast system reanalysis. *Am. Meteorol. Soc.* **2010**, *91*, 1015–1057. [CrossRef]

85. Funk, C.; Peterson, P.; Landsfeld, M.; Pedreros, D.; Verdin, J.; Shukla, S.; Husak, G.; Rowland, J.; Harrison, L.; Hoell, A.; et al. The climate hazards infrared precipitation with stations-a new environmental record for monitoring extremes. *Sci. Data* **2015**, *2*, 1–21. [CrossRef] [PubMed]

86. Di Gregorio, A. *UN Land Cover Classification System (LCCS)-Classification Concepts and User Manual for Software Version 2*; Food Agriculture Organization of the United Nation. Rome, Italy, 2005.

87. Doxsey-Whitfield, E.; MacManus, K.; Adamo, S.B.; Pistolesi, L.; Squires, J.; Borkovska, O.; Baptista, S.R. Taking advantage of the improved availability of census data: A first look at the gridded population of the world, version 4. *Pap. Appl. Geogr.* **2015**, *1*, 226–234. [CrossRef]

88. Kummu, M.; Taka, M.; Guillaume, J.H.A. Data descriptor: Gridded global datasets for gross domestic product and human development index over 1990–2015. *Sci. Data* **2018**, *5*, 1–15. [CrossRef] [PubMed]

89. Román, M.O.; Wang, Z.; Sun, Q.; Kalb, V.; Miller, S.D.; Molthan, A.; Schultz, L.; Bell, J.; Stokes, E.C.; Pandey, B.; et al. NASA's black marble nighttime lights product suite. *Remote Sens. Environ.* **2018**, *210*, 113–143. [CrossRef]

90. Voldoire, A.; Sanchez-Gomez, E.; Melia, D.S.; Decharme, B.; Cassou, C.; Senesi, S.; Valcke, S.; Beau, I.; Alias, A.; Chevallier, M.; et al. The CNRM-CM5.1 global climate model: Description and basic evaluation. *Clim. Dyn.* **2013**, *40*, 2091–2121. [CrossRef]

91. Kamworapan, S.; Surussavadee, C. Evaluation of CMIP5 global climate models for simulating climatological temperature and precipitation for Southeast Asia. *Adv. Meteorol.* **2019**. [CrossRef]

92. Watanabe, M.; Suzuki, T.; O'ishi, R.; Komuro, Y.; Watanabe, S.; Emori, S.; Takemura, T.; Chikira, M.; Ogura, T.; Sekiguchi, M.; et al. Improved climate simulation by MIROC5: Mean states, variability, and climate sensitivity. *J. Clim.* **2010**, *23*, 6312–6335. [CrossRef]

93. Monsef, H.A.; Ayman, S.H.A.; Smith, S.E. Locating suitable mangrove plantation sites along the Saudi Arabia Red Sea Coast. *J. Afr. Earth Sci.* **2013**, *83*, 1–9. [CrossRef]

94. Duke, C.N. Mangrove floristics and biogeography. *Trop. Mangrove Ecosyst.* **1992**, *41*, 63–100.

95. Leong, R.C.; Friess, D.A.; Crase, B.; Lee, W.K.; Webb, E.L. High-resolution pattern of mangrove species distribution is controlled by surface elevation. *Estuar. Coast. Shelf Sci.* **2017**, *202*, 185–192. [CrossRef]

96. Suprakto, B.; Soemarno, M.; Arfianti, D. Development of mangrove conservation area based on land suitability and environmental carrying capacity (case study Probolinggo Coastal Area, East Java, Indonesia). *Int. J. Ecosyst.* **2014**, *4*, 107–118.

97. Clarke, L.D.; Hannon, N.J. The mangrove swamp and salt marsh communities of the Sydney District: II the holocoenotic complex with particular reference to physiography. *J. Ecol.* **1969**, *57*, 213–234. [CrossRef]

98. Primavera, J.H.; Savaris, J.D.; Bajoyo, B.; Coching, J.D.; Curnick, D.J.; Golbeque, R.; Guzman, A.T.; Henderin, J.Q.; Joven, R.V.; Loma, R.A.; et al. *Manual on Community-Based Mangrove Rehabilitation-Mangrove Manual Series No.1*; ZSL: London, UK, 2012; Volume viii, 240p.

99. New Country Classifications by inCome Level: 2019–2020. Available online: https://blogs.worldbank.org/opendata/new-country-classifications-income-level-2019-2020 (accessed on 18 December 2019).

100. Caraka, R.E.; Lee, Y.; Kurniawan, R.; Herliansyah, R.; Kaban, P.A.; Nasution, B.I.; Gio, P.U.; Chen, R.C.; Toharudin, T.; Pardamean, B. Impact of COVID-19 large scale restriction on environment and economy in Indonesia. *Glob. J. Environ. Sci. Manag.* **2020**, *2*, 65–84.

101. Woolson, R.F. Wilcoxon signed-rank test. *Wiley Encycl. Clin. Trials* **2007**, *9*, 1–3.

102. Worthington, T.; Spalding, M. Mangrove Restoration Potential: A global map highlighting a critical opportunity. 2018. Available online: https://doi.org/10.17863/CAM.39153 (accessed on 24 June 2020).

103. McKee, K. Soil physiochemical patterns and mangrove species distribution-reciprocal effercts? *J. Ecol.* **1993**, *81*, 477–487. [CrossRef]

104. Ellison, J. South Pacific mangroves may respond to predicted climate change and sea level rise. In *Climate Change in the South Pacific: Impacts and Responses in Australia, New Zealand, and Small Islands States*; Gillespie, A., Burns, W., Eds.; Kluwer Academic Publishers: Dordrecht, The Netherlands, 2000; Chapter 15; pp. 289–301.

105. Snedaker, S. Mangroves and climate change in the Florida and Caribbean region: Scenario and hypotheses. *Hydrobiologia* **1995**, *295*, 43–49. [CrossRef]

106. Snedaker, S. Impact on mangroves. In *Climate Change in the Intra-American Seas: Implications of Future Climate Change on the Ecosystems and Socio-economic Structure of the Marine and Coastal Regimes of the Caribbean Sea, Gulf of Mexico, Bahamas and N. E. Coast of South America*; Maul, G.A., Ed.; Edward Arnold: London, UK, 1993; pp. 282–305.

107. Krauss, K.W.; McKee, K.L.; Lovelock, C.E.; Cahoon, D.R.; Saintilan, N.; Reef, R.; Chen, L. How mangrove forests adjust to rising sea level. *New Phytol.* **2013**, *202*, 19–34. [CrossRef]

108. UN High Commissioner for Refugees (UNHCR). The Sustainable Development Goals and Addressing Statelessness, March 2017. Available online: https://www.refworld.org/docid/58b6e3364.html (accessed on 16 May 2020).

109. United Nations Framework Convention on Climate Change (UNFCCC). Paris Agreement. 12 December 2015. Available online: https://unfccc.int/files/essential_background/convention/application/pdf/english_paris_agreement.pdf (accessed on 24 June 2020).

110. Sakti, A.D.; Fauzi, A.I.; Wilwantikta, F.N.; Rajagukguk, Y.S.; Sudhana, S.A.; Yayusman, L.F.; Syahid, L.N.; Sritarapipat, T.; Principe, J.A.; Trang, N.T.Q.; et al. Multi-source remote sensing data product analysis: Investigating anthropogenic and naturogenic impacts on mangroves in Southeast Asia. *Remote Sensing* **2020**, *12*, 2720. [CrossRef]

111. Zabel, F.; Delzeit, R.; Schneider, J.M.; Seppelt, R.; Mauser, W.; Václavík, T. Global impacts of future cropland expansion and intensification on agricultural markets and biodiversity. *Nat. Commun.* **2019**, *10*, 2844. [CrossRef]

112. Sakti, A.D.; Takeuchi, W. A Data-Intensive Approach to Address Food Sustainability: Integrating Optic and Microwave Satellite Imagery for Developing Long-Term Global Cropping Intensity and Sowing Month from 2001 to 2015. *Sustainability* **2020**, *12*, 3227. [CrossRef]

113. Zhou, Y.; Varquez, A.C.G.; Kanda, M. High-resolution global urban growth projection based on multiple applications of the SLEUTH urban growth model. *Sci. Data* **2019**, *6*, 34. [CrossRef] [PubMed]

114. Donchyts, G.; Baart, F.; Winsemius, H.; Gorelick, N.; Kwadijk, J.; Giesen, N. Earth's surface water change over the past 30 years. *Nat. Clim. Chang.* **2016**, *6*, 810–813. [CrossRef]

115. Sakti, A.D.; Tsuyuki, S. Spectral Mixture Analysis of Peatland Imagery for Land Cover Study of Highly Degraded Peatland in Indonesia. In *The International Archives of the Photogrammetry, Remote Sensing and Spatial Information Science*; Copernicus Publications: Göttingen, Germany, 2015; Volume XL-7/W3.

116. Peddle, D.R.; Brunke, S.P.; Hall, F.G. A Comparison of Spectral Mixture Analysis and Ten Vegetation Indices for Estimating Boreal Forest Biophysical Information from Airborne Data. *Can. J. Remote Sens.* **2001**, *27*, 627–635. [CrossRef]

117. Tveito, O.E. The Developments in Spatialization of Meteorological and Climatological Elements. In *Spatial Interpolation for Climate Data: The Use of GIS in Climatology and Meteorology*; Dobesch, H., Dumolard, P., Dyras, L., Eds.; ISTE Ltd.: London, UK, 2007; pp. 73–86.

118. Szentimrey, T.; Bihari, Z.; Szalai, S. Comparison of Geostatistical and Meteorological Interpolation Methods (What is What?). In *Spatial Interpolation for Climate Data: The Use of GIS in Climatology and Meteorology*; Dobesch, H., Dumolard, P., Dyras, L., Eds.; ISTE Ltd.: London, UK, 2007; pp. 45–56.

Publisher's Note: MDPI stays neutral with regard to jurisdictional claims in published maps and institutional affiliations.

 *remote sensing*

Article

Monitoring Bark Beetle Forest Damage in Central Europe. A Remote Sensing Approach Validated with Field Data

Angel Fernandez-Carrillo [1,*], Zdeněk Patočka [2], Lumír Dobrovolný [3], Antonio Franco-Nieto [1] and Beatriz Revilla-Romero [1]

[1] Remote Sensing and Geospatial Analytics Division, GMV, Isaac Newton 11, P.T.M. Tres Cantos, E-28760 Madrid, Spain; afranco@gmv.com (A.F.-N.); brevilla@gmv.com (B.R.-R.)

[2] Department of Forest Management and Applied Geoinformatics, Faculty of Forestry and Wood Technology, Mendel University in Brno, 613 00 Brno, Czech Republic; zdenek.patocka@mendelu.cz

[3] University Forest Enterprise Masaryk Forest Křtiny, Mendel University in Brno, 613 00 Brno, Czech Republic; lumir.dobrovolny@slpkrtiny.cz

* Correspondence: aafernandez@gmv.com; Tel.: +34-660-136-356

Received: 9 October 2020; Accepted: 30 October 2020; Published: 5 November 2020

Abstract: Over the last decades, climate change has triggered an increase in the frequency of spruce bark beetle (*Ips typographus* L.) in Central Europe. More than 50% of forests in the Czech Republic are seriously threatened by this pest, leading to high ecological and economic losses. The exponential increase of bark beetle infestation hinders the implementation of costly field campaigns to prevent and mitigate its effects. Remote sensing may help to overcome such limitations as it provides frequent and spatially continuous data on vegetation condition. Using Sentinel-2 images as main input, two models have been developed to test the ability of this data source to map bark beetle damage and severity. All models were based on a change detection approach, and required the generation of previous forest mask and dominant species maps. The first damage mapping model was developed for 2019 and 2020, and it was based on bi-temporal regressions in spruce areas to estimate forest vitality and bark beetle damage. A second model was developed for 2020 considering all forest area, but excluding clear-cuts and completely dead areas, in order to map only changes in stands dominated by alive trees. The three products were validated with in situ data. All the maps showed high accuracies (acc > 0.80). Accuracy was higher than 0.95 and F1-score was higher than 0.88 for areas with high severity, with omission errors under 0.09 in all cases. This confirmed the ability of all the models to detect bark beetle attack at the last phases. Areas with no damage or low severity showed more complex results. The no damage category yielded greater commission errors and relative bias (CEs − 0.30–0.42, relB = 0.42–0.51). The similar results obtained for 2020 leaving out clear-cuts and dead trees proved that the proposed methods could be used to help forest managers fight bark beetle pests. These biotic damage products based on Sentinel-2 can be set up for any location to derive regular forest vitality maps and inform of early damage.

Keywords: bark beetle; *Ips typographus* L.; pest; remote sensing; change detection; forest damage; spruce; Sentinel 2; damage mapping; multi-temporal regression

1. Introduction

In the last three centuries, the forests in the Czech Republic have changed dramatically. Triggered by economic reasons, natural mixed forests were replaced by even-aged spruce or pine monocultures. The spruce currently accounts for 49.2% of stands and pine does for 15.9% [1]. These stands are less stable and more threatened by abiotic factors (e.g., drought, wind, snow, glazed frost) and, more recently, by anthropogenic influences (e.g., air pollution) as well [2]. Over the last years, decrease of forest vitality, growth under stress, and greater susceptibility to spruce bark beetle (*Ips typographus* L.) infestation are being caused by global climate change, especially long-term droughts with extremely high temperatures in the vegetative period [2]. The common harvested volume per year in the Czech Republic is about 15 million m^3 in total and around 1 million m^3 of this amount is infested by insects. However, in the last 5 years, a rapid increase in wood infested by insects (WII) has been registered: 2015—16.1 million m^3 of harvested wood (2.3 million m^3 of WII); 2016—17. 6 million m^3 (4.4 million m^3 of WII); 2017—19.3 million m^3 (5.8 million m^3 of WII); 2018—25.6 million m^3 (13 million m^3 of WII); and 2019—32.5 million m^3 (22.8 million m^3 of WII, and 66,000 ha of damaged forests) [3]. For 2020, the estimate is ranking between 40 and 60 million m^3 of WII [4].

Permanent monitoring, early identification, and harvesting of affected trees are essential in forest protection. Although areas under bark beetle infestation can be easily determined through field work, there are currently no standard field methods in the Czech Republic for early identification of areas affected by outbreaks [5]. The situation becomes more serious as the infested area increases, hindering the possibility of collecting enough data. Moreover, conventional ground methods are time consuming and, in the case of large damaged area, they are usually expensive and not efficient, as they require the assessment of individual trees (e.g., recording the concentration of red-brown dust in bark crevices, holes or marks in the bark, and peeling the bark for more precise determination of infestation stage) in large regions [6]. This lack of control can contribute to an increase in bark beetle population levels with an increasingly negative impact on mortality and loss of forest yield. The increase of outbreaks in periods of extreme drought predicted in the context of climate change [7] could translate into a significant increase in the conditions for the spread of this pest [8–11]. Hence, more effective plans to control the pests are urgently needed. More specifically, in Central Europe, the area prone to be affected by pests such as bark beetle is growing [12–14]; therefore, the need for monitoring large areas is steadily increasing, thus hindering the design and implementation of field campaigns.

Remote sensing data are useful for detecting and monitoring areas infested by bark beetles [15–17], as they provide global, spatially continuous, and periodic data on vegetation condition [8]. Remote sensing data can also contribute to reduce costs associated with field campaigns, as there are a large number of freely available data sources with global coverage and regular revisit times (e.g., Landsat and Sentinel programmes).

Trees infested by bark beetle (Figure 1) go through three phases: (1) the green attack stage, when the tree needles remain green; (2) the red attack, when needles turn progressively from green to yellow and reddish tones; and (3) the grey attack, when dead needles fall and the bare tree stands grey [12,18]. Dead trees (i.e., in red or grey attack phases) are clearly visible using a wide range of satellite images [15]. Most studies have used multispectral high resolution data as main input [19–22], reaching accuracies around 90% for tree mortality (i.e., red and grey attack) detection associated with bark beetle infestation [20,23]. The results have been similar in studies that used very high resolution images (i.e., <1 m) [24,25]. Special interest has been given those studies based on very high resolution hyperspectral images [5,26–29], which have yielded accuracies close to 90% for bark beetle detection at red or grey attack phases. Synthetic Aperture Radar (SAR) data have also been tested to detect tree mortality induced by bark beetle [17], reaching mapping accuracies between 55% and 88% [30].

While detection at red and grey phases is valuable for assessing environmental and economic losses caused by bark beetles, it is not effective to stop the spread, as insects have already completed their lifecycle. Early alert systems are needed to curb the spread as well as to help foresters know the factors facilitating bark beetle attack [22]. The effects of bark beetle on leaf properties affect reflectance in the near-infrared (NIR) and shortwave infrared (SWIR) spectral domains (i.e., 730–1370 nm) [18], which are represented within Sentinel-2 and Landsat spectral bands. The use of satellite imagery to detect bark beetle infestation at green attack phase has not yielded satisfactory results in most cases [5,19,22,28]. Nevertheless, some correlation between forest health and later bark beetle attack have been reported, using both optical [5,19] and L-band SAR data [30]. Approaches based on multi-temporal vegetation indices have proven to be the most effective to detect bark beetle effects at the green attack phase. The correlation between satellite-derived vegetation indices and areas affected by bark beetle at early stages has been tested with Sentinel-2 and Landsat-8 [22] and Landsat-5 images [19]. These studies have proven that changes in photosynthetic activity, vegetation water content, and leaf area index (LAI) are correlated with bark beetle infestation at green attack phase, especially with Sentinel-2, yielding a user's accuracy of 67% for green attack mapping.

Figure 1. Trees infested by bark beetles (*Ips typographus* L.) in the study area. True colour orthophoto with 6 cm spatial resolution captured on 17/05/2020.

Since 1986, a regular forest health assessment has been performed in the Czech Republic using the systematic network of the International Cooperative Programme on Assessment and Monitoring of Air Pollution Effects in Forests (ICP Forests). This network is based on core monitoring sites of 16 × 16 km and selected areas of 8 × 8 km, with a total of 306 areas. At altitudes from 150 m to 1100 m, approximately 11,000 trees are evaluated every year, representing 28 species of trees in different age classes. In addition to ground-based evaluation, Landsat-5 satellite images have been used since the mid-1980s, allowing a uniform systematic evaluation unbiased by a subjective factor in ground visual assessment [31]. Since 2016, satellite images from the Sentinel-2 satellites have been analysed at one-year intervals. Only cloudless images of the Czech Republic are used, selected during the phenological summer (i.e., June to August). Leaf area index (LAI) maps are derived from Sentinel-2 images, which have been validated with LAI ground surveys and defoliation values from ICP Forests data. The health status of the stands is assessed on the basis of LAI changes, which significantly correlate with defoliation in the selected time interval [32,33]. For more detailed mapping of bark beetle outbreaks at higher spatial resolution and more frequent updating, a monitoring system to detect salvage cutting and standing dead wood using Planet images has been used since 2018 in the Czech Republic. The following main layers are used to build this system: (1) Areas dominated by spruce and higher than 12 m. The forest tree map is created by the classification of Sentinel-2 images, performed on the basis of the spectral response of tree species in different phenological phases using training data collected during the National Forest Inventory (NFI) ground survey. The Planet system also uses a digital surface model (DSM) derived from aerial images taken in 2016 and 2017. The DSM is normalized using digital terrain model derived from airborne laser scanning to obtain a canopy height model. (2) Map of LAI decrease, which detects either the clear-cuts or dead wood. Normalized difference vegetation index (NDVI) images at 3 m spatial resolution and two additional categories are detected: standing dead forest and newly established clear-cuts. These categories are distinguished based on the Triangular Greenness Index (TGI) [34].

Different Czech institutions are also conducting research into the use of multispectral and hyperspectral cameras carried by unmanned aerial vehicles (UAVs) [25,35,36]. These systems can provide images at local scales almost anytime with very high spatial resolution, which allows the evaluation of health at the level of individual trees. The disadvantage of UAVs compared with high resolution satellites such as Sentinel-2 or Landsat is the cost of data acquisition and the complexity of data processing, although this disadvantage is gradually disappearing with the development of cloud computing. Another shortcoming is that most multispectral cameras are used in UAV-based studies, which allow the calculation of only a few vegetation indices (e.g., NDVI or Normalized Difference Red Edge Index-NDRE) [37]. Hyperspectral cameras are more expensive, and data processing is more complicated. In contrast, Sentinel-2 provides free, easy to process data with a short revisit time, thus allowing more regular estimates about vegetation status.

In this study, a multi-temporal regression-based change detection method is proposed to map areas affected by bark beetle with different degree of severity at 10 m spatial resolution. Moreover, a comparison is carried out between forest vitality estimates and field records of forest damage caused by *Ips typographus* L. infestation.

The aim of this study is to test a methodology for the automatic mapping of bark beetle-induced damage using Sentinel-2 images. The specific objectives of this work are (a) to evaluate the performance of the damage detection algorithm, (b) to study the feasibility of discriminating areas affected with different intensity or level of mortality, and (c) to assess the influence of possible clear-cuts not related to bark beetle in the final damage maps.

2. Study Area and Data

2.1. Study Area

The study site was located in a forest area owned by Mendel University in Brno and managed by the University Forest Enterprise (UFE), in Křtiny, Czech Republic (49.170°N, 16.440°E. Figure 2). This property covers around 10,000 ha of forest stands. Mixed forests are prevailing where the proportion is 38% of conifers (19% spruce, 8% pine, 8% larch, and others) and 62% of broadleaves (33% beech, 15% oak, 8% hornbeam, and others). Altitude ranges between 200 and 570 m. Mean annual rainfall is 610 mm and mean annual temperature is 7.5 C. UFE forest management is focused on close-to-nature methods, using minimum clear-cuts and a high share of natural regeneration.

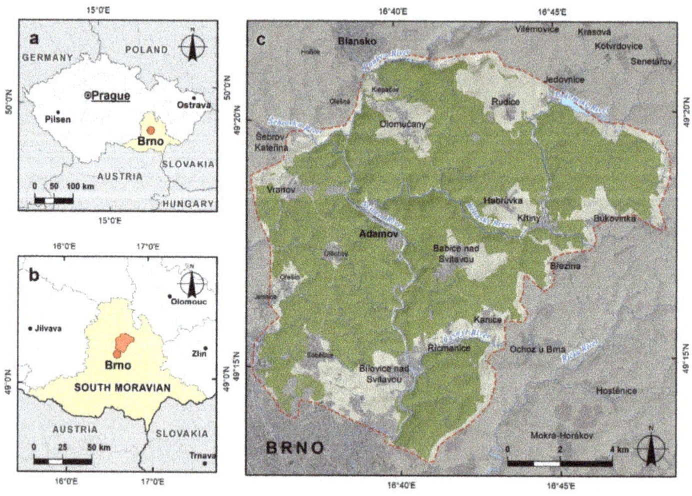

Figure 2. Location of study area. Overview maps of the (**a**) national and (**b**) regional context. (**c**) Forest area managed by University Forest Enterprise (UFE).

In the Czech Republic, salvage cutting has increased dramatically in the last three years. Figure 3 shows the total wood production in millions of cubic meters. Table 1 shows salvage cutting volume at the UFE forest.

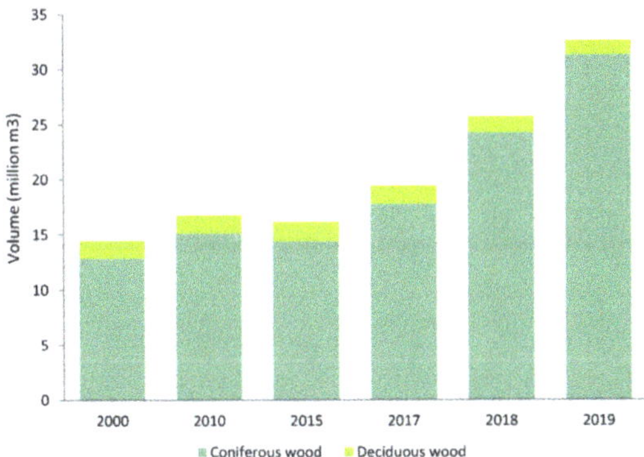

Figure 3. Progress of the volume of timber harvesting in millions of cubic meters, divided into coniferous and deciduous timber. Source: Ministry of Agriculture of the Czech Republic, 2020 [31].

Table 1. Summary of salvage cutting volumes (m^3) per species and year (until 15 July 2020) at the University Forest Enterprise (UFE) forest.

Species	2018	2019	2020	Total
Picea abies	30,564	59,958	33,897	124,419
Pinus sylvestris	2187	6735	4671	13,594
Larix decidua	4212	5711	1608	11,531
Fagus sylvatica	846	5167	1440	7452
Abies alba	1153	2563	503	4220
Quercus petraea	296	1020	471	1786
Fraxinus excelsior	93	377	473	943
Pseudotsuga menziesii	184	262	98	543
Abies grandis	35	191	8	234
Quercus robus	33	150	21	204
Carpinus betulus	64	116	23	203
Other	258	369	71	698
Total	39,925	82,619	43,284	165,828

A total of 32.58 million cubic meters of raw wood was harvested in the forests of the Czech Republic in 2019, which means a further increase of 6.89 million m^3 compared with the previous year. The salvage cutting of 30.94 million m^3 of wood contributed significantly to this volume. The share of salvage cutting in 2019 was 95%. Thus, the initial conditions for planned forest management continued to deteriorate. In terms of the composition of harvesting by tree species, the volume of coniferous wood harvested increased by 7.1 million m^3 compared with 2018 to a total of 31.31 million m^3. The share of coniferous wood harvesting in total harvesting was approximately 96%. The proportion of deciduous and coniferous wood harvesting is due to salvage cutting, especially the wood infested by bark beetle (Figure 4).

In 2019, 20.7 million m^3 of harvested spruce wood infested by bark beetle was registered in the Czech Republic, which represents an increase compared with 2018 by more than 70%, when approximately 12 million m^3 was recorded (2017—5.34 million m^3) (Figure 5). Harvested wood in 2019 was practically exclusively infested with European spruce bark beetle (*Ips typographus* L.), which is usually accompanied by six-toothed spruce bark beetle (*Pityogenes chalcographus* L.) and double-spined bark beetle (*Ips duplicatus* Sahlberg), especially in northern and central Moravia and Silesia, but locally often elsewhere.

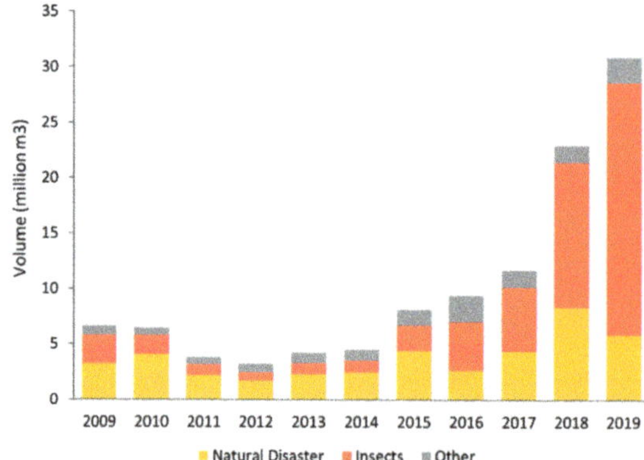

Figure 4. Progress of the volume of salvage cutting grouped by cause. Source: Ministry of Agriculture of the Czech Republic, 2020 [31].

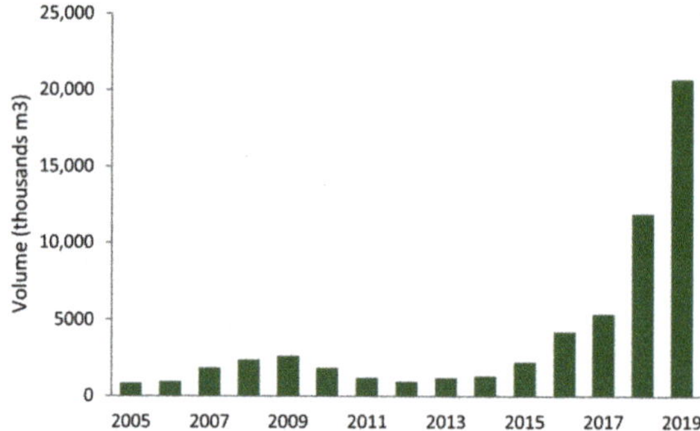

Figure 5. Recorded volume of infested spruce wood harvested in the years 2005 to 2019. Source: Ministry of Agriculture of the Czech Republic, 2020 [31].

The massive spruce infestation with bark beetles has been triggered by higher average temperatures and lower precipitation over the last five years compared with the long-term normal. Extreme drought prevailed, especially in 2015 and 2018 (Figure 6).

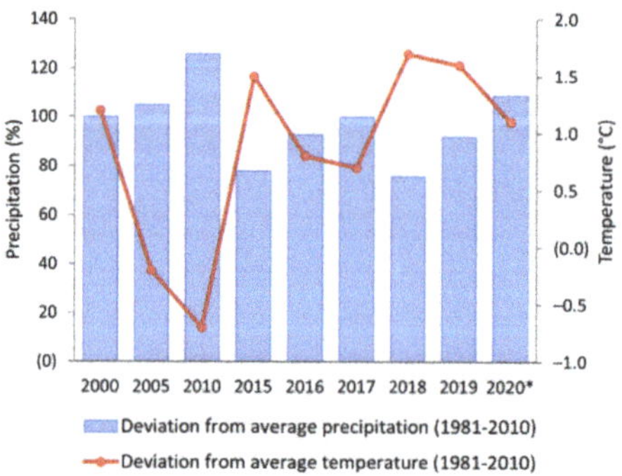

Figure 6. Deviations from the mean temperatures and precipitation (1981–2010) in the period 2000–2020. * Records for 2020 were acquired at the end of August. Source: Czech Hydrometeorological Institute, 2020 [38,39].

2.2. Sentinel-2 Satellite Images

In this study, we used Sentinel-2 images provided by Copernicus Open Access Hub to detect bark beetle damage. These images were acquired in format Level-2A (bottom of atmosphere reflectance). Sentinel-2 L2A images are generated by applying atmospheric and topographic correction algorithms to the Level-1C (top of atmosphere reflectance) images [40]. In order to focus the change detection in areas prone to be attacked by the spruce bark beetle, forest/non-forest and dominant species maps were produced for the study area (Figure 7). These first maps were also developed using Sentinel-2 images (Table 2).

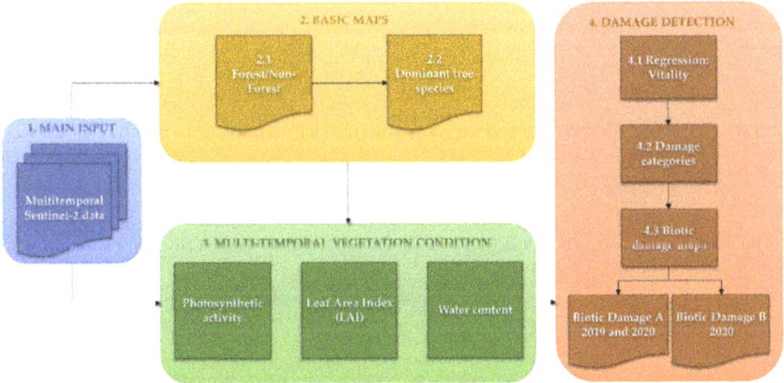

Figure 7. General workflow to obtain bark beetle damage from Sentinel-2 images.

Table 2. Sentinel-2 images used to produce each classification map.

Product	Timing	Forest/Non-Forest	Dominant Species	Biotic Damage
Biotic damage A	t_0	2018-03-22	2018-03-22	2018-09-10
2019	t_1	2018-08-29	2018-08-29	2019-06-30
Biotic damage A	t_0	2019-04-01	2019-04-01	2019-06-30
2020	t_1	2019-06-30	2019-06-30	2020-07-01
Biotic damage B	t_0	2020-04-02	2020-04-02	2019-06-30
2020	t_1	2020-07-01	2020-07-01	2020-07-01

Different Sentinel-2 bands (Table 3), vegetation indices, and texture indices were used for each process. The whole process, summarised in Figure 7, is explained in the subsequent sections.

Table 3. Sentinel-2 Level 2A bands. NIR, near-infrared; SWIR, shortwave infrared.

Sentinel-2 Bands	Central Wavelength (nm) *	Spatial Resolution (m)
Band 01-Coastal aerosol	442.7	60
Band 02-Blue	492.4	10
Band 03-Green	559.8	10
Band 04-Red	664.6	10
Band 05-Red Edge 1	704.1	20
Band 06-Red Edge 2	740.5	20
Band 07-Red Edge 3	782.8	20
Band 08-NIR	832.8	10
Band 8A-Narrow NIR	864.7	20
Band 09-Water vapour	945.1	60
Band 10-Cirrus	1373.5	60
Band 11-SWIR 1	1613.7	20
Band 12-SWIR 2	2202.4	20

* Central wavelength of Sentinel-2A MultiSpectral Instrument (MSI) bands.

2.3. Ground Truth Data

Records of salvage cutting and records of clear-cuts were used to build the ground truth dataset, together with a forest stand map derived from forest management plan (valid for the period 2013–2022). The records were collected continuously by foresters for the needs of the forest enterprise and state forest administration, and contained the ID of forest stands, month and year of the cutting, volume, type of cutting, cause, and area. A detailed explanation of ground truth generation can be found in Section 3.4.

3. Methods

Two algorithms were developed to map bark beetle damage: biotic damage A (BDA) and biotic damage B (BDB), considering as damage any negative deviation from the standard development of the forest (i.e., clear decrease of forest health). Both algorithms work by detecting changes between two dates. The BDA algorithm included masking Sentinel-2 surface reflectance images to areas dominated by spruce in the starting date of the pair. Hence, any area with dead trees or clear-cuts might be detected as bark beetle damage. In the case of BDB, satellite images were masked to the whole forest area in the final date of the pair (not in the initial one). Thus, all dead trees and clear-cuts could not be classified as damaged areas by the BDB algorithm. The objective of the difference between BDA and BDB was to test the ability of the algorithms to detect damage in non-cut areas and with most of the trees still alive. This allowed us to assess the true performance of our approach to detect the effects caused by bark beetle, as BDB leaves out possible over-detection caused by clear-cuts. Three different layers were ultimately developed (Table 2, Figure 7 Step 4): (1) the BDA layer using 2019 imagery (BDA19), (2) the BDA layer using 2020 imagery (BDA20), and (3) the BDB layer using 2020 imagery (BDB20) to map only changes in stands dominated by alive trees. The algorithms for BDA19 and BDA20 were identical, the only difference being the dates of the input imagery (2018–2019 and 2019–2020, respectively). The common steps for producing all damage maps from Sentinel-2 images are summarised in the flowchart (Figure 7) and explained in detail in the following subsections.

3.1. Forest/Non-Forest Classification

To derive any of the damage detection layers explained above (Figure 7, Step 4.), we first apply a classification algorithm to discard all the pixels belonging to non-forest cover (Figure 7, Step 2.1.) [41,42]. To represent phenological variations in the forest canopy, two Sentinel-2 images were used, one for winter and one for summer (Table 2). Green, red, NIR, and SWIR bands were selected. SWIR bands, originally at 20 m spatial resolution, were resampled to the resolution of the other bands, 10 m. Normalized difference vegetation index (NDVI, Equation (1)) [43] and modified soil adjusted vegetation index (MSAVI, Equation (2)) [44,45] were computed together with two texture indices: homogeneity (Equation (3)) and entropy (Equation (4)) [46]. A supervised classification method based on random forest algorithm [47] was used to classify pixels in forest and non-forest categories with the selected Sentinel-2 bands and the vegetation and texture indices as input. The output of the classification was a binary forest/non-forest map of the study area at 10 m spatial resolution. A validation protocol was specifically designed to carry out an independent validation of the forest mask [42], based on a stratified random sampling strategy. According to this validation, the forest mask had an accuracy of 0.97 for the study area in 2018. This process was repeated to update the forest mask to 2019 and 2020.

$$NDVI = \frac{(NIR - Red)}{(NIR + Red)} \tag{1}$$

$$MSAVI = \frac{\left(2{\cdot}NIR + 1 - \sqrt{(2{\cdot}NIR + 1)^2 - 8 \cdot (NIR - Red)}\right)}{2} \tag{2}$$

$$Homogeneity \; = \; \sum_{i=0}^{n-1} \sum_{j=0}^{n-1} P(i,j)(i-j)^2 \tag{3}$$

$$Entropy \; = \; \sum_{i=0}^{n-1} \sum_{j=0}^{n-1} P(i,j) \log P(i,\,j) \tag{4}$$

where n is the number of grey levels and $P(i,\,j)$ defines the entries of the grey-level co-occurrence matrix.

3.2. Species Identification

The next step (Figure 7, Step 2.2.) was to identify the areas dominated by spruce (i.e., those that are prone to be attacked by the *Ips typographus* L.), for which a dominant tree species classification was carried out [48]. The bi-temporal Sentinel-2 bands and indices used in previous steps were masked with the forest/non-forest map, selecting only the forest pixels. A random forest model was trained with polygons containing the main tree species present in the area, namely, *Carpinus betulus* L., *Fagus sylvatica* L., *Larix decidua* Mill., *Picea abies* (L.) Karst., *Pinus Sylvestris* L., and *Quercus robur* L. Training samples were obtained joining forest stands with information from the local forest management plan. For each of the species considered, stands with an abundance of 70% or higher were pre-selected. Finally, the smaller stands that actually contain the specific tree species were selected for training.

The species attacked by *Ips typographus* L. (i.e, *Picea abies* (L.) Karst) were classified with an accuracy of 0.91 in 2018 according to the cross-validation performed during the model training. This dominant species map was produced for 2018, 2019, and 2020, as it was made with the forest mask.

3.3. Anomaly Detection and Damage Detection

Once spruce areas were clearly delimited for each year, a new method (Figure 7, Step 3) was proposed to map the bark beetle damage suffered by the spruce forests during a year (i.e., between a pair of dates). The first step was to compute vegetation indices able to represent vegetation condition for the two dates (t_0 and t_1 for each year, in Table 2). Based on the literature review about the effects of bark beetle [49], the following vegetation indices were computed for each date: NDVI (Equation (1)), MSAVI (Equation (2)), normalized difference moisture index (NDMI, Equation (5)), and green leaf area index (LAI_{green}, Equation (6)) [50,51].

$$NDMI \; = \; \frac{(SWIR1 - NIR)}{(SWIR1 + NIR)} \tag{5}$$

$$LAI_{green} \; = \; 6.753 \cdot \frac{(RE1 - Red)}{(RE1 + Red)} \tag{6}$$

NDVI and SAVI are direct indicators of plant greenness and photosynthetic activity, NDMI is an indicator of vegetation water content, and LAI_{green} is an estimate of the proportion of green leaves per area. As defoliation is one of the main effects caused by bark beetles [15], LAI has been considered as one of the key variables in this study. Moreover, we proposed the use of Sentinel-2-derived LAI_{green}, as it considers only the area of green leaves, not only being an indicator of defoliation, but also of leaf browning [50,51].

To derive the BDA19 and BDA20 products, vegetation indices were masked with the spruce polygons identified in the dominant tree species map. For each index, a bi-temporal ordinary least squares (OLS) regression was carried out in order to model the typical behaviour of spruce forest patches between t_0 and t_1. The first index was used as the independent variable, and the second one as the dependent variable in each regression. The results were images of estimated NDVI, MSAVI, NDMI, and LAI_{green} for t_1 based on the average trend of spruce forests in the studied period. An error image was computed for each index using the estimated and real values of the indices in t_1. These images show the deviations from the expected condition of healthy spruce masses, as it was assumed that regression would be adjusted to healthy forests and disturbed areas would appear as deviations from that ideal status. Hence, these error images showed estimates of temporal decrease or increase of photosynthetic activity, vegetation water content, and number of green leaves.

In order to obtain a single image of vegetation vitality changes (Figure 7, Step 4.1.), all error images were standardised (i.e., pixel values expressing differences between real and estimated indices were converted to z-scores). Z-scores (Equation (7)) represent the number of standard deviations between a value and the population mean.

$$z = \frac{x - \mu}{\sigma} \tag{7}$$

where μ is the mean of the population and σ is the standard deviation of the population.

The mean of all the standardised images was computed to obtain a single image of average changes in vegetation vitality. Pixel values in this image may be interpreted as mean standard deviations from expected NDVI, MSAVI, NDMI, and LAI$_{green}$ of undisturbed spruce forest. Finally, these vitality changes were reclassified (Figure 7, Step 4.2.) to obtain a categorical map of bark beetle damage (Figure 7, final outputs, Step 4.3.), using the following rule based on z-scores: 0–1 no damage, 1–2 minor damage, 2–3 moderate damage, and >3 severe damage.

3.4. Ground Truth Data and Validation

To validate the damage products, a ground truth dataset was generated through a field campaign. An equalized stratified random sampling was used for data collection. As field works are costly, a limit of 200 points per product was established. To stratify the samples, 50 random points were generated per class and product, obtaining 200 points per product and 600 in total.

The collected records containing the ID of forest stands, month and year of the cutting, volume, type of cutting, cause, and area were joined to the forest stand map using the forest stand ID. Different classes of forest stands representing the degree and year of damage were created using the actual vitality of the stands based on the information of the records. This modified forest stand map was used to decide to which class each point belongs in terms of actual damage. This decision was made based on the occurrence of damage in the given forest stand, not on the occurrence on the exact position of the point, as the records were focused only at the level of the given forest stand. As dead trees with inactive bark beetle are no longer harvested in 2019 and 2020, in the case of doubt (no record of harvesting, but occurrence of damage detected), a field evaluation was also performed.

An accuracy assessment was carried out based on the analysis of the confusion matrix derived from the damage maps and ground truth points. The overall accuracy was computed for each product. In order to obtain performance metrics per class, each class was binarised against the rest, considering as positive the category of interest and negative the sum of all other classes. The following performance metrics were computed per product and damage category: accuracy (Equation (8)), precision (Equation (9)), recall (Equation (10)), F1-score (Equation (11)), omission error (Equation (12)), commission error (Equation (13)), and relative bias (Equation (14)) [42].

$$acc. = \frac{True\ positive\ + True\ negative}{sample\ size\ (n)} \tag{8}$$

$$prec. = \frac{True\ positive}{True\ positive + False\ positive} \tag{9}$$

$$rec. = \frac{True\ positive}{True\ positive + False\ negative} \tag{10}$$

$$F1score = \frac{2 \cdot True\ positive}{2 \cdot True\ Positive + False\ positive + False\ negative} \tag{11}$$

$$OE \;=\; \frac{False\ negative}{True\ positive + False\ negative} \qquad (12)$$

$$CE \;=\; \frac{False\ positive}{True\ positive + False\ positive} \qquad (13)$$

$$relB \;=\; \frac{(False\ positive - False\ negative)}{True\ positive + False\ negative} \qquad (14)$$

As an additional exercise, each point of the ground truth dataset was characterised with the vitality value of the corresponding year. The objective of this procedure was to study the statistical robustness of each class and detect possible overlaps or other inconsistences across the categories defined in the final maps.

4. Results

4.1. Bark Beetle Damage Maps

The BDA product was developed for 2019 (Figure 8) and 2020, while the BDB was produced only for 2020. Each product had two outputs: (1) a continuous map representing changes in forest vitality (Figure 8, left panel) and (2) a categorical map showing the areas of biotic damage (Figure 8, right panel), excluding cut areas for the BDB20 map.

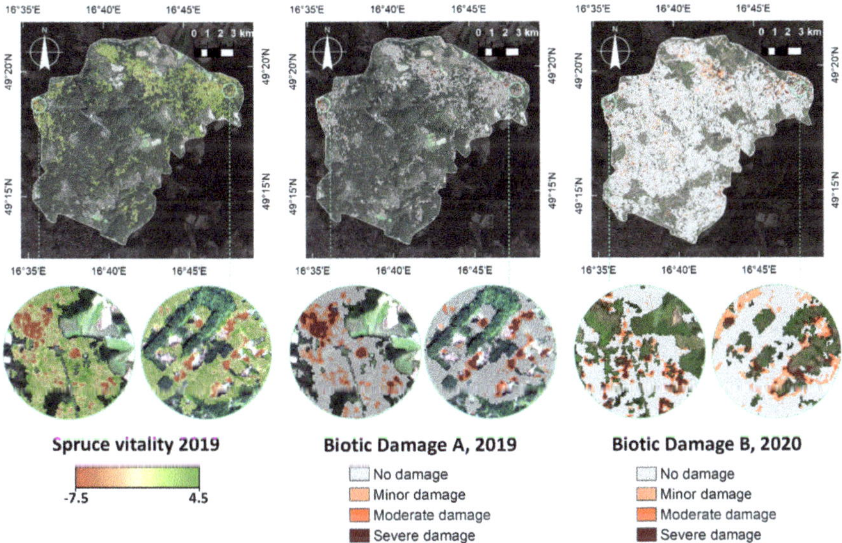

Figure 8. Maps of spruce vitality in 2019 (**left**), BDA19 (**middle**), and BDB20 (**right**). Images: Sentinel-2 acquired on 2019-08-29 (Spruce vitality 2019 and BDA19) and 2020-07-01 (BDB0) in true color composition (blue/green/red). BDA, biotic damage A; BDB, biotic damage B.

4.2. Validation Results

The three categorical maps were validated following the methods explained in Section 3.4. Confusion matrices (Figure 9) show that the algorithm performed as expected, as errors were low compared with the agreement between the classification maps and ground truth. Errors seemed to be slightly higher for the BDB20 product.

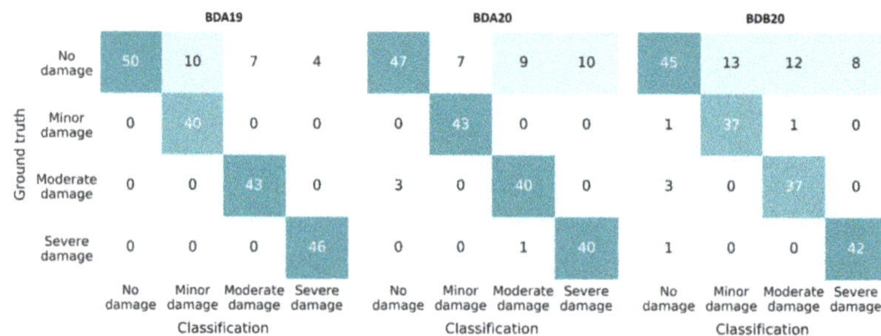

Figure 9. Confusion matrices of biotic damage (2019) (**left**), biotic damage (2020) (**middle**), and biotic damage (2020) excluding pixels of dead trees (**right**).

Overall accuracies were higher than 0.80 for all products (accBDA19 = 0.90, accBDA20 = 0.85, accBDB20 = 0.81). Metrics per class (Table 4) confirmed the good performance of the algorithms. BDA products showed a similar performance for different years, with accuracies increasing with the severity of the attack. Accuracy and F1-score were higher for the severe damage class (acc19 = 0.98, F1-19 = 0.96, acc20 = 0.95, F1-20 = 0.88), thus confirming that most affected areas were easier to detect. Nevertheless, cut areas might have influenced this result as they were classified into the severe damage class. On the other hand, the no damage class yielded lower accuracy and F1-score (acc19 = 0.90, F1-19 = 0.83, acc20 = 0.86, F1-20 = 0.76). The BDA20 product showed slightly worse results for all classes (acc$_{\text{no-damage}}$ = 0.81, F1$_{\text{no-damage}}$ = 0.70, acc$_{\text{severe-damage}}$ = 0.96, F1$_{\text{severe-damage}}$ = 0.90).

Table 4. Performance metrics by product and damage category. BDA, biotic damage A; BDB, biotic damage B; acc., accuracy; prec., precision; rec., recall; CE, commission error; OE, omission error; relB, relative bias.

Product	Class	acc.	prec.	rec.	F1	CE	OE	relB
	No damage	0.90	1.00	0.70	0.83	0.00	0.30	0.42
BDA19	Minor damage	0.95	0.80	1.00	0.89	0.20	0.00	−0.20
	Moderate damage	0.97	0.86	1.00	0.93	0.14	0.00	−0.14
	Severe damage	0.98	0.92	1.00	0.96	0.08	0.00	−0.08
	No damage	0.86	0.94	0.64	0.76	0.06	0.36	0.46
BDA20	Minor damage	0.97	0.86	1.00	0.93	0.14	0.00	−0.14
	Moderate damage	0.94	0.80	0.93	0.86	0.20	0.07	−0.14
	Severe damage	0.95	0.80	0.98	0.88	0.20	0.02	−0.18
	No damage	0.81	0.90	0.58	0.70	0.10	0.42	0.56
BDB20	Minor damage	0.93	0.74	0.95	0.83	0.26	0.05	−0.22
	Moderate damage	0.92	0.74	0.93	0.82	0.26	0.08	−0.20
	Severe damage	0.96	0.84	0.98	0.90	0.16	0.02	−0.14

Error metrics showed similar behaviour, being higher for BDB20 and lower for BDA19. Commission errors ranged from 0.08 (BDA19, severe damage) to 0.26 (BDB20, minor and moderate damage) and were higher than omission errors in all damage classes. The negative relative bias in these categories confirmed that the algorithm tends to over-estimate damaged areas, although values are not greater than 0.22. Omission errors were in general very low, between 0.00 and 0.08 for all damage classes in the three maps. The high recall in these classes (0.93–1.00) confirmed that the algorithms are able to detect affected areas, missing a small proportion of positive cases. On the other hand, omission errors were significantly higher than commission errors in the no damage class, leading to high positive relative bias ranging from 0.46 to 0.56. This was another signal of over-estimation of damaged areas.

Box plots of forest vitality per class (Figure 10) confirmed the clear correlation between the actual areas attacked by the bark beetle and the decrease of forest vitality, measured as a standardised mean of changes in NDVI, MSAVI, NDMI, and LAI_{green}. The interquartile range (IQR) represents the data dispersion and is calculated with the difference between 75th and 25th percentiles. IQR is greater in no damage (IQR_{BDA19} = 1.49, IQR_{BDA20} = 2.22, and IQR_{BDB20} = 2.63) than in the rest of the categories for the three products analysed. The overlap between categories decreases with the increase of severity, with moderate and severe damage not overlapping any other class in the three products. Outliers are present in the no damage, minor damage, and moderate damage categories, with severe damage being the only class without outliers in the three maps. This confirms that areas with severe damage are easier to detect and identify than the other classes, as inferred from performance metrics (Table 4).

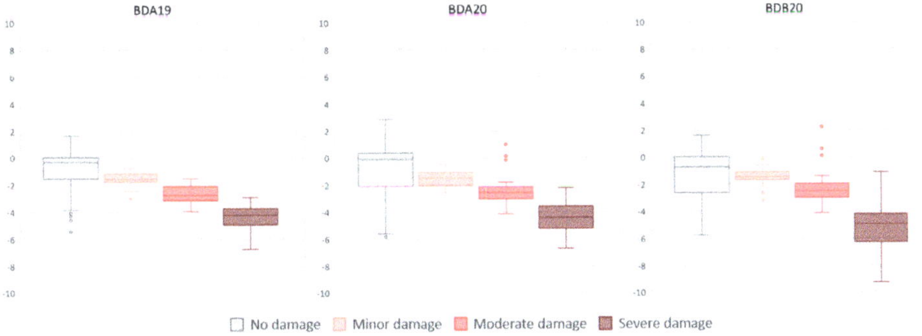

Figure 10. Box plots representing changes in vitality per ground truth class for each product.

Analysing the three products separately, some differences were noted. Overlap of no damage and minor damage between the 25th and 75th percentiles was clearer in the BDB20 product. In this product, there was also overlapping between no damage and moderate damage. However, the severe damage class was clearly discriminated from no damage and minor damage. This behaviour might suggest that a more precise discrimination could be achieved by reducing the number of classes.

5. Discussion

The results showed that the biotic damage algorithm worked consistently in different years. However, accuracy decreased by 5% between BDA19 and BDA20, while in these years, the proportion of spruce detected as damaged was very similar (7.2% and 7.4% of the total spruce area, respectively). Hence, although the results for both years were similar regarding performance metrics, a longer time series analysis would be needed to study the robustness of the method across time and quantify typical inter-annual variations. In the case of BDB20 (i.e., similar to BDA20, but considering areas dominated by alive trees of any species at the end of 2020, not only spruce areas), the lower accuracy ($accHD20$ = 0.81) may be explained by two main factors. Firstly, this product considers only alive trees, and minor and moderate changes have greater weight than severe changes compared with the other products. Hence, the overall metrics tend to be lower in those products where complex intermediate classes are more frequent than extremes that are easier to identify. Secondly, all species are considered within the BDB product, and not only spruce, subtle changes in areas dominated by deciduous species, which are also more sensitive to climate variations (e.g., drought), might have led to a greater confusion between the no damage and damage classes, as shown in the confusion matrices (Figure 9). However, the accuracy of the BDB map was high and, more importantly, it confirmed that the high accuracy obtained for the first two products was not due to the detection of cut areas, thus damage caused by bark beetle was clearly detected in most cases.

The three models presented higher commission than omission and negative relative bias in all damage classes, suggesting a certain degree of over-estimation of damaged areas, as confirmed by higher recall and lower precision. Assuming that a precise balance between commission and omission errors was hard to achieve, we prioritised the commission of damaged areas, as the aim of the products is to help forest managers to optimise the control of the pest. Although this led to high positive bias in the no damage class (i.e., there was a relatively high omission), accuracy and F1 were still high for this category. For BDB20, the no damage class presented the worst results with a recall of 0.58 and an omission error of 0.42. This confirmed that further improvements should be made to avoid the confusion between bark beetle damage and minor changes in forest vitality caused by other factors across the different forest types present in the study area. Different solutions might be applied in order to balance the errors of the no damage category, such as adding more spectral indices or change the temporal windows. Time series could allow carrying out the damage assessment using a combination of shorter periods, thus minimising the possible influence of slow changes in vegetation cover.

The results clearly showed that the minor damage class was often confused with no damage, as the limit between small regular changes in vegetation condition and small, but anomalous changes is hard to define. This suggests that the models would improve significantly if the no damage category was removed. Nevertheless, we decided to leave this class because it might be directly related to green-attack detection [22], which would be essential to design an early detection system [52].

The approach proposed in this article may improve previous attempts to detect bark beetle infestation with Sentinel-2 data. The ability of Sentinel-2 for this purpose has been tested by different authors with variable results. Abdullah et al. [22] reported a user's accuracy of 67% (i.e., agreement between pixels detected as infestation and ground truth) and a producer's accuracy of 71% for green attack detection, while our user's accuracies were not lower than 74% and our producer's accuracies not lower than 95% (CE = 0.26 and OE = 0.05 in the worst model, BDB20) for the minor damage category, which, as stated before, might correspond to green attack. Moreover, our approach was validated with field data, while Abdullah et al. [22] derived ground truth information from the visual analysis of aerial photography, which might have added a certain level of uncertainty into the accuracy assessment. Both studies coincide in the affirmation that changes in photosynthetic activity, green leaves, and humidity triggered by bark beetle can be detected from early infestation stages using multispectral Sentinel-2 data. The relationship between LAI maps derived from Sentinel-2 and defoliation caused by bark beetle infestation has also been confirmed by Barka et al. [33] (r^2 = 0.58). An advantage of the methods proposed in this study compared with the previously mentioned ones is that it does not require training data to detect bark beetle infestation, as the algorithms used unsupervised methods based on a statistical change detection approach. Zimmermann and Hoffmann [23] used a change detection approach to detect areas affected by *Ips typographus* in Germany and Switzerland. Although they obtained low commission errors for the positive class (CE = 0.03–0.12 versus CE = 0.08–0.26 of our algorithm), omission errors were substantially higher than in our study (OE = 0.48–0.60 versus OE = 0.00–0.08), leading to a high bias in the final results.

Regarding very high resolution, Worldview-2 showed potential to discriminate green attack from healthy trees [53]. However, the spectral differences were so small that the accuracy of random forest classification and logistic regression did not exceed 70%. RapidEye data have been used to map green attack, nearly equivalent to our minor severity class, reaching a kappa coefficient of 0.51 [54]. In comparison with these studies, our performance metrics yielded higher values, with an accuracy of 93% for the worst model. Nevertheless, it must be noted that the spatial resolutions of multispectral Worldview-2 (1.84 m at nadir) and RapidEye (6.5 m at nadir) allow the detection of smaller groups or even individual trees when compared with Sentinel-2. Although bark beetle damage mapping has been carried out using UAV very high resolution imagery with promising results [24,27–29], they rely on planned flights, which usually require high costs and cannot be repeated frequently. The value of using Sentinel-2 data is the free cost of imagery at relatively high spatial resolution (10 m), but mainly the revisit time [15] of 2–3 days at mid-latitudes. Relatively high performance metrics have also been reported using SAR data, with accuracies of 0.65–0.88 for L-band data [30] and kappa of 0.23 for green attack detection using X-band [55]. However, SAR-based detections were highly dependent on local environmental conditions [30]. The combination of very high resolution multispectral RapidEye and TerraSAR-X data yielded a kappa coefficient of 0.74, considerably improving the results based solely on optical and SAR data [55]. The results shown in this study are likely to improve if SAR data (e.g., Sentinel-1) are included in the models.

Using remote sensing products, we also need to consider that tree damage stages do not always correspond to the colour changes in the crown. In extreme cases, it occurs that the tree bark is totally peeled, the bark beetle has completed its development and flew out, while the crown is still green. This fact complicates the damage assessment with remote sensing optically derived products.

The next point of discussion focused on the local expertise of how the products developed in this study overcome some of the limitations of other datasets currently provided at the national level, as is the case of Czech Republic. Czech Hydrometeorological Institute provides a map of the sums of effective temperatures above 7.5 °C, knowing that the population dynamics of bark beetle are significantly affected by the weather, especially air temperatures, which determine the swarming and the number of generations. As the bark beetle swarms when temperatures above 7.5 °C sum up to 540 °C [56], this temperature map is intended to help forest managers in planning and prevention activities by estimating swarming dates. Even when the temperatures map is somewhat useful for managing spruce forests, there are other factors affecting the ecology of bark beetle that are not considered within this map. Moreover, it does not have the intention of being used for damage assessment purposes.

Next, also based on local expert knowledge, we assessed the monitoring system for the detection of salvage cutting and standing dead wood using Planet images established in the Czech Republic since 2018 (see Section 1) in comparison with the products produced and validated here. The Planet-based dataset is updated four times a year [57]. Although this product is useful to control dead trees and subsequent clear-cuts caused by bark beetle, it does not provide any information about the status of trees in intermediate infestation phases (i.e., it is equivalent to a binary classification detecting only dead trees). Contrarily, our bark beetle damage maps provide insights on the level of affection of different areas, even those dominated by trees that are still alive, as proven with BDB20. This suggests that the proposed method is able to improve Planet-based very high resolution maps by providing estimates of areas under red and green attack phase. Moreover, the costs of maps based on Sentinel-2 is lower because of the free availability of imagery. Nevertheless, the methods proposed here are limited to mapping relatively large areas, as the spatial resolution of 10 m does not allow detecting single trees affected by bark beetles. Hence, it is necessary that several trees within the pixel present enough spectral change to be detected as anomalous. The same limitation was found by other authors [23], still making these kind of methods useful for intermediate landscape scales.

Further research is still to be made on the field of bark beetle attack forecasting models, which are being highly demanded by forest managers [52]. Although the methods proposed here could be used for short-term forecasting because of their ability to detect areas with minor damage that might be in the first infestation stages, there is no evidence of a clear correlation between these areas and green attack phase. Moreover, future models should include climate data as input in order to perform mid-term forecasts allowing to mitigate the negative effects of this pest before its complete development.

6. Conclusions

In this study, a method was proposed to map bark beetle damage from satellite imagery. Using Sentinel-2 images as input, two multi-temporal regression models were built to detect and map the severity of bark beetle outbreaks on spruce forests in the Czech Republic at 10 m spatial resolution. The first model (BDA) was applied to map the damage occurred in years 2019 and 2020. The second model (BDB) was applied to map the bark beetle damage in 2020, in order to map only changes in stands dominated by alive trees. Both products were validated using a ground truth dataset generated through a field campaign and forest management plans. Different performance metrics were computed to assess the quality and errors of the three maps produced. Finally, forest vitality used to develop the biotic damage layers was compared against ground truth to study the overlaps between classes.

All products showed good performance, with accuracies higher than 0.80. The severe damage class yielded the best performance metrics in all products (acc > 0.95, F1 > 0.88), while the no damage yielded the worse metrics (acc > 0.81, F1 > 0.70). Commission errors were higher than omission errors in all positive damage classes, leading to a high relative bias in the no damage class (relB = 0.42–0.46). BDA products showed slightly better results than BDB (accBDA19 = 0.90, accBDA20 = 0.85, accBDB20 = 0.81) because of the easier identification of areas with clear-cuts or dominated by dead trees. However, the performance metrics yielded by BDB proved that the algorithm was able to identify areas affected with low severity unaffected by dead trees or clear-cuts, suggesting that some areas were correctly mapped at red and green attack phases. Changes in forest vitality grouped by ground truth classes confirmed that pixels were easier to classify in the correct class for the severe and moderate damage classes. Contrarily, the existing overlap of vitality values between classes of no damage and minor damage highlighted the difficulty of clearly discriminating infested areas with subtle signals of decay.

Comparing the proposed methods and outputs with the datasets currently used in the Czech Republic for bark beetle damage mapping, the products presented several advantages. Firstly, the cost is relatively low as it is based on freely available Sentinel-2 images. Secondly, the biotic damage maps provide information about damage intensity, suggesting that it might be used not only for damage assessment, but also to help forest managers in planning their prevention and mitigation activities. Biotic damage products can be set up for any location to monitor the forest vitality to derive regular maps as needed, for example, every month. Nevertheless, the presented methods are not valid to identify individual affected trees given their spatial resolution. Future research should be carried out to confirm the detection at the green attack phase and complement the existing studies with forecasting products.

Author Contributions: Conceptualization, A.F.-C. and Z.P.; Methodology, A.F.-C., Z.P., and L.D.; Software, A.F.-C.; Validation, Z.P., L.D., and A.F.-C.; Formal Analysis, A.F.-C.; Investigation, A.F.-C., Z.P., and L.D.; Resources, GMV and UFE; Data Curation, A.F.-C.; Writing—Original Draft Preparation, A.F.-C., Z.P., B.R.-R., and A.F.-N.; Writing—Review and Editing, B.R.-R.; Visualization, A.F.-N.; Supervision, B.R.-R.; Project Administration, MSF; Funding Acquisition, MSF. All authors have read and agreed to the published version of the manuscript.

Funding: This research was funded by European Union's Horizon 2020 research and innovation programme, under grant agreement 776045.

Conflicts of Interest: The authors declare no conflict of interest.

References

1. Forest Managent Institute of the Czech Republic. Information on the State of Forests from the Comprehensive Forest Management Plans for 2019. Available online: http://www.uhul.cz/ke-stazeni/informace-o-lese/slhp (accessed on 26 October 2020).

2. Zahradník, P.; Zahradníková, M. Salvage felling in the Czech Republic's forests during the last twenty years. *Cent. Eur. For. J.* **2019**, *65*, 12–20. [CrossRef]

3. Czech Statistical Office. Reports of Czech Republic Statistical Office from Forestry Sector. Available online: https://www.czso.cz/csu/czso/forestry-2019 (accessed on 28 August 2020).

4. Ebner, G. Up to 60 million m3 of Bark Beetle Damage This Year. Available online: https://www.timber-online.net/rundholz/2020/02/up-to-60-million-m--of-bark-beetle-damage-this-year.html (accessed on 24 August 2020).

5. Lausch, A.; Heurich, M.; Gordalla, D.; Dobner, H.-J.; Gwillym-Margianto, S.; Salbach, C. Forecasting potential bark beetle outbreaks based on spruce forest vitality using hyperspectral remote-sensing techniques at different scales. *For. Ecol. Manag.* **2013**, *308*, 76–89. [CrossRef]

6. Schlyter, F.; Birgersson, G.; Byers, J.A.; Löfqvist, J.; Bergström, G. Field response of spruce bark beetle, Ips typographus, to aggregation pheromone candidates. *J. Chem. Ecol.* **1987**, *13*, 701–716. [CrossRef] [PubMed]

7.   Biedermann, P.H.W.; Müller, J.; Grégoire, J.-C.; Gruppe, A.; Hagge, J.; Hammerbacher, A.; Hofstetter, R.W.; Kandasamy, D.; Kolarik, M.; Kostovcik, M.; et al. Bark Beetle Population Dynamics in the Anthropocene: Challenges and Solutions. *Trends Ecol. Evol.* **2019**, *34*, 914–924. [CrossRef]

8.   Trumbore, S.; Brando, P.; Hartmann, H. Forest health and global change. *Science* **2015**, *349*, 814–818. [CrossRef]

9.   Jönsson, A.M.; Appelberg, G.; Harding, S.; Bärring, L. Spatio-temporal impact of climate change on the activity and voltinism of the spruce bark beetle, Ips typographus. *Glob. Chang. Biol.* **2009**, *15*, 486–499. [CrossRef]

10. Seidl, R.; Rammer, W.; Jäger, D.; Lexer, M.J. Impact of bark beetle (*Ips typographus* L.) disturbance on timber production and carbon sequestration in different management strategies under climate change. *For. Ecol. Manag.* **2008**, *256*, 209–220. [CrossRef]

11. Seidl, R.; Rammer, W. Climate change amplifies the interactions between wind and bark beetle disturbances in forest landscapes. *Landsc. Ecol.* **2017**, *32*, 1485–1498. [CrossRef]

12. Wermelinger, B. Ecology and management of the spruce bark beetle Ips typographus—A review of recent research. *For. Ecol. Manag.* **2004**, *202*, 67–82. [CrossRef]

13. Krejzar, T. Bark Beetle Outbreak in the Czech Republic: Challenges and Solutions. European Network INTEGRATE Seminar on Managing bark Beetle Impacts on Forests. Białowieża, Poland. 2018. Available online: https://informar.eu/sites/default/files/pdf/Presentation%20CZ%20062018.pdf (accessed on 7 October 2020).

14. Stereńczak, K.; Mielcarek, M.; Kamińska, A.; Kraszewski, B.; Piasecka, Ż.; Miścicki, S.; Heurich, M. Influence of selected habitat and stand factors on bark beetle *Ips typographus* (L.) outbreak in the Białowieża Forest. *For. Ecol. Manag.* **2020**, *459*, 117826. [CrossRef]

15. Hall, R.J.; Castilla, G.; White, J.C.; Cooke, B.J.; Skakun, R.S. Remote sensing of forest pest damage: A review and lessons learned from a Canadian perspective. *Can. Entomol.* **2016**, *148*, S296–S356. [CrossRef]

16. Wulder, M.A.; Dymond, C.C.; White, J.C.; Leckie, D.G.; Carroll, A.L. Surveying mountain pine beetle damage of forests: A review of remote sensing opportunities. *For. Ecol. Manag.* **2006**, *221*, 27–41. [CrossRef]

17. Hollaus, M.; Vreugdenhil, M. Radar Satellite Imagery for Detecting Bark Beetle Outbreaks in Forests. *Curr. For. Rep.* **2019**, *5*, 240–250. [CrossRef]

18. Abdullah, H.; Darvishzadeh, R.; Skidmore, A.K.; Groen, T.A.; Heurich, M. European spruce bark beetle (*Ips typographus*, L.) green attack affects foliar reflectance and biochemical properties. *Int. J. Appl. Earth Obs. Geoinf.* **2018**, *64*, 199–209. [CrossRef]

19. Coops, N.C.; Waring, R.H.; Wulder, M.A.; White, J.C. Prediction and assessment of bark beetle-induced mortality of lodgepole pine using estimates of stand vigor derived from remotely sensed data. *Remote Sens. Environ.* **2009**, *113*, 1058–1066. [CrossRef]

20. Meddens, A.J.H.; Hicke, J.A.; Vierling, L.A.; Hudak, A.T. Evaluating methods to detect bark beetle-caused tree mortality using single-date and multi-date Landsat imagery. *Remote Sens. Environ.* **2013**, *132*, 49–58. [CrossRef]

21. Meigs, G.W.; Kennedy, R.E.; Cohen, W.B. A Landsat time series approach to characterize bark beetle and defoliator impacts on tree mortality and surface fuels in conifer forests. *Remote Sens. Environ.* **2011**, *115*, 3707–3718. [CrossRef]

22. Abdullah, H.; Skidmore, A.K.; Darvishzadeh, R.; Heurich, M. Sentinel-2 accurately maps green-attack stage of European spruce bark beetle (*Ips typographus*, L.) compared with Landsat-8. *Remote Sens. Ecol. Conserv.* **2019**, *5*, 87–106. [CrossRef]

23. Zimmermann, S.; Hoffmann, K. Evaluating the capabilities of Sentinel-2 data for large-area detection of bark beetle infestation in the Central German Uplands. *J. Appl. Remote Sens.* **2020**, *14*, 024515. [CrossRef]

24. Safonova, A.; Tabik, S.; Alcaraz-Segura, D.; Rubtsov, A.; Maglinets, Y.; Herrera, F. Detection of Fir Trees (*Abies sibirica*) Damaged by the Bark Beetle in Unmanned Aerial Vehicle Images with Deep Learning. *Remote Sens.* **2019**, *11*, 643. [CrossRef]

25. Klouček, T.; Komárek, J.; Surový, P.; Hrach, K.; Janata, P.; Vašíček, B. The use of UAV mounted sensors for precise detection of bark beetle infestation. *Remote Sens.* **2019**, *11*, 1561. [CrossRef]

26. Näsi, R.; Honkavaara, E.; Blomqvist, M.; Lyytikäinen-Saarenmaa, P.; Hakala, T.; Viljanen, N.; Kantola, T.; Holopainen, M. Remote sensing of bark beetle damage in urban forests at individual tree level using a novel hyperspectral camera from UAV and aircraft. *Urban For. Urban Green.* **2018**, *30*, 72–83. [CrossRef]

27. Näsi, R.; Honkavaara, E.; Lyytikäinen-Saarenmaa, P.; Blomqvist, M.; Litkey, P.; Hakala, T.; Viljanen, N.; Kantola, T.; Tanhuanpää, T.; Holopainen, M. Using UAV-based photogrammetry and hyperspectral imaging for mapping bark beetle damage at tree-level. *Remote Sens.* **2015**, *7*, 15467–15493. [CrossRef]

28. Fassnacht, F.E.; Latifi, H.; Ghosh, A.; Joshi, P.K.; Koch, B. Assessing the potential of hyperspectral imagery to map bark beetle-induced tree mortality. *Remote Sens. Environ.* **2014**, *140*, 533–548. [CrossRef]

29. Honkavaara, E.; Näsi, R.; Oliveira, R.; Viljanen, N.; Suomalainen, J.; Khoramshahi, E.; Hakala, T.; Nevalainen, O.; Markelin, L.; Vuorinen, M.; et al. Using multitemporal hyper-and multispectral UAV imaging for detecting bark beetle infestation on norway spruce. *Int. Arch. Photogramm. Remote Sens. Spat. Inf. Sci. ISPRS Arch.* **2020**, *43*, 429–434. [CrossRef]

30. Tanase, M.A.; Aponte, C.; Mermoz, S.; Bouvet, A.; Le Toan, T.; Heurich, M. Detection of windthrows and insect outbreaks by L-band SAR: A case study in the Bavarian Forest National Park. *Remote Sens. Environ.* **2018**, *209*, 700–711. [CrossRef]

31. Ministry of Agriculture of the Czech Republic. Information on Forests and Forestry in the Czech Republic by 2019. Available online: http://eagri.cz/public/web/file/658587/Zprava_o_stavu_lesa_2019.pdf (accessed on 7 October 2020).

32. Lukeš, P.; Strejček, R.; Křístek, Š.; Mlčoušek, M. *Forest Health Assessment in Czech Republic Using Sentinel-2 Satellite Data. Certified Methodology*; Forest Management Institute: Brandýs nad Labem, Czech Republic, 2018; ISBN 978-80-88184-21-8.

33. Barka, I.; Lukeš, P.; Bucha, T.; Hlásny, T.; Strejček, R.; Mlčoušek, M.; Křístek, Š. Remote sensing-based forest health monitoring systems—Case studies from Czechia and Slovakia. *Cent. Eur. For. J.* **2018**, *64*, 259–275. [CrossRef]

34. De Ocampo, A.L.P.; Bandala, A.A.; Dadios, E.P. Estimation of Triangular Greenness Index for Unknown PeakWavelength Sensitivity of CMOS-acquired Crop Images. In Proceedings of the 2019 IEEE 11th International Conference on Humanoid, Nanotechnology, Information Technology, Communication and Control, Environment, and Management (HNICEM), Laoag, Philippines, 29 November–1 December 2019; pp. 1–5.

35. Minařík, R.; Langhammer, J. Use of a multispectral UAV photogrammetry for detection and tracking of forest disturbance dynamics. *Int. Arch. Photogramm. Remote Sens. Spat. Inf. Sci.* **2016**, *XLI-B8*, 711–718. [CrossRef]

36. Brovkina, O.; Cienciala, E.; Surový, P.; Janata, P. Unmanned aerial vehicles (UAV) for assessment of qualitative classification of Norway spruce in temperate forest stands. *Geo-Spat. Inf. Sci.* **2018**, *21*, 12–20. [CrossRef]

37. Balková, M.; Bajer, A.; Patočka, Z.; Mikita, T. Visual Exposure of Rock Outcrops in the Context of a Forest Disease Outbreak Simulation Based on a Canopy Height Model and Spectral Information Acquired by an Unmanned Aerial Vehicle. *ISPRS Int. J. Geo-Inf.* **2020**, *9*, 325. [CrossRef]

38. Czech Hydrometeorological Institute. Territorial Air Temperature. Available online: http://portal.chmi.cz/historicka-data/pocasi/uzemni-teploty?l=en (accessed on 7 October 2020).

39. Czech Hydrometeorological Institute. Territorial Precipitation. Available online: http://portal.chmi.cz/historicka-data/pocasi/uzemni-srazky?l=en (accessed on 7 October 2020).

40. European Space Agency. *Sentinel-2 Level-2A Algorithm Theoretical Basis Document*; European Space Agency: Paris, France, 2020.

41. Fernandez-Carrillo, A.; de la Fuente, D.; Rivas-Gonzalez, F.W.; Franco-Nieto, A. A Sentinel-2 unsupervised forest mask for European sites. *Proc. SPIE* **2019**, *11156*, 111560Y.

42. Fernandez-Carrillo, A.; Franco-Nieto, A.; Pinto-Bañuls, E.; Basarte-Mena, M.; Revilla-Romero, B. Designing a Validation Protocol for Remote Sensing Based Operational Forest Masks Applications. Comparison of Products Across Europe. *Remote Sens.* **2020**, *12*, 3159. [CrossRef]

43. Tucker, C.J.; Elgin, J.H.; McMurtrey, J.E.; Fan, C.J. Monitoring corn and soybean crop development with hand-held radiometer spectral data. *Remote Sens. Environ.* **1979**, *8*, 237–248. [CrossRef]

44. Huete, A.R. A soil-adjusted vegetation index (SAVI). *Remote Sens. Environ.* **1988**, *25*, 295–309. [CrossRef]

45. Qi, J.; Chehbouni, A.; Huete, A.R.; Kerr, Y.H.; Sorooshian, S. A modified soil adjusted vegetation index. *Remote Sens. Environ.* **1994**, *48*, 119–126. [CrossRef]

46. Haralick, R.M. Statistical and structural approaches to texture. *Proc. IEEE* **1979**, *67*, 786–804. [CrossRef]

47. Breiman, L. Random Forests. *Mach. Learn.* **2001**, *45*, 5–32. [CrossRef]

48. Fernandez-Carrillo, A.; de la Fuente, D.; Rivas-Gonzalez, F.W.; Franco-Nieto, A. An automatic Sentinel-2 forest types classification over the Roncal Valley, Navarre: Spain. *Proc. SPIE* **2019**, *11156*, 111561N.

49. Havašová, M.; Bucha, T.; Ferenčík, J.; Jakuš, R. Applicability of a vegetation indices-based method to map bark beetle outbreaks in the High Tatra Mountains. *Ann. For. Res.* **2015**, *58*, 295–310. [CrossRef]

50. Pasqualotto, N.; Delegido, J.; Van Wittenberghe, S.; Rinaldi, M.; Moreno, J. Multi-Crop Green LAI Estimation with a New Simple Sentinel-2 LAI Index (SeLI). *Sensors* **2019**, *19*, 904. [CrossRef]

51. Delegido, J.; Verrelst, J.; Alonso, L.; Moreno, J. Evaluation of sentinel-2 red-edge bands for empirical estimation of green LAI and chlorophyll content. *Sensors* **2011**, *11*, 7063–7081. [CrossRef]

52. De Groot, M.; Ogris, N. Short-term forecasting of bark beetle outbreaks on two economically important conifer tree species. *For. Ecol. Manag.* **2019**, *450*, 117495. [CrossRef]

53. Immitzer, M.; Atzberger, C. Early detection of bark beetle infestation in Norway spruce (Picea abies, L.) using worldView-2 data frühzeitige erkennung von borkenkä ferbefall an fichten mittels worldView-2 satellitendaten. *Photogramm. Fernerkund. Geoinf.* **2014**, *2014*, 351–367. [CrossRef]

54. Marx, A. Detection and classification of bark beetle infestation in pure norway spruce stands with multi-temporal RapidEye imagery and data mining techniques. *Photogramm. Fernerkund. Geoinf.* **2010**, *2010*, 243–252. [CrossRef]

55. Ortiz, M.S.; Breidenbach, J.; Kändler, G. Early Detection of Bark Beetle Green Attack Using TerraSAR-X and RapidEye Data. *Remote Sens.* **2013**, *5*, 1912–1931. [CrossRef]

56. Czech Hydrometeorological Institute. Weather and Bark Beetle. Available online: https://portal.chmi.cz/aktualni-situace/aktualni-stav-pocasi/ceska-republika/pocasi-a-kurovec (accessed on 8 October 2020).

57. Hájek, F.; Lukeš, P.; Příhoda, J.; Křístek, Š.; Zahradník, P.; Kantorová, M.; Strejček, R.; Mlčoušek, M. Bark Beetle Map. Available online: https://www.kurovcovamapa.cz/o-projektu (accessed on 8 October 2020).

Publisher's Note: MDPI stays neutral with regard to jurisdictional claims in published maps and institutional affiliations.

 *remote sensing*

Article

The Use of Remotely Sensed Data and Polish NFI Plots for Prediction of Growing Stock Volume Using Different Predictive Methods

Paweł Hawryło [1,*], Saverio Francini [2], Gherardo Chirici [2], Francesca Giannetti [2], Karolina Parkitna [3], Grzegorz Krok [3], Krzysztof Mitelsztedt [3], Marek Lisańczuk [3], Krzysztof Stereńczak [3], Mariusz Ciesielski [3], Piotr Wężyk [1] and Jarosław Socha [1]

[1] Department of Forest Resources Management, Faculty of Forestry, University of Agriculture in Krakow, Al. 29 Listopada 46, 31-425 Kraków, Poland; piotr.wezyk@urk.edu.pl (P.W.); jaroslaw.socha@ur.krakow.pl (J.S.)

[2] Dipartimento di Scienze e Tecnologie Agrarie, Alimentari, Ambientali e Forestali, Università degli Studi di Firenze, 50145 Firenze, Italy; saverio.francini@unifi.it (S.F.); gherardo.chirici@unifi.it (G.C.); francesca.giannetti@unifi.it (F.G.)

[3] Department of Geomatics, Forest Research Institute, Braci Leśnej 3, 05-090 Sękocin Stary, Poland; k.parkitna@ibles.waw.pl (K.P.); g.krok@ibles.waw.pl (G.K.); k.mitelsztedt@ibles.waw.pl (K.M.); M.Lisanczuk@ibles.waw.pl (M.L.); k.sterenczak@ibles.waw.pl (K.S.); M.Ciesielski@ibles.waw.pl (M.C.)

* Correspondence: pawel.hawrylo@urk.edu.pl

Received: 12 September 2020; Accepted: 10 October 2020; Published: 13 October 2020

Abstract: Forest growing stock volume (GSV) is an important parameter in the context of forest resource management. National Forest Inventories (NFIs) are routinely used to estimate forest parameters, including GSV, for national or international reporting. Remotely sensed data are increasingly used as a source of auxiliary information for NFI data to improve the spatial precision of forest parameter estimates. In this study, we combine data from the NFI in Poland with satellite images of Landsat 7 and 3D point clouds collected with airborne laser scanning (ALS) technology to develop predictive models of GSV. We applied an area-based approach using 13,323 sample plots measured within the second cycle of the NFI in Poland (2010–2014) with poor positional accuracy from several to 15 m. Four different predictive approaches were evaluated: multiple linear regression, k-Nearest Neighbours, Random Forest and Deep Learning fully connected neural network. For each of these predictive methods, three sets of predictors were tested: ALS-derived, Landsat-derived and a combination of both. The developed models were validated at the stand level using field measurements from 360 reference forest stands. The best accuracy (RMSE% = 24.2%) and lowest systematic error (bias% = −2.2%) were obtained with a deep learning approach when both ALS- and Landsat-derived predictors were used. However, the differences between the evaluated predictive approaches were marginal when using the same set of predictor variables. Only a slight increase in model performance was observed when adding the Landsat-derived predictors to the ALS-derived ones. The obtained results showed that GSV can be predicted at the stand level with relatively low bias and reasonable accuracy for coniferous species, even using field sample plots with poor positional accuracy for model development. Our findings are especially important in the context of GSV prediction in areas where NFI data are available but the collection of accurate positions of field plots is not possible or justified because of economic reasons.

Keywords: airborne laser scanning; deep learning; Landsat; national forest inventory; stand volume

1. Introduction

Information about forests is collected at many spatial scales and with many different methods to deliver the information required for local, strategic and operational purposes. The forest growing stock volume (GSV) is an important variable for forest resource management. GSV provides the foundation for monitoring silvicultural treatments and changes in the forest ecosystem structure and functions. GSV can be based on data collected in the field or based on remote sensing-derived information. Ground-based National Forest Inventory (NFI) programs are usually carried out to obtain information at the strategic level. Information gathered on NFI field sample plots is subsequently used for national or regional forest management and planning, sustainability assessments or reporting to international conventions [1]. At the national or regional scale, forest volume is most commonly estimated on the basis of NFI data [2,3]. However, at the local scale, remote sensing data are increasingly used to obtain information on the smallest parts of the forest, particularly forest stands [4,5]. At present, remotely sensed data, especially Airborne Laser Scanning (ALS) point clouds, are used in forest inventories, where they are designed to support short-term forest management decisions at the local (stand) level related to harvest planning, the assessment of GSV and the planning of silvicultural activities [6]. GSV is one of the forest parameters with the highest interest for forest inventory activities. Detailed information about stand volume is crucial for making reasonable decisions concerning forest management. Accurate estimates of GSV are very important in the context of planning silvicultural activities and modelling forest productivity [7]. GSV is also the most important variable in carbon budget modelling [8].

The most common method for the estimation of GSV based on remotely sensed data is the area-based approach (ABA [9]). In this method, different remotely sensed data—most often ALS point clouds—are used for calculating various metrics that are then used as predictors in models developed using co-located ground plot measurements. This approach requires in situ data, such as field sample plots, with accurate information about their localization [10]. Unfortunately, precisely accurate measurements of sample plot positions require an expensive and time-consuming process because of the necessity to use, for example, advanced Global Navigation Satellite System (GNSS) receivers and perform post-processing procedures with data from the base-stations [11]. Only permanent sample plots do not generate additional costs after their establishment. On the other hand, relaxing requirements for the positioning accuracy of field plots may reduce the costs of the forest inventory but may also decrease the accuracy of the biophysical parameter estimates [4]. Accurate measurements of the plot coordinates and sizes of the sample plots are important factors that influence the accuracy of the forest inventory based on ALS data [12]. To some extent, a larger plot size can compensate for inaccurate measurements of sample plot coordinates [13,14] but the measurement costs will increase significantly.

The use of NFI field sample plots seems to be a reasonable logical compromise for optimizing the costs of the RS data used in national scale inventories, which provide operational and strategic information about the forest. The first attempts at integrating NFI plots with Landsat images began 30 years ago in Finland [15]. There are also other more recent positive experiences in the integration of remotely sensed data and NFI for descriptions of forest characteristics [16], specifically using ALS point clouds [17–19]. Hollaus et al. [18] used NFI plots and ALS point clouds for creating GSV model based on ALS data in Austria. Nord-Larsen et al. [19] developed models for selected forest parameters combing Danish NFI plots with ALS point clouds and the same was done for Sweden [17]. All these studies assumed that accurate measurements of sample plot positions are required for creating GSV predictive models. However, in some countries, such as Poland, NFI plots do not offer accurate information about plot coordinates. The positions of Polish NFI plots were measured using low-class GPS receivers; thus, their positional accuracy may vary from several to about 15 m. Socha et al. [7] proposed a method for using NFI data together with ALS point clouds even without any information about the plot coordinates. However, that method was concentrated on pure Scots pine-dominated stands and was tested in only one forest district. In this research, we aimed to evaluate Polish NFI plots as a source for the development of predictive models for GSV using ALS and Landsat data in varying types of stands located in different parts of Poland. We assessed whether the accuracy of GSV determination at the stand scale based on Polish NFI plots is sufficient for forest management purposes.

The main objectives of this study were (i) to evaluate the possibility of predicting the growing stock volume at the stand level using an area-based approach with NFI plots without accurate information about the plot coordinates; (ii) to compare the performance of different predictive approaches; (iii) to evaluate the possible differences in model performance using different types of predictor variables (e.g., Landsat, ALS, Landsat and ALS metrics); and (iv) to evaluate the model performance within stands dominated by different tree species (Scots pine, European beech, sessile oak and silver fir).

2. Materials and Methods

2.1. Study Area

Poland is located in Central Europe (Figure 1a) and covers a total area of 312,679 km^2. According to the Central Statistical Office [20], forest covers 29.6% of the country's area, which corresponds to 92,420 km^2. Poor and moderately rich forest habitats predominate and cover 57% of the forested area which is reflected in the dominance of coniferous tree species, which dominate over 68% of the area of Polish forests [21]. According to the National Forest Inventory [22], Scots pine (*Pinus sylvestris* L.) is predominant in most Polish tree stands. Among the deciduous species, oaks (*Quercus* spp.), silver birch (*Betula pendula* Roth.) and black alder (*Alnus glutinosa* Gaertn) are the species with the largest share. In mountainous areas, the share of Norway spruce (*Picea abies* Karst.), silver fir (*Abies alba* Milland) and European beech (*Fagus sylvatica* L.) is more significant. Forest ownership also has an impact on its management. In Poland, public forests are predominant (80.7%), with most of them under the administration of the State Forests. The age structure is mainly represented by III and IV age classes (41–60 and 61–80 years old, respectively). As reported by the National Forest Inventory at the end of 2016, the timber resources in Polish forests reached 2587 million m^3 of gross merchantable timber, out of which almost 80% is in the State Forests [21]. The mean growing stock volume of Polish forests is 269 m^3/ha, which is much higher than the average in European forests (163 m^3/ha) [23].

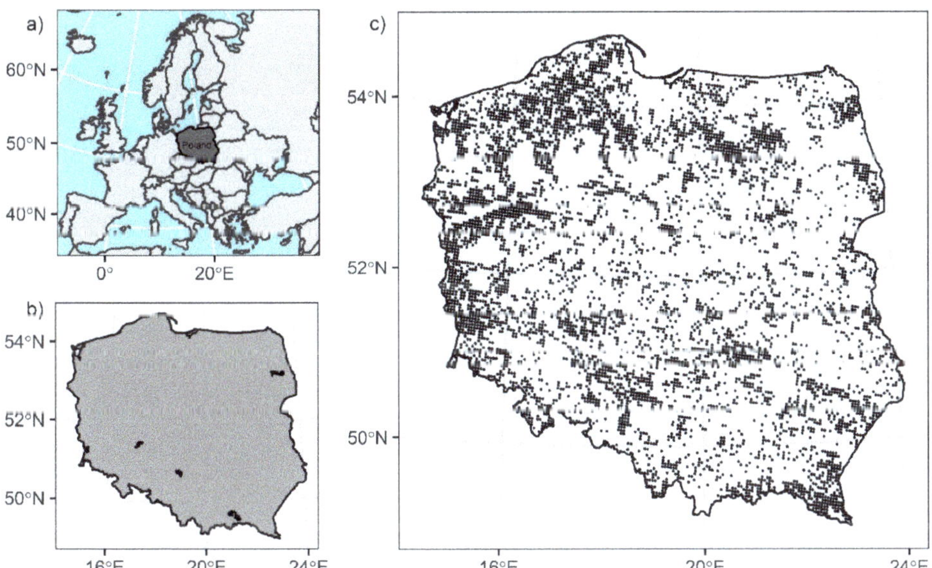

Figure 1. Localization of Poland in Europe (**a**) with the distribution of the reference forest stands ((**b**); black polygons) and the distribution of NFI (National Forest Inventory) plots in Poland ((**c**); black dots).

2.2. Polish National Forest Inventory Data

The National Forest Inventory started in Poland in 2005. This system uses a systematic scheme within a 4 km square grid of permanent sample plots. The 4 km grid is based on the pan-European 16 × 16 km forest monitoring network of the International Co-operative Programme on the Assessment and Monitoring of Air Pollution Effects on Forests [24]. Measurements of the NFI plots are performed in a 5-year inventory cycle measuring approximately 20% of the plots each year. In the first two cycles of the NFI, the sample plots were circular with 7.98 m (200 m^2), 11.28 m (400 m^2) and 12.62 m (500 m^2) radii depending on the forest stand age. In the last recent cycle (2015–2019), a radius of 11.28 m (400 m^2) was used for all plots. At each NFI sample plot, many tree and stand characteristics were measured. The diameter at breast height (DBH) was measured for all trees with a DBH ≥ 7 cm. The heights of the selected trees were measured to estimate the height curve. To obtain the GSV for a sample plot, first, the allometric models are used to predict individual tree volumes. Then, after aggregation from single trees, the plot-level GSV is calculated [25,26]. In this research, measurements from the second cycle of the NFI (2010–2014) were used since these measurements guaranteed the smallest difference between the time of the field measurements and the time of the ALS point cloud acquisition. For the analysis, 13,323 NFI sample plots were used, where the full area of the sample plot was located within the borders of a single forest stand and for which the ALS point clouds were available. The basic characteristics of the GSV at the plot level are presented in Table 1. The species share at the plot level was calculated using the volume of single trees.

Table 1. Characteristics of NFI plots used for the development of the GSV (growing stock volume) models (*n* = 13,323).

Dominant Tree Species	Number of Plots	Percentage of Plots (%)	Minimum GSV (m^3/ha)	Mean GSV (m^3/ha)	Maximum GSV (m^3/ha)	Standard Deviation of GSV (m^3/ha)
Scots pine	8334	62.6	0.2	300.7	935.1	139.6
silver birch	904	6.8	0.3	192.8	708.4	119.8
Norway spruce	731	5.5	0.4	309.3	1069.0	200.8
European beech	708	5.3	0.1	351.8	1452.8	220.6
sessile oak	680	5.1	0.2	296.0	1237.3	186.0
other species	663	5.0	0.2	243.4	1347.1	174.0
common alder	577	4.3	0.7	270.9	908.3	165.5
silver fir	459	3.4	2.7	368.0	1338.0	217.5
European larch	267	2.0	0.3	253.0	795.1	181.0

2.3. Landsat Images

After considering the possibility of using other forms of optical satellite imagery, we chose the imagery of Landsat 7 ETM+ to cover the study area because the scenes are freely available at a moderate resolution (30 × 30 m) compared to other types of satellite data that require a fee (e.g., Quickbirds and Ikonos) or are not available for the years of interest (e.g., Sentinel-2). In total, Poland is covered by 37 Landsat scenes (Figure 2), so we used the Google Earth Engine Platform (GEE) to obtain a cloud-free composite image. GEE provides a full repository of Landsat data and offers the possibility to select and process images with a specific threshold of cloud cover and a specific time range, as well as quickly obtain free cloud composite images that are produced combining different images of the same scenes [27]. Based on the cloud cover thresholds (i.e., less than 10%), the solar zenith angle (i.e., less than 76°) and the specific acquisition period (i.e., 1 January 2010–31 December 2014), we found that 420 images were available for all of Poland's area, with an average of 11 images for each Landsat scene (Figure 2). Among the 420 images selected, 24 were acquired between November and February.

For all the selected images, we used the bottom of the atmosphere reflectance values (BOA) calculated using the LEDAPS (Landsat ecosystem disturbance adaptive processing system) algorithm [28]. The images were subsequently masked for clouds using the Fmask (Function of the mask) algorithm [29]. Based on the processed images for reflectance and cloud masking, using GEE, we produced a cloud-free composite image for the whole of Poland. The spectral value of each pixel of the composite image was calculated as the *'median'* of all Landsat images available for the specific pixel. More details on GEE and composite images can be found in Gorelick et al. [27].

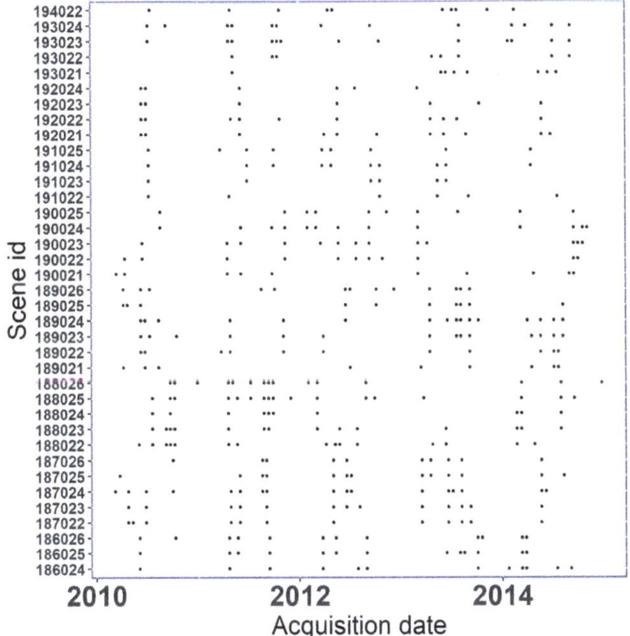

Figure 2. Numbers of the images available for each of the Landsat Scenes (black dots represent the available images).

2.4. Airborne Laser Scanning Point Clouds

The primary dataset for developing the GSV models in this research consisted of ALS point clouds acquired during the country-wide projects aimed mainly at creating a detail Digital Terrain Model for the whole of Poland. In the years 2011–2015, the Main Office of Geodesy and Cartography in Poland commissioned the development of altitude data using ALS technology for an area encompassing 92% of the country—i.e., 288,806 km^2 of Poland. In the following years (2015–2017), data collection for other areas of the country continued. ALS data were obtained in two standards: Standard I (mainly forest and rural areas outside cities), where the point density is at least four pulses per square meter (ppsm) and Standard II (94 cities in most without forests), with a density of at least 12 ppsm. The acquisition of ALS point clouds took place from mid-October to April, that is, in the leaf-off period, which at the transverse scan angle (allowed up to 25 degrees) guaranteed good penetration of the laser beams through the stand hood to the ground. Since the second cycle of the NFI (2010–2014) took place at nearly the same time as ALS data acquisition with the ISOK project (2011–2015), the absolute time difference between the two datasets was, in most cases (90.4%), less than three years (Table 2).

The second set of ALS data used in this study included point clouds obtained in 2015 and 2016 for the REMBIOFOR project. These data were used for validation of the developed predictive models on the validation stands. The data acquisition area included 6 forest districts where the reference forest stands were established (Figure 1b). Every single validation stand was scanned with the Riegl LiteMapper LMSQ680i laser scanning system. The scanner's operating frequencies were in the range of 300 to 400 kHz. The aircraft was flying at around 550 m above the mean ground level altitude, maintaining 30% tile-side overlap with a 60° FOV. The above mission parameters enabled the acquisition of spatially homogeneous multi-echo point clouds with average densities in the range from 10 to 14 ppsm.

Table 2. Maximum absolute time difference between the time of the field measurements at the NFI plots and the time of the ALS (airborne laser scanning) point cloud acquisition.

Maximum Absolute Difference between NFI and ALS (Years)	Number of Plots	Percentage of Plots (%)
1	5209	39.1
2	4246	31.9
3	2588	19.4
4	1092	8.2
5	188	1.4

2.5. Extraction of Predictor Variables

Two different sets of predictor variables were extracted from the remote sensing data associated with each NFI plot and each stand: optical metrics from Landsat cloud-free composites and ALS normalized point cloud-derived metrics.

In particular, we extracted the median value of each band of the Landsat 7 ETM+ composite image from the pixels associated with the area of each NFI plot and each reference forest stand. Based on the normalized ALS point clouds, the sets of metrics were calculated using the lidR package for R (Table 3 [30]).

Table 3. Predictor variables calculated for the NFI plots and used for the development of the GSV (growing stock volume) models.

Description of Predictor Variables	Acronyms
Bottom of atmosphere reflectance of Landsat 7 ETM+ spectral bands	B1_30, B2_30, B_3_30, B4_30, B5_30, B6_30, B7_30
Mean value of point heights (m)	zmean
Maximum value of point heights (m)	zmax
Standard deviation of point heights (m)	zsd
Skewness of point heights	Zskew
Kurtosis of point heights	Zkurt
Percentile values of point heights: 5th, 10th, 15th, ... , 95th (m)	zq5, zq10, zq15, ... , zq95
Entropy calculated as a normalized Shannon vertical complexity index	zentropy
Percentage of all returns above 2 m (%)	pzabove2
Percentage of all returns above *zmean* (%)	pzabovezmean
Cumulative percentage of returns from nine height layers. The height measurements were divided into 10 equal intervals according to Reference [31] (%)	zpcum1, zpcum2, zcum3, ... , zpcum9

2.6. Validation Data

The 360 forest stands located in 6 forest districts across Poland with a mean area of 1.1 ha were used as validation data (Figure 1b; Table 4). Field measurements were conducted in summer and autumn 2015 (four districts) and 2016 (two districts). In each stand, only trees with a DBH of at least 7 cm were measured. Moreover, in each stand, several heights for a dominant tree species were measured (minimum 20 trees). The measured trees were distributed evenly across the DBH range and forest stand layers. Height measurements were made using a Haglof Vertex IV ultrasonic hypsometer. Then, the volume for single trees was calculated using the allometric models created for Poland [25]. To calculate the GSV, the volumes of individual trees in the stand were summed and divided by the stand area. The ALS-derived metrics listed in Table 3 were calculated for each forest stand within a

30×30 m raster grid corresponding to the Landsat pixel size. The predicted GSV for each stand was calculated using GEE as the weighted mean from the raster cells overlapping with the stand borders. The overlapping area was used as the weight. See the GEE documentation for more in-depth information (https://developers.google.com/earth-engine/reducers_reduce_region#pixels-in-the-region).

Table 4. Characteristics of stands used for model validation (n = 360; mean stand area = 1.1 ha; GSV—growing stock volume).

Dominant Tree Species	Number of Stands	Percentage of Stands (%)	Minimum GSV (m³/ha)	Mean GSV (m³/ha)	Maximum GSV (m³/ha)	Standard Deviation of GSV (m³/ha)
Scots pine	270	75.0	111.0	332.6	593.0	97.4
silver fir	29	8.1	223.0	387.7	629.0	97.4
European beech	22	6.1	158.0	336.0	590.0	116.3
sessile oak	20	5.6	174.0	348.0	473.0	86.8
other species	19	5.3	190.0	277.6	479.0	77.7

2.7. Predictive Methods

We evaluated four methods for predicting GSV: three nonparametric approaches—Random Forest (RF), k-Nearest Neighbours (k-NN) and deep learning neural network (DL)—and one parametric approach—the multiple linear regression (LM) model.

The training set for developing the predictive models consisted of the GSV values measured in the field at the National Forest inventory plots (n = 13,323), while the validation set consisted of 360 stands with a mean area of 1.1 ha, ranging between 0.44 ha and 3.78 ha. For both the training set and validation set, the predictor variables consist of (i) Landsat metrics and (ii) ALS metrics (Table 3).

The predictive approaches were tested using three different sets of predictor variables (i.e., Landsat; ALS; and combined Landsat and ALS metrics) to compare the performance of each predictive approach using different types of predictors. Therefore, each imputation-predictive approach was optimized three times using the three different sets of predictors. In the following paragraphs, we detail how the predictive models were parametrized.

For each set of predictors (i.e., Landsat, ALS and Landsat+ALS), we calculated the variables' importance rankings using Random Forests and calculating the percentage increase in the mean square error (IncMSE%) by removing all variables one at a time. We calculated the IncMSE% through a 5-fold-cross-validation procedure that associates the relative standard error (MSE) to each variable.

Among the 7 Landsat predictors, the 41 ALS predictors and the 48 predictors for both Landsat and ALS, we identified the relevant and irrelevant predictors by iteratively removing the variables that were identified as less important [32]. At the end of each 5-fold cross-validation, we selected the set of predictor variables that had the lowest Mean Squared Error (MSE). Therefore, we selected a set of predictor variables for Landsat, a set of predictor variables for ALS and a set of predictor variables combining the Landsat and ALS variables. Then, the three sets of predictors were used to optimize the four predictive approaches.

Each predictive approach (i.e., Random Forest, k-NN, Neural Network and Multiple Linear Regression) was optimized three times using the three sets of predictor variables selected with the procedure described above. In this section, we will describe the different predictive approaches used.

2.7.1. Random Forests

RF is a decision tree algorithm introduced by Breiman [33]. It is commonly used for the spatial prediction of forest variables using remotely sensed data [34–38]. RF grows a set of regression trees (n_{tree}) with a certain depth (t_{depth}), using a randomly chosen subset of predictors for each one (m_{try}). To build trees, the out-of-bag samples (OOB) procedure is applied, where each tree is built independently based on bootstrap samples from the training dataset, while the remaining one-third of the samples are randomly left out. More details on the RF method can be found in the review of Belgiu and Drăgu [39] and in the article of Li et al. [32].

The hyperparameter optimization was performed using a random search tuning approach, through which we tested 50 randomly extracted combinations of n_{tree}, $m_{try\ and}$ t_{depth} (Table 5), searching for the combination that minimizes the Root Mean Square Error:

$$\text{RMSE} = \sqrt{\frac{\sum_{i=1}^{n}(y_i - \hat{y}_i)^2}{n}}, \tag{1}$$

where y_i and $\hat{y}_i$ denote the reference and predicted values, respectively, for the i-th sample plot and n is the number of plots.

Table 5. Set of tested parameters for each nonparametric predictive approach (KNN—k-nearest neighbours, RF—random forests and DL—deep learning).

Model	Hyperparameter	Values	Tuning Method
KNN	K Minkowski distance Distance metrics	1–40 Euclidean distance, Manhattan distance Unweighted, Weighted, Inverse, Reciprocal	Random Search
RF	n_{tree} m_{try} t_{depth}	100–500 2—number of predictors 2–10	Random Search
DL		Hidden layers Nodes Optimizer Learning Rate Dropout layers Dropout percentage Activation functions Regularization layers L1 regularization L2 regularization	Trial-and-error

The *RMSE* was evaluated using k-fold cross-validation. For each hyperparameter combination, we randomly divided the NFI plots into 5 reference set units and each one was deleted in sequence and predicted using the remaining plots. Thus, each sample plot was never used for both training and testing.

2.7.2. k-Nearest Neighbours

The k-NN imputation approach is well known method often used for developing predictive models in context of applications of remote sensing data in forestry [40]. With the k-Nearest Neighbours (k-NN) technique, predictions are calculated as linear combinations of observations for sample units that are the nearest to the population units for which the predictions are desired with respect to a selected distance metric in the space of the feature (auxiliary) variables. The optimization of k-NN parameters was performed using the same procedure used for RF but optimizing k (the number of nearest neighbours), the *Minkowski distance* parameter (among Euclidean or Manhattan distance) and the *distance metric*. A full list of the tested hyperparameters is provided in Table 5.

2.7.3. Deep Learning Fully Connected Neural Network

The Deep Learning model (DL) consists of N stacked layers composed of M nodes that facilitate learning through successive representations of the input data [41]. The DL model that we used was a Fully Connected Neural Network (FCNN), in which all nodes or neurons, in one layer are connected to the nodes in the next layer. The data are transformed in each layer using weights, which are specific parameters that link the nodes of subsequent layers [42]. During training, the optimum value of the weight of each node was optimized based on the loss function, which measures the difference between the observed and predicted values. In this study, modelling was done using the TensorFlow backend.

DL models require several hyperparameters to be set: (i) the number of hidden layers, (ii) the number of nodes, (iii) the optimizer and relative Learning Rate, (iv) the number of dropout layers, (v) the percentage of dropped out units in each layer, (vi) the different activation function configurations, (vii) the different values of lambda for L1 and L2 regularization strategies, (viii) the number of epochs and (ix) the activation function.

These hyperparameters are too numerous to use the same automatic random search optimization procedure used for k-NN and RF. Therefore, we used a trial-and-error approach, which, compared to random search, is more time consuming and requires more in-depth knowledge of the model and the effects of the hyperparameters.

As for RF and k-NN, we chose the best model configuration by minimizing the RMSE resulting from the 5-fold-cross validation procedure.

2.7.4. Multiple Linear Regression

Multiple linear regression (LM) method can be defined in the form of

$$y_i = \beta_0 + \beta_1 \cdot x_{1i} + \cdots + \beta_p \cdot x_{pi} + \varepsilon_i, \tag{2}$$

where i indexes the sample units, y_i represents the single response variable, $p \geq 1$ denotes the number of explanatory variables, $j = \{1, \ldots, p\}$ indexes the explanatory variables, β_j is the respective regression coefficient and ε_i denotes a random residual term with the distribution of $N(0, \sigma_i^2)$. LM does not have hyperparameters and does not require an optimization phase. However, we calculated the performance of LM with the same 5-fold-cross validation procedure used for RF, k-NN and DL.

2.8. Models Training and Validation

As a result of the previous processes, we identified the model configurations that ensure the best performance in terms of the RMSE. Then, we ultimately trained the k-NN, RF, DL and LM models with the identified configuration using all 13,323 NFI plots. We tested the best twelve models (i.e., the four predictive approaches using the three different sets of predictors) using a bootstrapping approach based on the reference stands ($n = 360$) to evaluate the performance using an independent dataset (never-seen-before data).

Resampling methods such as the bootstrap resampling technique can be applied to assess uncertainty for predictive approaches such as parametric and nonparametric approaches [43,44]. Bootstrapping is based on the notion of a bootstrap sample and the bootstrapping pair approach was used to construct bootstrap samples for this study [45]. We created 100,000 bootstrap samples (nboot = 100,000) that were considered to be consistent [43,45]. Each sample consisted of a sample drawn with a replacement from the original reference set with the same dimension of the original stand dataset ($n = 360$).

Based on each 100,000 bootstrap sample, we applied the models and calculated the RMSE as done in the optimization phase (Equation (1)). In this case, the n value was equal to the number of stands ($n = 360$). Moreover, we calculated the relative RMSE% as the percentage of the average ground reference value of the bootstrap samples.

For each model, at the end of the bootstrapping procedure, we calculated the mean $RMSE$ (μ) and the RMSE% (μ%) as the mean of the RMSE. The mean of the RMSE% achieved for each bootstrap sample was calculated as

$$\mu = \frac{\sum_{i=1}^{n} \text{RMSE}_i}{b}$$

$$\mu_{\%} = \frac{\sum_{i=1}^{n} \text{RMSE}\%_i}{b},$$

where RMSE$_i$ and *RMSE%$_i$* are the results achieved for the *i*-th bootstrap sample and *b* is the number of bootstrap samples; in our case, *b* = 100,000. Using the same procedure, we calculated the R^2, the bias and the relative bias for the bootstrap samples.

Moreover, we calculated confidence intervals of 95% for each parameter of performance as

$$\mu \pm t_{1-\alpha/2}\,\sigma(\mu)$$

$$\mu_\% \pm t_{1-\alpha/2}\,\sigma(\mu_\%),$$

where α is the significant value, $t_{1-\alpha/2}$ is the $1 - \frac{\alpha}{2}$ percentile of Student's t-distribution (in our case 1.96) and σ is the standard deviation.

Additionally, the R^2, RMSE% and bias% were calculated and analysed for the dominant tree species with at least 20 validation stands available.

3. Results

3.1. Set of Predictor Variables

Importance ranking using Random Forests and calculation of the percentage increase in the mean square error (IncMSE%) by removing all variables one at a time identified the best sets of predictor variables for the three sets of predictors (i.e., Landsat, ALS and Landsat+ALS).

The sets of predictor variables for the Landsat predictors consisted of all the bands (except for Band 6) being removed (Figure 3). However, for the ALS predictors, we removed zq10 and zq5 (Figure 3). For the set of a predictor that combined Landsat and ALS, we found that the same variables removed from the other two sets were also excluded (i.e., B6, zq10 and zq5; Figure 3).

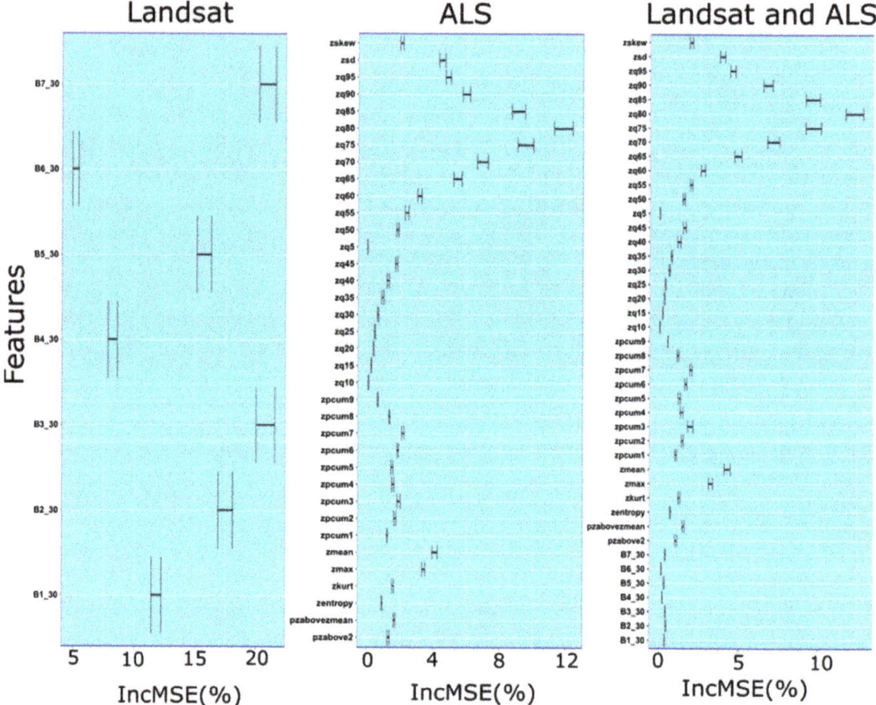

Figure 3. Importance ranking of Random Forests for each type of predictor: Landsat, ALS and Landsat and ALS. The graph reports the percentage increase in the mean square error (IncMSE%) of each type of predictor.

3.2. Optimization Results

Each predictive approach was optimized using the three sets of predictors to search for the hyperparameters that can achieve the lowest RMSE. The results of the calibration of the RF and KNN models are reported in Table 6. Using the 5-fold-cross validation procedure, we found that under the same approach, the configurations of the hyperparameters (Table 6) remain similar if different sets of predictors are used.

Table 6. The best hyperparameter configurations selected for each predictive approach and each set of predictors (RF—random forest, k-NN—k-nearest neighbours).

Predictive Approach	Predictors	Configurations		
		m_{try}	t_{depth}	n_{tree}
RF	Landsat	2	9	500
	ALS	2	6	500
	Landsat+ALS	2	3	500
		kmax	distance	kernel
k-NN	Landsat	36	Manhattan	unweighted
	ALS	37	Manhattan	inverse
	Landsat+ALS	35	Manhattan	inverse

In case of the neural network configurations for the DL models we matched the number of nodes in the input layer with the number of predictors. The number of hidden layers was two with 70 nodes each and LeakyReLU activation functions [46]. Moreover, the number of dropout layers was two, with the percentage of nodes randomly dropped out equal to 30%. The L1 and L2 regularization layers were not necessary and the selected optimizer was rmsprop [47].

3.3. Performance Assessment

The results of the bootstrapping procedure were comparable among different predictive approaches with the same sets of predictors. As shown in Figure 4, the results are consistent and comparable in terms of the confidence interval and distribution of the errors. Using the Landsat predictors, we found that all predictive approaches achieved results for an RMSE between 102.10 and 106.24 m^3/ha and an RMSE% between 30.45% and 31.69% with the RF that achieved the highest accuracy (Table 7). Using the ALS predictors, we found that all the predictive methods achieved RMSE results between 81.43 and 83.90 m^3/ha and RMSE% results between 24.30% and 25.03% with the DL that achieved the highest accuracy. Using the Landsat and ALS predictors, the evaluated predictive approaches achieved results for an RMSE between 80.98 and 84.41 m^3/ha and an RMSE% between 24.16% and 25.18% with the RF that achieved the lowest accuracy and the DL the highest accuracy.

Table 7. Performance of the developed growing stock volume models (DL—deep learning, KNN—k-nearest neighbours, LM—multiple linear regression and RF—random forests) assessed at the stand level (n = 360).

Model	Predictors	R^2	RMSE (m^3/ha)	RMSE%	Bias (m^3/ha)	Bias%
DL	Landsat	0.01	106.24	31.69	40.39	12.03
DL	ALS	0.39	81.43	24.30	−11.64	−3.48
DL	Landsat+ALS	0.38	80.98	24.16	−7.31	−2.2
KNN	Landsat	0.04	104.54	31.19	39.94	11.9
KNN	ALS	0.38	83.24	24.85	−16.77	−5.02
KNN	Landsat+ALS	0.38	83.19	24.83	−16.94	−5.06
LM	Landsat	0.05	104.92	31.3	38	11.3
LM	ALS	0.37	82.72	24.69	−11.7	−3.51
LM	Landsat+ALS	0.37	82.57	24.64	−12.01	−3.6
RF	Landsat	0.07	102.10	30.45	36.56	10.89
RF	ALS	0.38	83.90	25.03	−16.51	−4.93
RF	Landsat+ALS	0.39	84.41	25.18	−17.81	−5.32

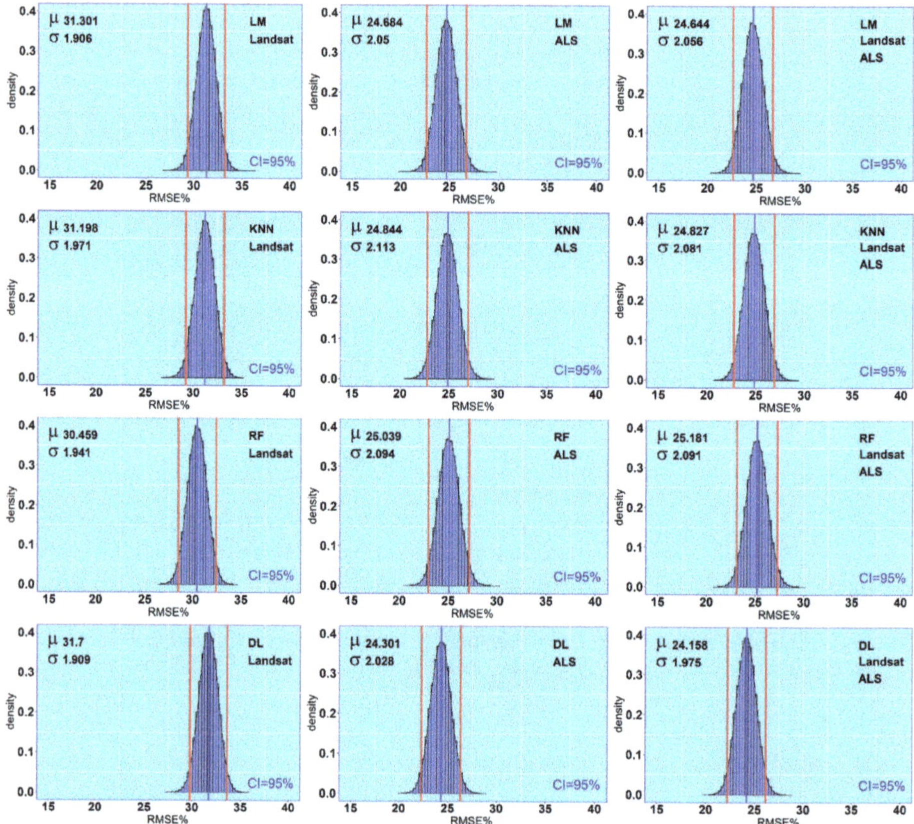

Figure 4. Results of the bootstrapping procedure for each considered approach (LM—multiple linear regression, KNN—k-nearest neighbours, RF—random forests and DL—deep learning) using different sets of predictor variables (Landsat, ALS and Landsat+ALS). The red line identifies the confidence interval.

All predictive approaches have similar results for R^2 (Figure 5). Using Landsat, only the R^2 dropped to 0.01. Similarly, using only Landsat, the bias was large, ranging between 40.39 and 36.56 m^3/ha. On the other hand, Landsat made an important contribution to DL when used together with the ALS. Indeed, when using DL with just the ALS, the bias was −11.64 m^3/ha but while using both the Landsat and the ALS, the bias dropped to −7.31 m^3/ha.

3.4. Performance Assessment per Dominant Tree Species

From the set of 360 validation stands, there were four dominant tree species for which at least 20 stands were available: European beech, Scots pine, sessile oak and silver fir. The other species were joined to one group of 19 stands. For these groups of stands, the performance of the models developed based on Landsat- and ALS-derived predictors was analysed (Figure 6, Table 8). The RMSE was relatively low for oak, pine and fir regardless of the predictive method, varying from 72.61 m^3/ha (20.99%) for oak-dominated stands with the DL method to 80.00 m^3/ha (24.06%) for pine-dominated stands predicted with the RF model. A considerably larger RMSE was obtained for the beech-dominated stand varying from 106.94 m^3/ha (DL) to 110.19 m^3/ha (LM). For beech-dominated stands, large systematic errors (bias) were also observed for all evaluated predictive methods: from −20.19% (−66.62 m^3/ha) for the DL model up to −23.71% (−78.55 m^3/ha) for LM. The relative bias for pine, oak and fir varied from 0.70% (fir; RF) to −6.73% (oak, LM). The R^2 averaged through all predictive methods was the highest for beech-dominated stands (0.50) and the lowest for fir-dominated stands (0.30). The R^2 was also low for the group of 19 stands dominated by "other species" (0.13).

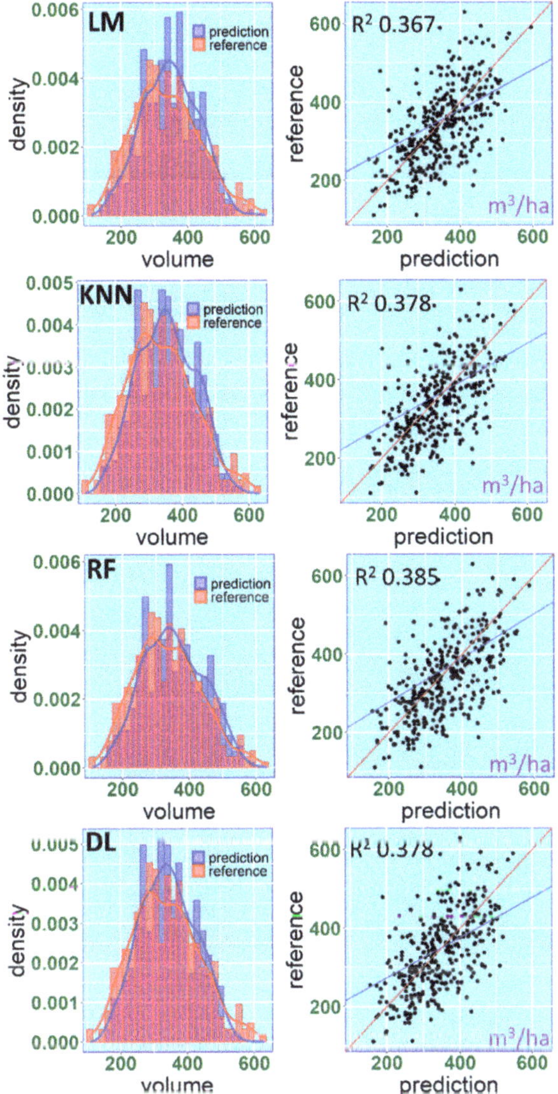

Figure 5. Observed vs. predicted values of GSV (growing stock volume) for 360 validation stands for each considered predictive approach (LM—multiple linear regression, KNN—k-nearest neighbours, RF—random forests and DL—deep learning) for the best set of predictors (Landsat+ALS).

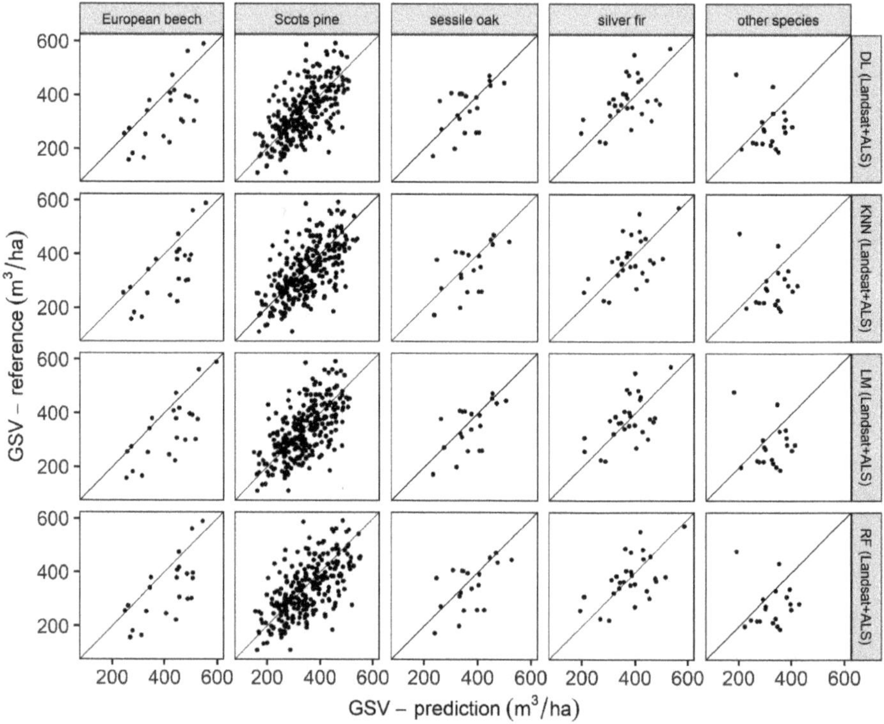

Figure 6. Observed vs. predicted values of GSV (growing stock volume) for validation stands obtained with the best set of predictors (Landsat+ALS) and grouped by dominant tree species and predictive approach (DL—deep learning, KNN—k-nearest neighbours, LM—multiple linear regression and RF—random forests).

Table 8. Performance of the developed growing stock volume models (DL—deep learning, KNN—k-nearest neighbours, LM—multiple linear regression and RF—random forests) assessed at the stand level for different dominant tree species (n = 360).

Species	Number of Stands	Method	R^2	RMSE (m³/ha)	RMSE%	Bias (m³/ha)	Bias%
European beech	22	DL	0.46	106.94	32.13	−66.62	−20.19
		KNN	0.51	108.39	32.59	−74.39	−22.5
		LM	0.53	110.19	33.11	−78.55	−23.71
		RF	0.49	109.92	33.03	−74.21	−22.45
Scots pine	270	DL	0.42	76.02	22.86	−2.80	−0.86
		KNN	0.42	78.6	23.64	−12.08	−3.65
		LM	0.40	77.79	23.40	−6.13	−1.86
		RF	0.42	80.00	24.06	−13.18	−3.98
sessile oak	20	DL	0.33	72.61	20.99	−10.91	−3.32
		KNN	0.31	77.22	22.34	−20.19	−5.99
		LM	0.36	73.38	21.24	−22.78	−6.73
		RF	0.32	77.62	22.45	−20.27	−6.01
silver fir	29	DL	0.28	87.74	22.63	18.01	4.53
		KNN	0.3	83.32	21.49	6.78	1.63
		LM	0.29	84.08	21.68	12.81	3.18
		RF	0.32	84.53	21.82	3.15	0.70
other species	19	DL	0.13	103.23	37.11	−37.98	−14.17
		KNN	0.12	109.53	39.45	−51.86	−19.19
		LM	0.14	108.51	39.04	−45.55	−16.91
		RF	0.12	109.09	39.27	−47.89	−17.74

4. Discussion

In this article, four predictive approaches (LM, KNN, RF and DL) trained based on NFI plots combined with ALS point clouds and Landsat images were evaluated for predicting GSV at the stand level. Three sets of predictor variables were tested (ALS, Landsat and Landsat+ALS) for all methods. Validation of the developed models was performed on 360 forest stands measured in the field.

When creating the Landsat mosaic, we did not divide the Landsat images into leaf-on (396 images) and leaf-off (24 images) datasets. Although we are aware that mixing leaf-on and leaf-off images is not a perfect approach, we assumed that since the final reflectance value was calculated as the median, the rare leaf-off images would not affect the final image consistently. On the other hand, rare high quality leaf-off images (cloud cover < 10% and solar zenith angle smaller than 76°) selected in winter can be informative for evergreen trees and areas when there are no images acquired during the leaf-on season, which may occur when algorithms are applied over huge areas.

The obtained results showed that there are no considerable differences in accuracy between the evaluated predictive approaches (Table 7, Figures 4 and 5). Generally, the DL method provided the lowest RMSE and bias; however, the differences between models were minimal. Larger differences between the predictive models were observed when validating within groups of dominant tree species. The results show that multiple linear regression (LM) can provide satisfactory results but there are no or only small, benefits from using more advanced methods, including machine learning (KNN and RF) or deep learning (DL). The linear model was used by Nilsson et al. [17] for country-level forest variables estimation in Sweden. The RMSE of 24.69% obtained in our study (LM) regardless of tress species is larger than that obtained by Nilsson et al. ([17]; RMSE: 17.2%–22.0%); however, this might be related to the higher species and structural variability of Polish forests. Parametric and nonparametric methods each have their own pros and cons. Previous works [9,48] pointed to the undoubted advantage of classic multiple linear regression, highlighting the ease of interpreting the model. The advantage of LM is also its ability to extrapolate extreme values. The construction of such models, however, requires more experience and more work for the selection of appropriate predictors. This problem disappears with the nonparametric methods of KNN, RF and DL used in this paper. These models can also provide good performance in the case of non-linear relationships between variables. However, the possibility of more frequent model overfitting and poor results in the case of overly small observation sets are potential disadvantages of these approaches [49,50].

The obtained results showed differences in the performance of the developed predictive models between dominant tree species. Nilsson et al. [17] reported for Sweden that when the proportion of broadleaved trees increased to 76%, the RMSE% of the GSV estimation increased to 27.9%. For our study, a similar relationship was observed. The model performance was relatively good for pine and fir, moderate for oak and the worst for beech. However, the obtained differences in RMSE% for coniferous and broadleaved stands were not so obvious. For example, the RMSE% for oak-dominated stands was even lower than that for pine-dominated stands. We observed that the systemic error (bias) is higher for stands dominated by broadleaved trees. In the case of beech-dominated stands, the relative bias increased to more than −20%, meaning that the developed models are not appropriate for these kinds of stands. It can be expected that creating species-specific models will improve the accuracy and reduce model bias [19]. However, the large overestimation of GSV for broadleaved stands is likely the result of using the ALS point clouds of different characteristics for model validation, and for model development. In cases when the ALS point clouds are collected with different pulse densities and at different times throughout the year, the ALS metrics obtained for broadleaved trees might not be comparable between two acquisitions. Our results suggest that such an approach can be acceptable for the prediction of GSV in coniferous forests but not in areas dominated by broadleaved species. To deal with this issue, predictive models for broadleaved trees could be developed based on metrics derived from the Canopy Height Model instead of using the point cloud-derived metrics.

In our research, we tested a deep learning method as a fully connected neural network using the same predictor variables applied in the other three predictive methods. When validating the results regardless of the dominant tree species, the best model accuracy among all tested predictive approaches and sets of predictor variables was obtained with the DL utilizing Landsat+ALS variables (RMSE% = 24.2; bias% = −2.2%) but this performance was only slightly better than that of the other tested methods. However, to derive reliable conclusions about the performance of DL for GSV predictions, different neural network architecture types must be evaluated, including the Convolutional Neural Network (CNN). Deep learning with CNN architecture—especially 3D CNN—can interpret raw data (e.g., ALS point clouds) without the need for human-derived explanatory variables such as height percentiles or density metrics, which are usually used as predictor variables in other imputation-predictive methods [51,52]. Ayrey et al. [51] showed that 3D CNN can considerably outperform the random forest method in predicting forest parameters based on ALS and satellite data.

In similar studies on using NFI plots and ALS point clouds for GSV estimation [17–19], the detailed selection of training plots and high accuracy measurements of the plot centres were used, while in our study, accurate coordinates of the plot centres were not available. This is likely the main reason for the relatively low R^2 values that we obtained. Nevertheless, our study shows that for coniferous species (Scots pine and silver fir), relatively low RMSE% (22%–23%) and bias% (1%) can be obtained for stand-level GSV predictions when combining ALS point clouds and Landsat images with NFI data, even when the positional errors of the field plots vary from several to about 15 m. It can be assumed that collecting accurate information about the plot centre coordinates in the next cycles of the Polish NFI may increase the accuracy of the predictive models of GSV. However, McRoberts et al. [53] observed only a small decrease in the standard error of the mean aboveground biomass in ALS-assisted estimates of aboveground biomass when using survey-grade GPS receivers with sub-meter accuracy compared to field grade GPS receivers with a 5–10 m accuracy. The aforementioned authors concluded that the high costs of acquiring a survey-grade GPS receiver are not justified in the case of ALS-assisted estimates of aboveground biomass at the national scale level in the USA.

Notably, in our study, for some of the NFI plots, there was up to a 5-year difference between the year of laser scanning and the year of the field survey. This difference may provide additional noise in the data used for training the predictive models because of tree growth, thinning activities, clear cutting or the occurrence of forest disturbances. As we are aware of these limitations, we assumed that the influence of these imperfections (positional inaccuracy and time lag) in the NFI data might be mitigated by using many training plots (13,323) through averaging the model parameters. We also maintain that a few-year difference between field measurements on NFI plots and the available ALS point clouds (or image-derived point clouds) may often be the case in real-world applications. Thus, we decided to use all NFI plots for which ALS data were available. The influence of time lag on the accuracy of GSV predictive models should be analysed in future studies.

When considering the forest management inventory at a local scale, an absolute bias of about 0%–3% (like that obtained for coniferous tree species) can be acceptable. However, this level of systematic error might be problematic for large areas of ALS-assisted estimates with GSV. The obtained bias might partially be the result of the abovementioned inaccuracies in the NFI data but it might also be the result of resolution difference between the differently sized training field plots (200 m^2, 400 m^2 and 500 m^2) and grids used for the model predictions (30 × 30 m; 900 m^2). Nevertheless, we decided to use a 30 × 30 m resolution grid for prediction to avoid resampling and average the Landsat-derived predictors. Packalen et al. [54] reported that a resolution mismatch between the field plot size and grid cell size used for predictions caused only a small increase of bias in ABA. The authors indicated that a higher prediction resolution compared to training resolution caused an underestimation of AGB. However, in our study, overestimation of GSV was observed in most cases, even though the prediction grid was larger than the training sample size.

Future research should evaluate how the time lag between NFI and ALS data influences the predictions of growing stock volume at the national level. It would also be valuable to evaluate more complex DL models based on raw ALS point cloud data including convolutional layers. On the other hand, simpler multiple regression models based on Canopy Height Models may be potentially more transferable compared to point cloud-based models, especially for stands dominated by broadleaved trees. The use of predictors from Sentinel-2 + ALS instead of Landsat+ALS also seems worth exploring.

5. Conclusions

Several conclusions were drawn from this study. First, using Landsat-derived predictors together with ALS only slightly increased the accuracy of GSV predictions compared to using only ALS-derived predictors. Second, the deep learning method in the form of fully connected layers does not provide considerably more accurate predictions of GSV compared to the LM, KNN and RF approaches. To benefit more from the deep learning approach, larger datasets and network architectures including convolutional layers are likely needed. Third, NFI plots with positional errors in the order of several to about 15 m combined with ALS point clouds can be used for the development of relatively accurate predictive models of the growing stock volume. In cases when a relatively large number of NFI plots is available for model training, the influence of positional inaccuracies and the time lag of NFI plots is likely mitigated partially by averaging the model parameters. The demonstrated approach is applicable for coniferous stands even when the ALS point clouds of different pulse densities are used for model training and prediction. However, when applied to forest stands dominated by broadleaved trees, large systematic errors may occur. Further research aimed at improving the integration methods of NFI data with remote sensing data is desirable, especially in the context of model transferability in broadleaved forests. Integrating NFI with remotely sensed data could provide substantial cost savings in the forest management inventory at a local scale.

Author Contributions: Conceptualization, P.H., G.C. and J.S.; methodology, P.H., G.C., S.F. and F.G.; formal analysis, P.H., S.F. and F.G.; data curation, P.H., J.S., K.S., K.P., G.K., K.M., M.L., M.C. and P.W.; writing—original draft preparation, P.H., S.F., G.C., F.G. and J.S.; writing—review and editing, K.S., K.P., G.K., K.M., M.L., M.C. and P.W.; funding acquisition, J.S. and K.S. All authors have read and agreed to the published version of the manuscript.

Funding: This research was supported by the projects REMBIOFOR and I-MAESTRO. The project "REMBIOFOR" was financed by The National Centre for Research and Development in Poland under the programme BIOSTRATEG, agreement no. BIOSTRATEG1/267755/4/NCBR/2015. The Project Innovative forest MAnagEment STrategies for a Resilient biOeconomy under climate change and disturbances (I-MAESTRO) is supported under the umbrella of ForestValue ERA-NET co-funded by the National Science Centre, Poland and the French Ministry of Agriculture, Agrifood and Forestry, as well as the French Ministry of Higher Education, Research and Innovation, German Federal Ministry of Food and Agriculture (BMEL) via the Agency for Renewable Resources (FNR), the Slovenian Ministry of Education, Science and Sport (MIZS). ForestValue received funding from the European Union's Horizon 2020 research and innovation programme under grant agreement N° 773324.

Conflicts of Interest: The authors declare no conflict of interest.

References

1. Kangas, A.; Astrup, R.; Breidenbach, J.; Fridman, J.; Gobakken, T.; Korhonen, K.T.; Maltamo, M.; Nilsson, M.; Nord-Larsen, T.; Næsset, E.; et al. Remote sensing and forest inventories in Nordic countries—Roadmap for the future. *Scand. J. For. Res.* **2018**, *7581*, 1–16. [CrossRef]

2. Tomppo, E.; Gschwantner, T.; Lawrence, M.; McRoberts, R.E. *National Forest Inventories: Pathways for Common Reporting*; Springer: Dordrecht, The Netherlands, 2010; ISBN 9789048132324.

3. Gschwantner, T.; Lanz, A.; Vidal, C.; Bosela, M.; Di Cosmo, L.; Fridman, J.; Gasparini, P.; Kuliešis, A.; Tomter, S.; Schadauer, K. Comparison of methods used in European National Forest Inventories for the estimation of volume increment: Towards harmonisation. *Ann. For. Sci.* **2016**. [CrossRef]

4. Næsset, E. Area-Based Inventory in Norway—From Innovation to an Operational Reality. In *Forestry Applications of Airborne Laser Scanning: Concepts and Case Studies*; Springer: Dordrecht, The Netherlands, 2014; pp. 215–240.

5. White, J.C.; Coops, N.C.; Wulder, M.A.; Vastaranta, M.; Hilker, T.; Tompalski, P. Remote Sensing Technologies for Enhancing Forest Inventories: A Review. *Can. J. Remote Sens.* **2016**, *42*, 619–641. [CrossRef]

6. Wulder, M.A.; Bater, C.W.; Coops, N.C.; Hilker, T.; White, J.C. The role of LiDAR in sustainable forest management. *For. Chron.* **2008**, *84*, 807–826. [CrossRef]

7. Socha, J.; Hawryło, P.; Pierzchalski, M.; Stereńczak, K.; Krok, G.; Wężyk, P.; Tymińska-Czabańska, L. An allometric area-based approach—A cost-effective method for stand volume estimation based on ALS and NFI data. *For. Int. J. For. Res.* **2019**, *93*, 344–358. [CrossRef]

8. Kurz, W.A.; Dymond, C.C.; White, T.M.; Stinson, G.; Shaw, C.H.; Rampley, G.J.; Smyth, C.; Simpson, B.N.; Neilson, E.T.; Trofymow, J.A.; et al. CBM-CFS3: A model of carbon-dynamics in forestry and land-use change implementing IPCC standards. *Ecol. Model.* **2009**, *220*, 480–504. [CrossRef]

9. White, J.C.; Wulder, M.A.; Varhola, A.; Vastaranta, M.; Coops, N.C.; Cook, B.D.; Pitt, D.G.; Woods, M. *A Best Practices Guide for Generating Forest Inventory Attributes from Airborne Laser Scanning Data Using an Area-Based Approach*; Technical Report FI-X-010; The Canadian Wood Fibre Centre: Victoria, BC, Canada, 2013.

10. Næsset, E. Estimating timber volume of forest stands using airborne laser scanner data. *Remote Sens. Environ.* **1997**, *61*, 246–253. [CrossRef]

11. Brach, M.; Stereńczak, K.; Bolibok, L.; Kwaśny, Ł.; Krok, G.; Laszkowski, M. Impacts of forest spatial structure on variation of the multipath phenomenon of navigation satellite signals. *Folia For. Pol.* **2019**, *61*, 3–21. [CrossRef]

12. Stereńczak, K.; Lisańczuk, M.; Parkitna, K.; Mitelsztedt, K.; Mroczek, P.; Miścicki, S. The influence of number and size of sample plots on modelling growing stock volume based on airborne laser scanning. *Drewno* **2018**, *61*. [CrossRef]

13. Gobakken, T.; Næsset, E. Assessing effects of positioning errors and sample plot size on biophysical stand properties derived from airborne laser scanner data. *Can. J. For. Res.* **2009**, *39*, 1036–1052. [CrossRef]

14. Mauro, F.; Valbuena, R.; Manzanera, J.A.; García-Abril, A. Influence of global navigation satellite system errors in positioning inventory plots for treeheight distribution studies. *Can. J. For. Res.* **2011**, *41*, 11–23. [CrossRef]

15. Tomppo, E. Satellite image-based National Forest Inventory of Finland. In Proceedings of the ISPRS, COMISSION VII, Mid-Term Symposium Global and Environmental Monitoring, Techniques and Impacts, Victoria, BC, Canada, 17–21 September 1990; pp. 419–424.

16. Chirici, G.; Giannetti, F.; McRoberts, R.E.; Travaglini, D.; Pecchi, M.; Maselli, F.; Chiesi, M.; Corona, P. Wall-to-wall spatial prediction of growing stock volume based on Italian National Forest Inventory plots and remotely sensed data. *Int. J. Appl. Earth Obs. Geoinf.* **2020**, *84*, 101959. [CrossRef]

17. Nilsson, M.; Nordkvist, K.; Jonzén, J.; Lindgren, N.; Axensten, P.; Wallerman, J.; Egberth, M.; Larsson, S.; Nilsson, L.; Eriksson, J.; et al. A nationwide forest attribute map of Sweden predicted using airborne laser scanning data and field data from the National Forest Inventory. *Remote Sens. Environ.* **2017**, *194*, 447–454. [CrossRef]

18. Hollaus, M.; Dorigo, W.; Wagner, W.; Schadauer, K.; Höfle, B.; Maier, B. Operational wide-area stem volume estimation based on airborne laser scanning and national forest inventory data. *Int. J. Remote Sens.* **2009**, *30*, 5159–5175. [CrossRef]

19. Nord-Larsen, T.; Schumacher, J. Estimation of forest resources from a country wide laser scanning survey and national forest inventory data. *Remote Sens. Environ.* **2012**, *119*, 148–157. [CrossRef]

20. CSO. *Statistical Yearbook of Forestry*; Statistics Poland: Warsaw, Poland, 2019. (In Polish)

21. SFIC. *Forests in Poland*; The State Forests Information Centre: Warsaw, Poland, 2018.

22. NFI National Forest Inventory. *Wielkoobszarowa Inwentaryzacja Stanu Lasów, Wyniki za Okres 2014–2018*; Biuro Urządzania Lasu i Geodezji Leśnej: Warsaw, Poland, 2018. (In Polish)

23. Forest Europe. State of Europe's Forests. In Proceedings of the Ministerial Conference on the Protection of Forests in Europe, Madrid, Spain, 20–21 October 2015.

24. ICP ICP Forests Manual. *Manual on Methods and Criteria for Harmonized Sampling, Assessment, Monitoring and Analysis of the Effects of Air Pollution on Forests*; Thünen Institute Forest Ecosystems: Eberswalde, Germany, 2016.

25. Bruchwald, A.; Rymer-Dudzińska, T.; Dudek, A.; Michalak, K.; Wróblewski, L.; Zasada, M. Wzory empiryczne do określania wysokości i pierśnicowej liczby kształtu grubizny drzewa. *Sylwan* **2000**, *144*, 5–12.

26. Gschwantner, T.; Alberdi, I.; Balázs, A.; Bauwens, S.; Bender, S.; Borota, D.; Bosela, M.; Bouriaud, O.; Cañellas, I.; Donis, J.; et al. Harmonisation of stem volume estimates in European National Forest Inventories. *Ann. For. Sci.* **2019**, *76*. [CrossRef]

27. Gorelick, N.; Hancher, M.; Dixon, M.; Ilyushchenko, S.; Thau, D.; Moore, R. Google Earth Engine: Planetary-scale geospatial analysis for everyone. *Remote Sens. Environ.* **2017**, *202*, 18–27. [CrossRef]

28. Schmidt, G.; Jenkerson, C.B.; Masek, J.; Vermote, E.; Gao, F. *Landsat Ecosystem Disturbance Adaptive Processing System (LEDAPS) Algorithm Description*; U.S. Geological Survey: Reston, VA, USA, 2013.

29. Zhu, Z.; Woodcock, C.E. Object-based cloud and cloud shadow detection in Landsat imagery. *Remote Sens. Environ.* **2012**, *118*, 83–94. [CrossRef]

30. Roussel, J.-R.; Auty, D. lidR: Airborne LiDAR Data Manipulation and Visualization for Forestry Applications. R Package Version 2.2.2. Available online: https://CRAN.R-project.org/package=lidR2020 (accessed on 28 January 2020).

31. Woods, M.; Lim, K.; Treitz, P. Predicting forest stand variables from LiDAR data in the Great Lakes—St. Lawrence forest of Ontario. *For. Chron.* **2008**, *84*, 827–839. [CrossRef]

32. Li, I.Z.W.; Xin, X.P.; Huan, T.A.N.G.; Fan, Y.A.N.G.; Chen, B.R.; Zhang, B.H. Estimating grassland LAI using the Random Forests approach and Landsat imagery in the meadow steppe of Hulunber, China. *J. Integr. Agric.* **2017**, *16*, 286–297. [CrossRef]

33. Breiman, L. Random Forests. *Mach. Learn.* **2001**, *45*, 5–32. [CrossRef]

34. Baccini, A.; Goetz, S.J.; Walker, W.S.; Laporte, N.T.; Sun, M.; Sulla-Menashe, D.; Hackler, J.; Beck, P.S.A.; Dubayah, R.; Friedl, M.A.; et al. Estimated carbon dioxide emissions from tropical deforestation improved by carbon-density maps. *Nat. Clim. Chang.* **2012**, *2*, 182–185. [CrossRef]

35. Evans, J.S.; Cushman, S.A. Gradient modeling of conifer species using random forests. *Landsc. Ecol.* **2009**, *24*, 673–683. [CrossRef]

36. Falkowski, M.J.; Evans, J.S.; Martinuzzi, S.; Gessler, P.E.; Hudak, A.T. Characterizing forest succession with lidar data: An evaluation for the Inland Northwest, USA. *Remote Sens. Environ.* **2009**, *113*, 946–956. [CrossRef]

37. Houghton, R.A. Balancing the Global Carbon Budget. *Annu. Rev. Earth Planet. Sci.* **2007**, *35*, 313–347. [CrossRef]

38. Stumpf, A.; Kerle, N. Object-oriented mapping of landslides using Random Forests. *Remote Sens. Environ.* **2011**, *115*, 2564–2577. [CrossRef]

39. Belgiu, M.; Drăgu, L. Random forest in remote sensing: A review of applications and future directions. *ISPRS J. Photogramm. Remote Sens.* **2016**, *114*, 24–31. [CrossRef]

40. Chirici, G.; Mura, M.; McInerney, D.; Py, N.; Tomppo, E.O.; Waser, L.T.; Travaglini, D.; McRoberts, R.E. A meta-analysis and review of the literature on the k-Nearest Neighbors technique for forestry applications that use remotely sensed data. *Remote Sens. Environ.* **2016**, *176*, 282–294. [CrossRef]

41. Goodfellow, I.; Bengio, Y.; Courville, A. *Deep Learning*; MIT Press: Cambridge, MA, USA, 2016.

42. Chollet, F. *Deep Learning with Python*; Manning Publications Co.: New York, NY, USA, 2018.

43. Mura, M.; McRoberts, R.E.; Chirici, G.; Marchetti, M. Statistical inference for forest structural diversity indices using airborne laser scanning data and the k-Nearest Neighbors technique. *Remote Sens. Environ.* **2016**, *186*, 678–686. [CrossRef]

44. McRoberts, R.E. Estimating forest attribute parameters for small areas using nearest neighbors techniques. *For. Ecol. Manag.* **2012**, *272*, 3–12. [CrossRef]

45. Efron, B.; Tibshirani, R.J. *An Introduction to the Bootstrap*; Chapman & Hall/CRC: Boca Raton, FL, USA, 1994.

46. Jebadurai, J.; Jebadurai, I.J.; Paulraj, G.J.L.; Samuel, N.E. Super-resolution of digital images using CNN with leaky ReLU. *Int. J. Recent Technol. Eng.* **2019**, *8*, 210–212. [CrossRef]

47. Huk, M. *Stochastic Optimization of Contextual Neural Networks with RMSprop BT—Intelligent Information and Database Systems*; Nguyen, N.T., Jearanaitanakij, K., Selamat, A., Trawiński, B., Chittayasothorn, S., Eds.; Springer International Publishing: Cham, Switzerland, 2020; pp. 343–352.

48. Miścicki, S.; Stereńczak, K. Określanie miąższości i zagęszczenia drzew w drzewostanach centralnej Polski na podstawie danych lotniczego skanowania laserowego w dwufazowej metodzie inwentaryzacji zasobów drzewnych. *Leśne Pr. Badaw.* **2013**, *74*, 127–136. [CrossRef]

49. Guyon, I.; Elisseeff, A. An introduction to variable and feature selection. *J. Mach. Learn. Res.* **2003**, *3*, 1157–1182. [CrossRef]

50. Straub, C.; Weinacker, H.; Koch, B. A comparison of different methods for forest resource estimation using information from airborne laser scanning and CIR orthophotos. *Eur. J. For. Res.* **2010**, *129*, 1069–1080. [CrossRef]

51. Ayrey, E.; Hayes, D.J.; Kilbride, J.B.; Fraver, S.; Kershaw, J.A.; Cook, B.D.; Weiskittel, A.R. Synthesizing Disparate LiDAR and Satellite Datasets through Deep Learning to Generate Wall-to-Wall Regional Forest Inventories. *BioRxiv* **2019**, 580514. [CrossRef]
52. Ayrey, E.; Hayes, D.J. The use of three-dimensional convolutional neural networks to interpret LiDAR for forest inventory. *Remote Sens.* **2018**, *10*, 649. [CrossRef]
53. McRoberts, R.E.; Chen, Q.; Walters, B.F.; Kaisershot, D.J. The effects of global positioning system receiver accuracy on airborne laser scanning-assisted estimates of aboveground biomass. *Remote Sens. Environ.* **2018**, *207*, 42–49. [CrossRef]
54. Packalen, P.; Strunk, J.; Packalen, T.; Maltamo, M.; Mehtätalo, L. Resolution dependence in an area-based approach to forest inventory with airborne laser scanning. *Remote Sens. Environ.* **2019**, *224*, 192–201. [CrossRef]

Article

The Importance of High Resolution Digital Elevation Models for Improved Hydrological Simulations of a Mediterranean Forested Catchment

João Rocha [1,*], André Duarte [1], Margarida Silva [1], Sérgio Fabres [1], José Vasques [2], Beatriz Revilla-Romero [3] and Ana Quintela [1]

[1] RAIZ—Forest and Paper Research Institute, Quinta de S. Francisco, Rua José Estevão (EN 230-1), Eixo, 3800-783 Aveiro, Portugal; andre.duarte@thenavigatorcompany.com (A.D.); margarida.silva@thenavigatorcompany.com (M.S.); sergio.fabres@thenavigatorcompany.com (S.F.); ana.quintela@thenavigatorcompany.com (A.Q.)

[2] Navigator Forest Portugal, Zona Industrial da Mitrena, 2910-738 Setúbal, Portugal; jose.vasques@thenavigatorcompany.com

[3] Remote Sensing and Geospatial Analytics Division, GMV, Isaac Newton 11, P.T.M. Tres Cantos, E-28760 Madrid, Spain; brevilla@gmv.com

* Correspondence: joao.rocha@thenavigatorcompany.com

Received: 27 August 2020; Accepted: 8 October 2020; Published: 10 October 2020

Abstract: Eco-hydrological models can be used to support effective land management and planning of forest resources. These models require a Digital Elevation Model (DEM), in order to accurately represent the morphological surface and to simulate catchment responses. This is particularly relevant on low altimetry catchments, where a high resolution DEM can result in a more accurate representation of terrain morphology (e.g., slope, flow direction), and therefore a better prediction of hydrological responses. This work intended to use Soil and Water Assessment Tool (SWAT) to assess the influence of DEM resolutions (1 m, 10 m and 30 m) on the accuracy of catchment representations and hydrological responses on a low relief forest catchment with a dry and hot summer Mediterranean climate. The catchment responses were simulated using independent SWAT models built up using three DEMs. These resolutions resulted in marked differences regarding the total number of channels, their length as well as the hierarchy. Model performance was increasingly improved using fine resolutions DEM, revealing a bR^2 (0.87, 0.85 and 0.85), NSE (0.84, 0.67 and 0.60) and Pbias (-14.1, -27.0 and -38.7), respectively, for 1 m, 10 m and 30 m resolutions. This translates into a better timing of the flow, improved volume simulation and significantly less underestimation of the flow.

Keywords: forested catchment; forestry; hydrological modeling; SWAT model; DEM

---

1. Introduction

Forests play a significant role on the hydrological cycle and have a key importance on ecosystems regulation. Planted forest productivity is highly dependent on water availability [1]. Therefore, the evaluation of watersheds' hydrological response is important especially in regions that, under present day conditions, register water restrictions such as the Mediterranean bioclimatic zones [2,3]. In Portugal, forests cover about 36% of the total mainland area [4], holding an important role on the environment and on the national economy.

In order to obtain insight on hydrological responses and on water availability at a catchment scale, hydrological models may be used. These models require, as an input, a morphological surface representation, and the digital elevation model (DEM) is one of the most common. DEM is a numeric representation of a surface arranged in a set of regular grids each containing the three-dimensional (3D) (x, y, z) coordinate records [5]. A DEM may be produced based on three arrangements: contour lines (x, y data on equal elevation lines); triangular irregular network (non-overlapped linked triplets of nodes with x, y, z data); or mass points (x, y, z data regularly or irregularly distributed) [6].

DEM data sources are typically obtained by using: (i) remote sensing techniques on which the sources may be Interferometric Synthetic Aperture Radar (IfSAR), either from the Shuttle Radar Topography Mission (SRTM), Advanced Spaceborne Thermal Emission or Reflection Radiometer; (ii) photogrammetry with stereo pairs of aerial imagery (e.g., satellite, airborne and Unmanned Aerial Vehicles); (iii) laser scanning by Light Detection and Ranging (LiDAR); (iv) interpolated global positioning system and total stations; and (v) cartographic digitalization on pre-existing topographic (contour) maps [7,8]. Photogrammetry, IfSAR and LiDAR are preferential sources to produce DEMs [9,10].

From the DEM it can be extracted geomorphological and hydrological data (e.g., aspect, slope, stream network, watershed delineation, flow direction and accumulation patch) and, therefore, the horizontal and vertical resolutions of the DEM will impact on the accuracy of the linked extract products. As concluded by Sørensen et al. [11] and Vaze et al. [12], the input DEM (accuracy and resolution) will impact the ranges of hydrological and topographic indices derived from the DEM. Following Tan et al. [13,14] and Xu et al. [15], DEM uncertainties are linked with the input resolution (provided by the data sources) and the applied resampling techniques.

Physically based semi-distributed and distributed hydrological models are powerful tools to simulate catchment processes and linked responses. Moreover, hydrological models may be used to support effective land management and planning of forest resources. They can be used as effective decision support tools [16,17] to set site-specific forestry management practices.

The Soil and Water Assessment Tool (SWAT) is a continuous time, physical based and semi-distributed ecohydrological model that simulates on a daily basis landscape processes and watershed responses [18,19]. The model has been widely used for three decades and it has been applied to assess different inputs and scenarios for multiple variety of catchments and objectives (e.g., climate change impacts, land use and land cover changes, water quality and quantity, sediment exportation, management practice effects) resulting in more than 3900 published research papers [20]. The SWAT model has been successfully used to address the impacts of different resolution input data on catchments and hydrological responses [21–23] and some studies have focused on the use of low to high resolution input DEM to assess SWAT performance and simulation results [24–27]. The links between different input DEM resolutions and the uncertainty on model predictions have been discussed in other research studies [28–30].

SWAT procedures that are DEM-dependent (e.g., watershed delineation, stream network definition, sub-basins) will be largely constrained by the DEM resolution and uncertainty. Additionally, as stressed by Goulden et al. [31], the lack of knowledge on DEM uncertainty and the potential undefined cascade impact on hydrological model simulations may result in inaccurate assumptions and under or overestimated conclusions for the modeler. Wu et al. [32] applied the SWAT to 10 catchments and used three DEM resolutions (250 m, 500 m and 1000 m) to assess the links between automatic watershed extraction (from DEM), watershed parameters and area threshold values. They concluded that a DEM resolution increase results in a more refined flow direction and accumulation calculation, meaning a more realistic representation of the stream network.

Tan et al. [13] found that SWAT generated stream network was set to be more sensitive with regard to the DEM resolution rather than DEM source or the applied resampling technique. Furthermore, Tan et al. [14] obtained better streamflow simulations using 20 m and 60 m resolution DEMs as well as improved DEM sensitivity analysis upon the integration of the smallest area threshold (1000 ha) value.

Camargos et al. [33] debated the use of fine (regional data) and coarse (global data) detail inputs to improve the robustness of SWAT model simulations. Zhang et al. [29] and Xu et al. [15] analyzed the influence of contrasting DEM resolutions on SWAT simulations for hydrological responses and nutrient exports. Goulden et al. [24] presented a comprehensive analysis on DEM resolution and sensitivity of the SWAT outputs, namely the increase of slope class with higher DEM resolutions and the linked increase of sediments loads as a consequence of the reduction on watershed areas using higher resolutions.

The great majority of SWAT modeling studies explicitly fails to consider the simulations on forest dominated catchments despite the relevance of these ecosystems and provided services. In addition, there is also a gap on studies that focus on forest dominated catchments on water scarce regions under the influence of Mediterranean-type climate.

This study focuses on the use of DEMs with different resolutions to improve the accuracy of hydrological modeling in a forest dominated catchment in a water scarce region in Alentejo (Portugal). To this end, the SWAT model was applied on relative flat catchment with three DEM cell sizes (1 m, 10 m and 30 m) in order to simulate the impacts of different DEMs on the representation of watershed characteristics, processes and responses.

There are two major research questions that we intend to address: What is the impact of different DEM resolutions on surface representation and watershed properties in a flat area? Is it worthwhile to use very high resolution DEMs to simulate catchment processes and responses in a forest dominated catchment under the influence of Mediterranean climate?

2. Material and Methods

2.1. The Study Area

The Caniceira catchment (is located in Alcolobre river basin (in the context of the wider Tagus transboundary river basin), in the Alentejo region in mainland Portugal. The catchment is under the influence of a Mediterranean climate and registers an average annual precipitation of about 720 mm with a dry warm summer and a mild winter and it is characterized by a temperate with dry and hot summer climate (Csa) according to the Köppen classification [34] (Figure 1). In addition, it is categorized as a sub-humid mesothermal, registering a severe water-deficit, especially during summer time (C2 B′2) [35,36]. The catchment is located in a relatively flatten area with elevations ranging from 99 to 164 m (above sea level) and covers an area of about 267 ha.

Regarding the geological background, the catchment is located in the context of the Lower Tagus Tertiary basin, whose genesis is related to Pyrenean compression [37]. The catchment is dominated by Miocene and Pliocene fluvial sedimentary deposits, conglomerates and reddish to yellowish clay sandstones and some Quaternary sands and gravels beds. The Tertiary deposits overlay in discordance with the gneiss and migmatite formations from Precambrian [38]. The hydrogeological context of the area is associated with an extensive multilayered aquifer system, shaped by variable in depth layers of sandstones alternating with impervious clay layers [39,40].

Following the Portuguese soil classification [41,42] and the FAO-UNESCO classification [43,44], the catchment soils are mainly Arenic Endoleptic Regosols and Arenic Epileptic Regosols, with little or no profile differentiation, on which bedrock is located between 25 cm and 50 cm (Epileptic) or between 50 cm and 100 cm (Endoleptic) with sandy or coarser texture.

This is a forest dominated catchment largely occupied by evergreen broadleaf species (*Eucalyptus globulus*, *Quercus suber*—cork oak, *Quercus faginea*—green oak, *Pinus pinaster*—maritime pine; Table 1) and by sclerophyllous vegetation (Mediterranean scrubland). The catchment is linked to a riparian gallery of high nature conservation value, including alders and narrow-leaved ash as the most common tree cover and otter, polecat and Eurasian sparrowhawk as faunistic representatives. The land use and land cover data (1:10 scale) was obtained with field observation and detailed mapping survey validated with pan-sharped World-View-2 orthoimage from 8 August 2019.

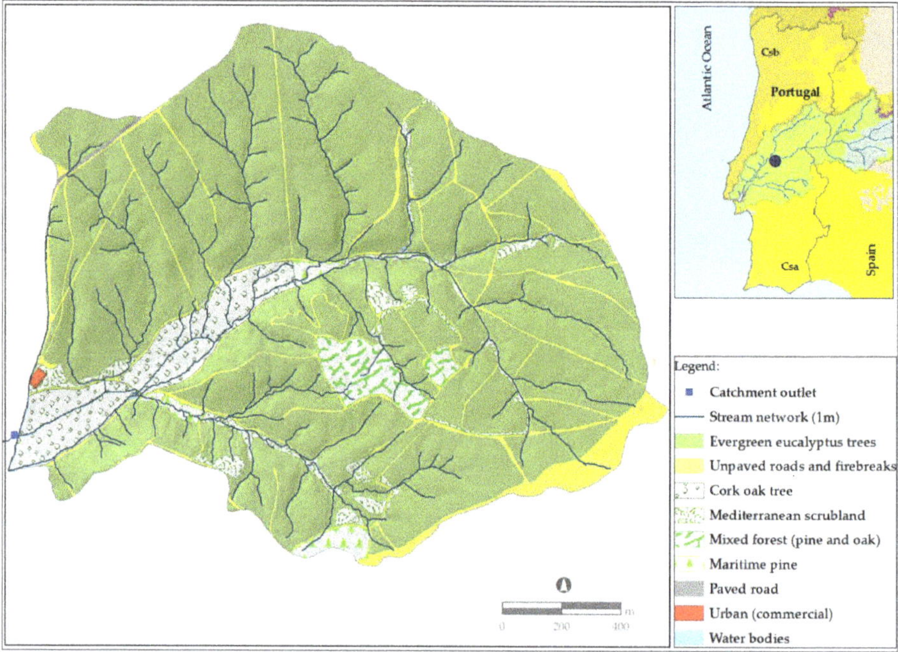

Figure 1. Land cover and stream network (1 m resolution) at Caniceira catchment, in the context of the Tagus transboundary river basin with the Köppen classification (top right map).

Table 1. Land use and land cover (LULC) distribution at Caniceira catchment.

LULC	Area ha	Area %
Eucalyptus	210.9	78.9
Unpaved roads and firebreaks	19.9	7.4
Cork oaks	16.4	6.1
Mediterranean scrubland	9.2	3.4
Mixed forest (maritime pine, green oak)	6.3	2.4
Maritime pine	3.7	1.4
Paved road	0.4	0.2
Urban (commercial)	0.2	0.1
Water bodies	0.1	0.0

2.2. DEM Input Data

The DEMs used as input in the SWAT model were obtained using: Synthetic Aperture Radar interferometric data, GPS and digitalization surveys and Airborne Laser Scanning surveying data.

The SRTM from 2000 in combination with the United States Geological Survey's GTOPO30 data set, provided about 80% worldwide coverage of the SAR interferometric data, which were resampled at 1 arc-second resolution (~30 m × ~30 m) [45]. The SRTM-DEM was validated for Portugal [46,47] and is freely available. In addition, it has been used worldwide as an input DEM in the SWAT model [28,48,49].

The 10 m resolution DEM was based on the contour lines data (National Cartography Series—1:10,000 scale), produced by the Portuguese Geographical Institute and by the Geographical Institute of the Military, with the European Terrestrial Reference System coordinated system

(PT-TM06/ETRS89). This is the official data for Portugal and included, among others, the National Geodetic Network and GPS Permanent Network to provide an expected vertical accuracy ranging from 1 to 1.5 m [46]. A triangulated irregular network surface was generated from contour lines through the Delaunay triangulation and was converted to raster with a 10 m × 10 m cell size, in order to respect the contour lines equidistance.

The 1 m resolution DEM was provided by a LiDAR survey (April 2019) with a flight performed at approximately 2792 m with a 30% overlap between sweeps. The point clouds with an average point density of 2 points m^{-2} were captured using Leica ALS80HP, a high performance airborne scanner operating at a pulse rate 704 kHz and scan rate of 73.5 Hz. The airborne laser scanning point clouds were processed and classified using CloudCompare and FUSION/LDV 3.80 software [50,51]. The 1 m gridded surface model was interpolated using the GridSurfaceCreate tool in FUSION/LDV 3.60+. Wooded areas may induce positional errors on DEMs. However, as the planted forest in the catchment follow the same compartments (tree plantation plots) since the late 1950s, it is possible to get several sectors with no vegetation, namely during the periods between clear cuts and new coppice cycle (each 12 years). The operations included regular management of the understory vegetation in-between plantation lines. As a result, the airborne laser scan was able to get the surface information without the canopy interference. Additionally, the canopy density on this planted forest, in association with the unpaved roads and firebreaks networks account for several gaps of vegetation and therefore, exposing the topographic surface, which results in an improved accuracy of the laser scan on the actual surface. The official available input data for the 10 m DEM also benefits from this reducing the eventual elevation errors provided on wooded areas.

The three DEMs represent the topographic surface for an almost 20-year period, as the data began to be collected in the late 1990s (for the 10 m DEM), in the earlier 2000 (for the 30 m DEM) and in 2019 (for the 1 m DEM). As a result, there is a temporal gap in the actual representation of the topographic surface. Nevertheless, no significant geomorphological changes are expected to have occurred in the catchment.

2.3. SWAT Modeling Approach and Calibration Routines

The catchment responses were simulated using the SWAT model, through the ArcGIS interface (ArcSWAT version 2012). In order to assess the influence of the DEMs at different resolutions (1 m, 10 m and 30 m), three independent SWAT models were built up, as each new input DEM requires a single new SWAT model.

In the SWAT model, the stream network definition can be done via DEM by "forcing" the model to define the stream network (DEM-based) or by using a pre-defined shapefile on which each line represents a reach with a unique identification code as well as the flow direction in the sub-basins. Similarly, the watershed delineation may be achieved based on the DEM or by using a pre defined polygon shapefile. The stream networks were calculated with a D8 hydrological algorithm [52], within the Terrain Analysis Using Digital Elevation Models (TauDEM), which is a set of tools for the analysis of terrain using DEMs. The pre-defined watershed limit was calculated with the D8 algorithm based on the 1 m DEM.

To assess the accuracy of the stream network three metrics were used: drainage density (Dd = average total stream length/catchment area; a higher value represents a more agglutination of the channels, [53]); sinuosity (S = length of meandering/straight-line distance; ranging from one for a straight line to zero for a curvy line, [54]); vertices index (Vi = number of vertices/total stream length, wherein a close to zero value indicates lower geometry complexity, with no curvy sectors).

As the catchment presents little altimetry variation, the use of a pre-defined stream network and watershed shapefiles were considered rather than using the DEM-based option in the SWAT model.

In addition, back in the early 2000s, a water management strategy was implemented in a flat sector at the Caniceira catchment outlet, to prevent the occurrence of soil waterlogging during the raining season and as to be used as water storage pond in the driest period. As a result, a small retention pond was created and two stream lines were merged, which only left one channel free flowing. On this flat area, the unmistakable geomorphological evidence of the two parallel thalwegs was captured by the fine detailed DEM (1 m resolution) and included in the extracted stream network. These channel lines are linked to the topographic surface but present no relation with the actual flow direction and altered the actual location of the catchment outlet. On the TauDEM-based stream network, there is a partial a gap on the linkage to the man-made pond (oval marker on Figure 2) and a missed representation of the actual stream lines nearby the catchment outlet (highlighted by the rectangle marker on Figure 2), which impact the outlet location. As a consequence, the stream lines shapefile (.shp) was edited using the ArcGIS Editing toolbox, and the polyline vertices rectified and some vertex deleted in order eliminate these incongruences and to redefine the correct flow direction. Additionally, a field survey was done with a Real Time Kinematic GNSS high-accuracy antenna, to provide a set of 225 ground survey-points in order to assess the positional accuracy of the 1 m DEM based stream network, in particular z (elevation) values. The survey-points were acquired with a triple-frequency GNSS receiver (Arrow Gold antenna) linked to the Portuguese active GNSS network (RENEP). The reference ellipsoid is the GRS80, the coordinate reference system is the ETRS89/PT-TM06 (EPSG:3763) and the vertical heights are based on the GeodPT08 geoid model.

This validation procedure insures a more realistic representation of the flow direction on this flat terminal sector and the definition of a more precise location of the gauged catchment outlet, ultimately leading to an accurate representation of the catchments responses and an improved robustness of the SWAT model. It is worth stressing that the extracted stream network represents channel lines according to the topographic surface, but does not necessarily mean free flow on all channels, as for present day conditions the catchment is facing water scarcity constrains. Figure 2 presents the TauDEM based stream network (dotted line) and the rectified (edited) stream network for the flat terminal sector of the catchment.

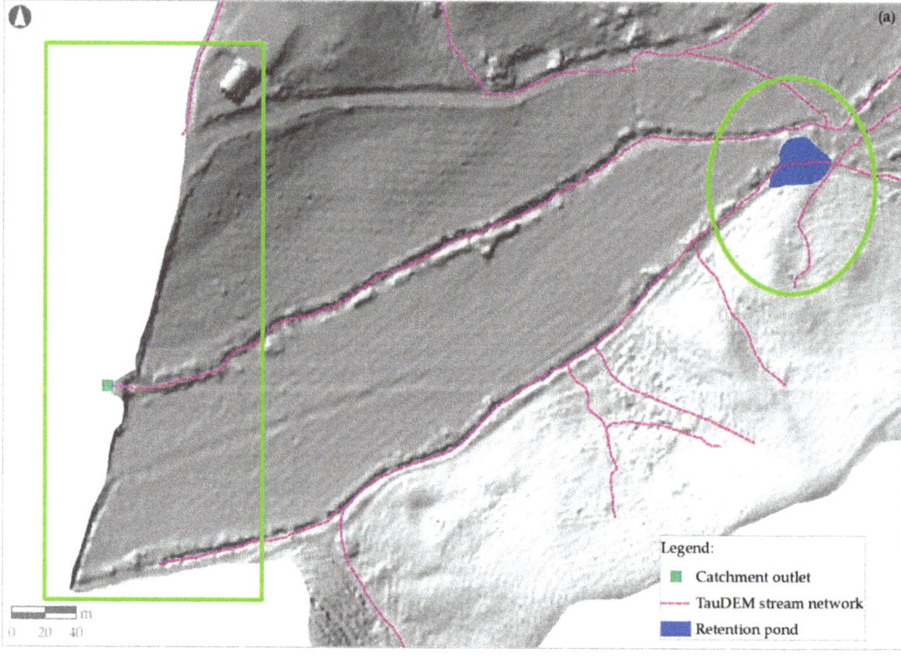

Figure 2. *Cont.*

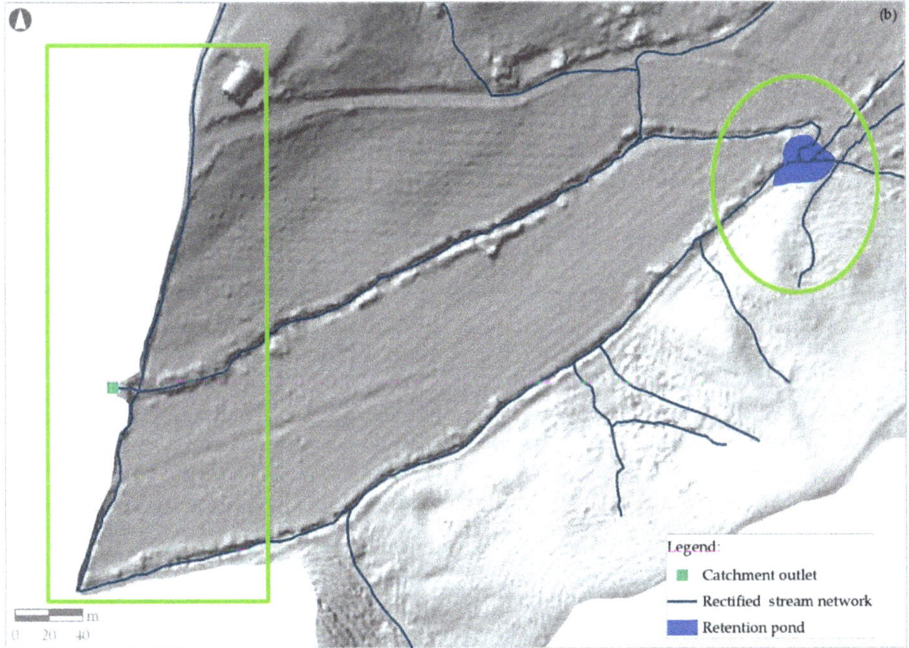

Figure 2. TauDEM generated (**a**) and rectified stream network (**b**), on the flat terminal sector of the catchment for the 1 m DEM resolution, with major differences highlighted in green.

SWAT defines sets of hydrological response units (HRUs) as a unique combination of overlaying land use, soil type and slope for a sub-basin. In the SWAT+ (a revised SWAT version), it is possible to separate water and other land areas per sub-basin (in contrary to the previous SWAT model) and, from the landscape units (LSUs), it is possible to define the HRUs [55].

SWAT is mainly divided in two phases: land phase and routing (water) phase. Within the land phase the amount of water and sediments is controlled, as well as nutrients loads in the main channel for each sub-basin; on the routing phase the movement of water, sediments and nutrients in the stream network throughout the watershed catchment is controlled [56].

To run the model, some input data are mandatory: soil, land cover and crop/plant parameters and climate records.

Soil data resulted of a sampling survey of a minimum of five soil samples per hectare. All samples were analyzed in the laboratory in order to define a comprehensive database of soil physical and chemical characteristics (namely texture, pH, exchangeable cations).

Specific-site parameters for eucalyptus trees were provided by forest inventories, Scholander pressure chamber, leaf area index surveys, sap flow sensors and dendrometers. Eco-physiological data for the other land uses were adapted from the SWAT database and also following previous research [57–60].

The meteorological sub-daily data (precipitation, air temperature, humidity, solar radiation and wind), were provided by two automatic meteorological stations installed in the catchment area since 2012. Sub-daily hydrological records were provided by an automatic stream flow gauge installed in 2019 in a weir open channel at the catchment outlet.

The SWAT management operations database included information on several operations (e.g., soil tillage, planting and maintenance fertilizations, harvesting procedures) that may have influenced the plant growing cycle and the catchment hydrological responses. Plant growth was calculated using a modified version of the Environmental Policy Integrated Calculator (EPIC) crop model [61]. The model simulates the evolution of leaf area index, biomass accumulation as well as the yield for different plants. A detailed operation schedule (for eucalyptus) was defined by date (adapted to local meteorological conditions) rather than by heat units. In addition, refined SWAT parameters (e.g., plantation dates, tillage operations, planting and maintenance fertilizations, weed control, plant growth cycle, leaf area index) were included in the model database based on the long term (50 years) management operations held at Caniceira and also following previous research [57–60]. The model was run on a daily basis from 2012 to 2020 with a warm-up period of five years to minimize the effect of the initial systems variables ensuring proper model dynamic equilibrium.

Calibration routines were performed with Calibration and Uncertainty Analysis Program (SWAT-CUP [62,63]). SWAT-CUP is an iterative calibration procedure that allows the selection of five optimization algorithms and different objective functions (10 in total) and on which multiple variables (parameters) can be assumed and simultaneously calibrated, whether meant for a specific HRU or a set of HRUs or sub-basins. For the Caniceira catchment, a total of 12 sub-basins were considered. The Sequential Uncertainty Fitting algorithm (SUFI2) produces a set of good solutions within the considered parameters and their ranges, and for each iteration, the arrays were gradually narrowed through several iterations. The algorithm accounts for potential sources of uncertainty and helps to determine the most impacting parameters. The calibration was performed for each SWAT model (1 m, 10 m and 30 m resolution DEMs) on two iterations, each with 450 simulations.

Model performance (observed versus calibrated data) was evaluated using three indicators: (i) the Coefficient of Determination (R^2) multiplied by the coefficient of the regression line (bR^2); (ii) the Nash-Sutcliffe model efficiency index (NSE); (iii) the average percent model error (Pbias—percentage of bias) [64–67] (Table 2). The calibration results are expressed from the unsatisfactory to the very good performance rating, following the goodness-of-fit indicators proposed by Moriasi et al. [66].

Table 2. Ratings of the goodness-of-fit indicators for model performance (adapted from Moriasi et al. [66].

Performance Rating	bR^2	NSE	Pbias%
Very good	$0.75 < bR^2 \leq 1.00$	$0.75 < NSE \leq 1.00$	Pbias $< \pm10$
Good	$0.65 < bR^2 \leq 0.75$	$0.65 < NSE \leq 0.75$	$\pm10 \leq$ Pbias $< \pm15$
Satisfactory	$0.50 < bR^2 \leq 0.65$	$0.50 < NSE \leq 0.65$	$\pm15 \leq$ Pbias $< \pm25$
Unsatisfactory	$bR^2 \leq 0.50$	$NSE \leq 0.50$	Pbias $\geq \pm25$

3. Results and Discussion

3.1. Impacts of DEM Resolution on Surface Representation and Watershed Properties

The watershed limit was calculated in order to assess the impact of increasing resolution on products extracted from different DEMs. The watershed delineation resulted in a total area value ranging from 267 ha (1 m) to 224 ha (30 m). The watershed area on the 30 m DEM results from a non-realistic generalization of watershed limit due to a super pixel cell size (30 m × 30 m) that accounts for a higher area within the influence of each pixel. Charrier et al. [68] pointed out that the use of very high resolution DEMs could better suit for more realistic watershed delineation, which is true for the 1 m and 10 m resolutions DEMs. In order to avoid these constrains imposed by the DEM resolution on the watershed area, a pre-defined watershed shapefile (calculated with D8 algorithm based on the 1 m DEM) was used in the model. Luo et al. [69] used a pre-defined set of watershed and stream shapefiles for a plain study-area and obtained an improved representation of the actual surface and also increased model accuracy, when compared with a DEM "forcing" procedure.

This resulted in a more realistic representation of the catchment (Figure 3), especially as it is located on a relatively flatten area with low amplitude on elevation values, in line with Luo et al. [69]. A reduction in total watershed area and linked geometry can be substantial, especially on low to medium size watersheds, as the proper representation of the surface and catchment responses is, therefore, affected. A closer analysis points to some incongruence whether due to the lack of detail on catchment headwaters or due to erroneous flow patch and direction calculations on more flat sectors. In addition, loss in detail from altimetry values for coarse resolutions DEMs resulted in an underestimation of the maximum value and an overestimation of the lower catchment altimetry values (Table 3), in line with results reported by Lin et al. [48], Zhang et al. [29] and Reddy et al. [25].

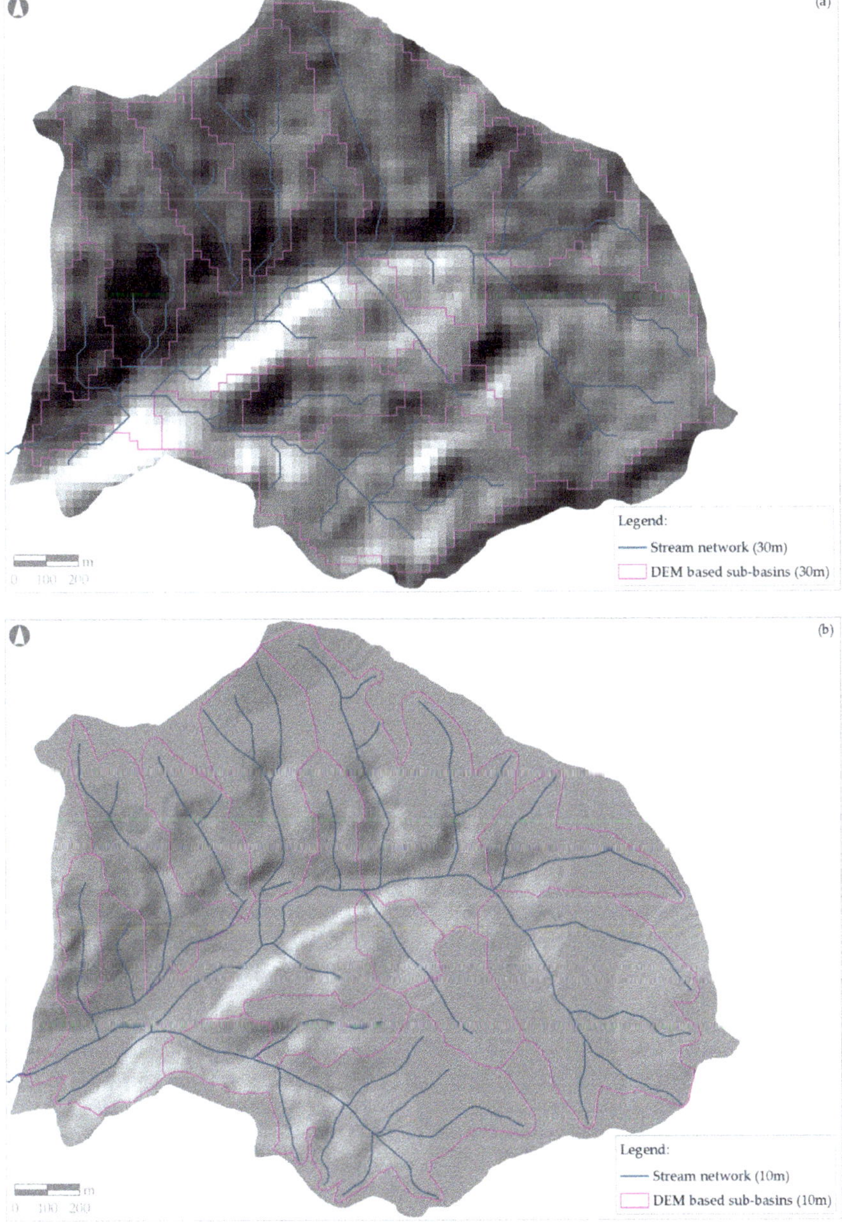

Figure 3. *Cont.*

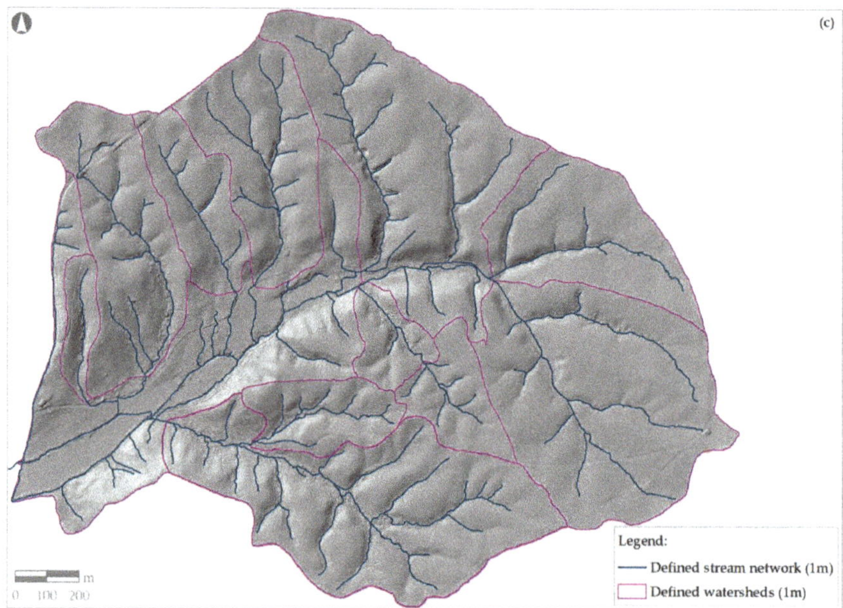

Figure 3. Pre-defined watershed, stream network and DEM-based sub-basins with (**a**) 30 m, (**b**) 10 m and (**c**) 1 m DEM resolution.

Table 3. Caniceira watershed characteristics extracted from 30 m, 10 m and 1 m DEM resolutions.

Watershed Characteristics	DEM Resolution		
	30 m	10 m	1 m
Number of channels	312	735	16,321
Channels length	14,955 m	16,274 m	26,317 m
Strahler order	4	5	7
Elevation (minimum)	105 m	102 m	99 m
Elevation (maximum)	169 m	160 m	164 m
Elevation (Std. deviation)	13.7	11.3	12.8
Drainage density	0.55 mm^{-2}	0.60 mm^{-2}	0.98 mm^{-2}
Sinuosity	0.45	0.12	0.06
Vertices index	0.04	0.09	1.24
Hydrological response units	263	297	402

The accurate three-dimensional representation of a surface is closely related to the resolutions of the input data. In fact, all the 12 sub-basins registered a decrease (from the 1 m to the 30 m resolution) on the number of channels and their length, the Strahler stream order numbers [70] and the slope class values (Table 3).

The 1 m resolution DEM resulted in a significantly higher number of HRUs. The SWAT capability to better simulate surface runoff will be increased as curve numbers will be better adjusted to slope values, based on HRUs slopes, although forest catchments may present a potential higher capability for water storage, as mentioned by Pang et al. [71].

3.2. Impacts of DEM Resolution on Catchment Processes and Responses

The analyses prior to the model calibration intended to obtain insight on the SWAT parameterization to highlight the impact of DEM resolutions, as the three models were built up with the same parameters. The baseline simulation (uncalibrated) on the 1 m DEM embodied a more reduced gap between the observed and simulated data, providing therefore a more realistic representation of the catchment response (bR^2 = 0.74, NSE = 0.58 and Pbias = −40.7). Moreover, observed, simulated (uncalibrated) and calibrated data presented a reduced difference, thus denoting a good overall trend on the model performance (following Moriasi et al. [64]—Table 2), and improved the goodness-of-fit indicators after calibration (bR^2 = 0.87, NSE = 0.84 and Pbias = −14.1). This translates into a better timing of the flow, improved volume simulation and significantly less underestimation of the flow.

The baseline simulations from the 10 m and 30 m resolutions denoted a gradual increase on the gap between observed and simulated. The overall model performance before calibration, based on the bR^2 (0.71 and 0.60, respectively) and NSE index (0.51 and 0.49, respectively) are nevertheless good (Table 4). The Pbias for 10 m and 30 m DEMs indicates an overall flow underestimation (−44.1 and −50.6, respectively) that may be related to the challenges of capturing the total rainfall amounts on extreme precipitation events. As expected, after calibration these values improved.

Table 4. Goodness-of-fit indicators for daily discharge at the Caniceira catchment outlet considering DEM at different resolutions before and after SWAT model calibration. bR^2—ranging from the optimal value (one) to zero; NSE—ranging from the optimal value (one) to zero (below zero indicates unacceptable model simulation); —ranging from the optimal value (zero) to positive or negative values representing model underestimation and overestimation, respectively.

	bR^2			NSE			Pbias		
DEM Resolution	**30 m**	**10 m**	**1 m**	**30 m**	**10 m**	**1 m**	**30 m**	**10 m**	**1 m**
Uncalibrated	0.60	0.71	0.74	0.49	0.51	0.58	−50.6	−44.1	−40.7
Calibrated	0.85	0.85	0.87	0.60	0.67	0.84	−38.7	−27.0	−14.1

Model simulations showed a gradual deterioration on model performance as DEM moved from 1 m, 10 m to 30 m resolution. This is translated in a gradual loss on accuracy for streamflow at the watershed outlet (gauged) (Figure 4). This analysis was performed in two stages: upon model parameterization (uncalibrated flow) and upon model calibration (calibrated flow). From 1 June to 30 September (2019), the catchment received a total of precipitation as little as 17.8 mm. Upon the first rain events on mid-October, the responses are showed at the catchment outlet where the river gauge and model outlet are located. The extreme precipitation event in December 2019 affecting Portugal ("Elsa storm") resulted in the highest flow peak visible on both stages hydrographs. The hydrological response provided by the flow peak denotes a higher peak representation for the 1 m DEM and a gradual decrease on peak representation for coarser resolutions. The catchment meteorological stations registered 114.4 mm of rain, respectively, with 68.4 mm on the 19th of December and 46.0 mm on the 20th of December (2019).

On this forest dominated Mediterranean catchment with low elevation range, the SWAT modeling of the responses to intense precipitation events have clearly showed a better fit using very high resolution DEM, either uncalibrated or calibrated, and can, therefore, be considered a very good model performance based on bR^2 and NSE, and a satisfactory performance based on Pbias [66,67].

Reddy et al. [25] found similar relation with different resolution DEM and stressed that accuracy of estimated runoff and sediment yield decreases with coarser resolutions. The rational for using coarse DEM resolutions has been largely discussed by several authors [6,72]. They concluded that despite poor accuracy, they present a reasonable compromise between available (freely) data and low computation demands and a suitable representation of hydrological processes and catchment responses, in the context of hydrological model simulations. The findings of the present work contrast some previous studies [28–31,73] that showed that high resolutions DEM have a negligible effect on streamflow reduction. As suggested by Tan el al. [13], such studies rely on coarser DEM resolutions. In addition, most models are calibrated using monthly to yearly records, which may uncover some short-term discharge tendencies, event-based responses and peak flow representations. Finally, fine DEM resolutions may not add any significant advantage in flow representation of large watersheds and in mountain areas [29].

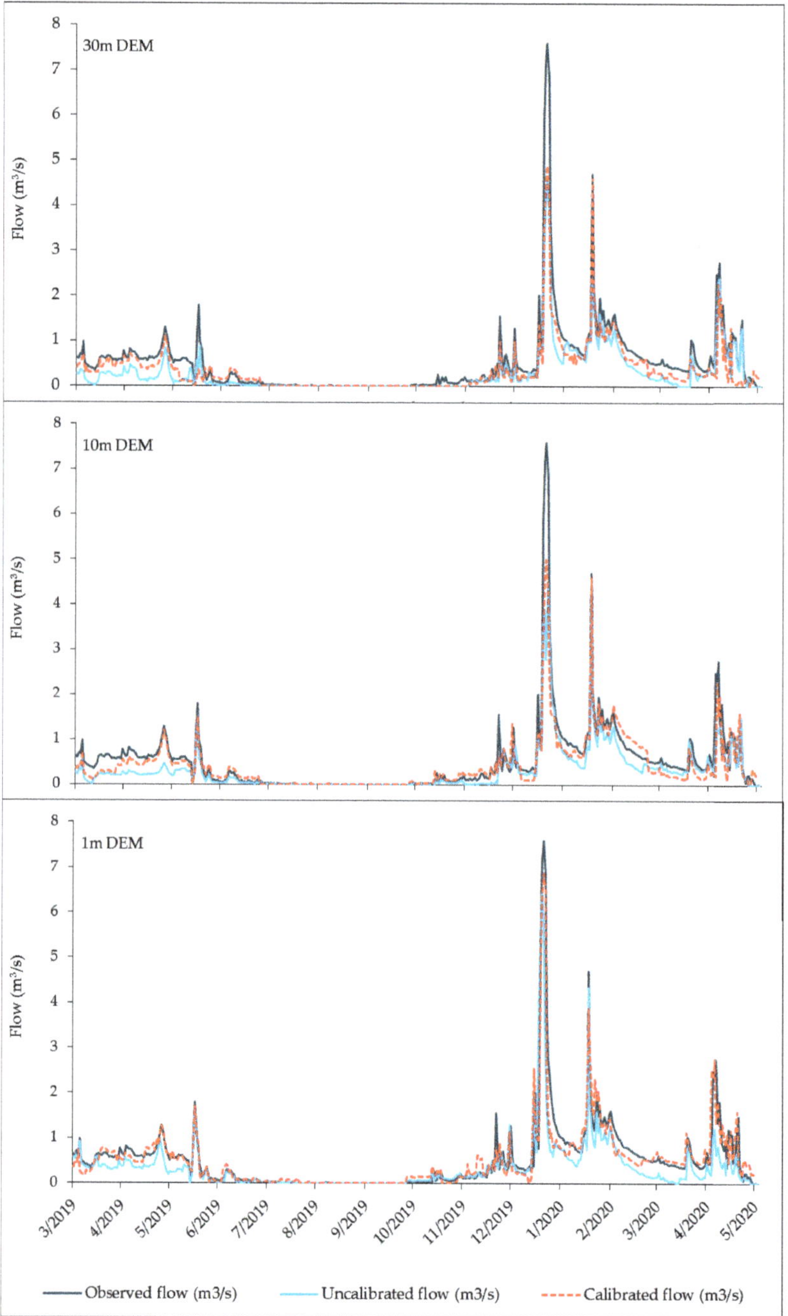

Figure 4. SWAT model performance for uncalibrated and calibrated flows for Caniceira catchment, for the three DEM resolutions.

A more detailed representation of the terrain supports the definition of site-specific forestry operations, better planning of LULC, stands location site-indices and stocking and harvesting operations. High resolution DEM can also be used in support the planning of forest roads and drainage infrastructures, in order to prevent biotic, waterlogging and forest fire risks.

On one hand, future work must be conducted in order to assess the real value for forest planning of modeling with fine resolution DEM, since fine detail DEM are often not freely available. On the other hand, future research may consider the use of SWAT model to simulate adaptive land management strategies to improve the resilience of stands to water scarcity, particularly in the context of climate change scenarios.

4. Conclusions

The study looked into two major research questions: (i) What is the impact of different DEM resolutions on surface representation and watershed properties in a flat area? (ii) Is it worthwhile to use very high resolution DEMs to simulate catchment processes and responses in a forest dominated catchment under the influence of Mediterranean climate?

In the low-relief Caniceira forest catchment with a low altimetry variation, the use of high resolution DEM resulted in an improved representation of the surface with a more accurate report of watershed characteristics, namely, the number and length of channels and the Strahler order. At the flat end sector of the catchment, the high detail DEM (1 m) generated two parallel channels with significance on the morphological representation but with no linkage to the actual water direction. The 1 m DEM was able to capture water bodies (ponds) and the tree plantation lines. Moreover, the coarse resolutions DEMs fail to capture these topographic evidences. The detailed input from 1 m DEM is better suited to model the hydrological response of this catchment, located in a water scarce region of Portugal (under the influence of Mediterranean climate). In fact, the model performance was increasingly improved as fine resolutions DEMs were implemented allowing a better representation of actual processes and events, such as better timing of the flow, improved volume simulation and significantly less underestimation of the flow.

Author Contributions: Conceptualization, J.R. and A.Q.; Data curation, J.R.; Formal analysis, J.R. and A.D.; Investigation, J.R. and A.Q.; Methodology, J.R. and A.Q.; Software, J.R. and A.D.; Validation, J.R.; Visualization, J.R.; Resources, M.S., J.V. and A.Q.; Project administration S.F. and A.Q.; Funding acquisition, M.S.; Supervision, S.F. and A.Q.; Writing—original draft, J.R. and A.Q.; Writing—review and editing, J.R., A.D., S.F., J.A., B.R.-R. and A.Q. All authors have read and agreed to the published version of the manuscript.

Funding: This study has been developed within the context of MySustainableForest project, which has received funding from the European Union's Horizon 2020 research and innovation program under grant agreement No 776045.

Acknowledgments: The authors would like to express appreciation to the efforts from Bruna Scarparo, Célia Fernandes, Fernando Arede and Filipe Louro on data collection and hydrological survey. The authors would like to express appreciation to Nuno Borralho for the revisions and English language correction.

Conflicts of Interest: The authors declare no conflict of interest.

References

1. Stape, J.L.; Binkley, D.; Ryan, M.G.; Fonseca, S.; Loos, R.A.; Takahashi, E.N.; Silva, C.R.; Silva, S.R.; Hakamada, R.E.; Ferreira, J.M.A.; et al. The Brazil Eucalyptus Potential Productivity Project: Influence of water, nutrients and stand uniformity on wood production. *For. Ecol. Manage.* **2010**, *259*, 1684–1694. [CrossRef]
2. Peel, M.C.; Finlayson, B.L.; McMahon, T.A. Updated world map of the Köppen-Geiger climate classification. *Hydrol. Earth Syst. Sci.* **2007**, *11*, 1633–1644. [CrossRef]
3. Spinoni, J.; Vogt, J.; Naumann, G.; Carrao, H.; Barbosa, P. Towards identifying areas at climatological risk of desertification using the Köppen-Geiger classification and FAO aridity index. *Int. J. Climatol.* **2015**, *35*, 2210–2222. [CrossRef]

4.   IFN6–*6° Inventário Florestal Nacional*; Instituto da Conservação da Natureza e das Florestas: Lisboa, Portugal, 2019. (In Portuguese)

5.   Sofia, G. Combining Geomorphometry, Feature Extraction Techniques and Earth-Surface Processes Research: The Way Forward. *Geomorphology* **2020**, *355*, 107055. [CrossRef]

6.   Zhang, W.; Montgomery, D.R. Digital elevation model grid size, landscape representation, and hydrologic simulations. *Water Resour. Res.* **1994**, *30*, 1019–1028. [CrossRef]

7.   Burrough, P.A.; Macdonnell, R.A.; Lloyd, C.D. *Principles of Geographical Information Systems*, 3rd ed.; Oxford University Press: New York, NY, USA, 2015; p. 352.

8.   Florinsky, I.V. *Digital Terrain Analysis in Soil Science and Geology*, 2nd ed.; Elsevier Academic Press: Amsterdam, The Netherlands, 2016; p. 379.

9.   Ouédraogo, M.; Degré, A.; Debouche, C.; Lisein, J. The evaluation of unmanned aerial system-based photogrammetry and terrestrial laser scanning to generate DEMs of agricultural watersheds. *Geomorphology* **2014**, *214*, 339–355. [CrossRef]

10.   Muhadi, N.A.; Abdullah, A.F.; Bejo, S.K.; Mahadi, M.R.; Mijic, A. The Use of LiDAR-Derived DEM in Flood Applications: A Review. *Remote Sens.* **2020**, *12*, 2308. [CrossRef]

11.   Sørensen, R.; Seibert, J. Effects of DEM resolution on the calculation of topographical indices: TWI and its components. *J. Hydrol.* **2007**, *347*, 79–89. [CrossRef]

12.   Vaze, J.; Teng, T.; Spencer, G. Impact of DEM accuracy and resolution on topographic indices. *Environ. Model. Softw.* **2010**, *25*, 1086–1098. [CrossRef]

13.   Tan, M.L.; Ficklin, D.L.; Dixon, B.; Ibrahim, A.L.; Yusop, Z.; Chaplot, V. Impacts of DEM resolution, source, and resampling technique on SWAT-simulated streamflow. *Appl. Geogr.* **2015**, *63*, 357–368. [CrossRef]

14.   Tan, M.L.; Ramli, H.P.; Tam, T.H. Effect of DEM resolution, source, resampling technique and area threshold on SWAT outputs. *Water Resour. Manag.* **2018**, *32*, 4591–4606. [CrossRef]

15.   Xu, F.; Dong, G.; Wang, Q.; Liu, L.; Yu, W.; Men, C.; Liu, R. Impacts of DEM uncertainties on critical source areas identification for non-point source pollution control based on SWAT model. *J. Hydrol.* **2016**, *540*, 355–367. [CrossRef]

16.   McDonald, S.; Mohammed, I.N.; Bolten, J.D.; Pulla, S.; Meechaiya, C.; Markert, A.W.; Srinivasang, R.; Nelson, J.; Lakshmi, V. A web-based decision support system tools: The Soil and Water Assessment Tool online visualization and analyses (SWATOnline) and NASA earth observation data downloading and reformatting tool (NASAaccess). *Environ. Model. Softw.* **2019**, *120*, 104499. [CrossRef] [PubMed]

17.   Ferraz, S.F.D.B.; Rodrigues, C.B.; Garcia, L.G.; Alvares, C.A.; Lima, W.D.P. Effects of Eucalyptus plantation on streamflow in Brazil: Moving beyond the water use debate. *For. Ecol. Manage.* **2019**, *453*, 1–10. [CrossRef]

18.   Di Luzio, M.; Srinivasan, R.; Arnold, J.G.; Neitsch, S.L. *Soil and Water Assessment Tool–ArcView GIS Interface Manual–Version 2000*; Grassland, Soil and Water Research Laboratory, Agricultural Research Service and Blackland Research Center, Texas Agricultural Experiment Station: Temple, TX, USA, 2001.

19.   Arnold, J.G.; Moriasi, D.N.; Gassman, P.W.; Abbaspour, K.C.; White, M.J.; Srinivasan, R.; Santhi, C.; Harmel, R.D.; van Griensven, A.; Van Liew, N.W.; et al. SWAT: Model use, calibration, and validation. *Trans. ASABE* **2012**, *55*, 1491–1508. [CrossRef]

20.   CARD 2020. *SWAT Literature Database for Peer-Reviewed Journal Articles*; Center for Agricultural and Rural Development, Iowa State University: Ames, IA, USA, 2019; Available online: https://www.card.iastate.edu/swat_articles/ (accessed on 4 May 2020).

21.   Cotter, A.S.; Chaubey, I.; Costello, T.A.; Soerens, T.S.; Nelson, M.A. Water quality model output uncertainty as affected by spatial resolution of input data. *J. Am. Water Resour. Assoc.* **2003**, *39*, 977–986. [CrossRef]

22.   Dixon, B.; Earls, J. Resample or not?! Effects of resolution of DEMs in watershed modeling. *Hydrol. Process.* **2009**, *23*, 1714–1724. [CrossRef]

23.   Duru, U.; Arabi, M.; Wohl, E.E. Modeling stream flow and sediment yield using the SWAT model: A case study of Ankara River basin, Turkey. *Phys. Geogr.* **2017**, *39*, 264–289. [CrossRef]

24.   Goulden, T.; Jamieson, R.; Hopkinson, C.; Sterling, S.; Sinclair, A.; Hebb, D. Sensitivity of hydrological outputs from SWAT to DEM spatial resolution. *Photogramm. Eng. Remote Sens.* **2014**, *80*, 639–652. [CrossRef]

25.   Reddy, A.S.; Reddy, M.J. Evaluating the influence of spatial resolutions of DEM on watershed runoff and sediment yield using SWAT. *J. Earth Syst. Sci.* **2015**, *124*, 1517–1529. [CrossRef]

26.   Nagaveni, C.; Kumar, K.P.; Ravibabu, M.V. Evaluation of TanDEMx and SRTM DEM on watershed simulated runoff estimation. *J. Earth Syst. Sci.* **2018**, *128*, 2. [CrossRef]

27. Munoth, P.; Goyal, R. Effects of DEM source, spatial resolution and drainage area threshold values on hydrological modeling. *Water Resour. Manag.* **2019**, *33*, 3303–3319. [CrossRef]

28. Lin, S.; Jing, C.; Coles, N.A.; Chaplot, V.; Moore, N.J.; Wu, J. Evaluating DEM source and resolution uncertainties in the Soil and Water Assessment Tool. *Stoch. Environ. Res. Risk Assess.* **2012**, *27*, 209–221. [CrossRef]

29. Zhang, P.; Liu, R.; Bao, Y.; Wang, J.; Yu, W.; Shen, Z. Uncertainty of SWAT model at different DEM resolutions in a large mountainous watershed. *Water Res.* **2014**, *53*, 132–144. [CrossRef]

30. Kumar, B.; Lakshmi, V.; Patra, K.C. Evaluating the Uncertainties in the SWAT Model Outputs due to DEM Grid Size and Resampling Techniques in a Large Himalayan River Basin. *J. Hydrol. Eng.* **2017**, *22*, 04017039. [CrossRef]

31. Goulden, T.; Hopkinson, C.; Jamieson, R.; Sterling, S. Sensitivity of DEM, slope, aspect and watershed attributes to LiDAR measurement uncertainty. *Remote Sens. Environ.* **2016**, *179*, 23–35. [CrossRef]

32. Wu, M.; Shi, P.; Chen, A.; Shen, C.; Wang, P. Impacts of DEM resolution and area threshold value uncertainty on the drainage network derived using SWAT. *Water SA* **2017**, *43*, 450. [CrossRef]

33. Camargos, C.; Julich, S.; Houska, T.; Bach, M.; Breuer, L. Effects of Input Data Content on the Uncertainty of Simulating Water Resources. *Water* **2018**, *10*, 621. [CrossRef]

34. Köppen, W. *Grundriss der Klimakunde*; Walter de Gruyter: Berlin, Germany, 1931; p. 388.

35. Thornthwaite, C.W. An approach toward a rational classification of climate. *Geogr. Rev.* **1948**, *38*, 55–94. [CrossRef]

36. Morais, J. Divisão climática de Portugal. In *Memórias e Notícias*; Publicações do Museu Mineralógico e Geológico da Universidade de Coimbra: Coimbra, Portugal, 1959; p. 27. (In Portuguese)

37. Dias, R.; Araújo, A.; Terrinha, P.; Kullberg, J.C. (Eds.) *Geologia de Portugal. Volume II Geologia Meso-Cenozóica de Portugal*; Escolar Editora: Lisboa, Portugal, 2013; p. 798. (In Portuguese)

38. Gonçalves, F.; Zbyzewski, G.; Carvalhosa, A.; Coelho, A.P. *Notícia Explicativa da Carta Geológica de Portugal na Escala 1/50 000, Folha 27-D*; Serviços Geológicos de Portugal: Lisboa, Portugal, 1979; p. 75. (In Portuguese)

39. Almeida, C.; Jesus, M.R.; Mendonça, J.J.L.; Gomes, A.J. *Sistemas Aquíferos de Portugal Continental-Sistema Aquífero das Aluviões do Tejo*; INAG: Lisboa, Portugal, 2000; p. 12.

40. Matias, M.J.; Marques, J.M.; Figueiredo, P.; Basto, M.J.; Abreu, M.M.; Carreira, P.M.; Ribeiro, C.; Flambo, A.; Feliciano, J.; Vicente, E.M. Assessment of pollution risk ascribed to Santa Margarida Military Camp activities (Portugal). *Environ. Geol.* **2008**, *56*, 1227–1235. [CrossRef]

41. SROA. *Carta dos Solos de Portugal. I Vol: Classificação e Caracterização Morfológica dos Solos*; Ministério da Economia, Secretaria de Estado da Agricultura, Serviço de Reconhecimento e Ordenamento Agrário: Lisboa, Portugal, 1970. (In Portuguese)

42. Cardoso, J.V.J.C. A Classificação dos Solos de Portugal-Nova Versão. *Boletim de Solos (SROA)* **1974**, *17*, 14–46. (In Portuguese)

43. FAO–UNESCO. *Soil Map of the World. Vol. I-Legend*; UNESCO: Paris, France, 1974.

44. FAO–UNESCO. *Soil Map of the World. Revised Legend*; World Soil Resources Report No. 60: Rome, Italy, 1988; p. 59.

45. Rabus, B.; Eineder, M.; Roth, A.; Bamler, R. The Shuttle Radar Topography Mission-A new class of digital elevation models acquired by spaceborne radar. *ISPRS J. Photogramm. Remote Sens.* **2003**, *57*, 241–262. [CrossRef]

46. Gonçalves, J.A.; Fernandes, J.C. *Assessment of SRTM 3 DEM in Portugal with Topographic Map Data*; EARScL 3D Remote Sensing Workshop: Porto, Portugal, 2005.

47. Gonçalves, J.A.; Morgado, A. Use of the SRTM DEM as a geo-referencing tool by elevation matching. *Int. Arch. Photogramm. Remote Sens. Spat. Inf. Sci.* **2008**, *37*, 879–883.

48. Lin, S.; Jing, C.; Chaplot, X.; Yu, X.; Zhang, Z.; Moore, N.; Wu, J. Effect of DEM resolution on SWAT outputs of runoff, sediment and nutrients. *Hydrol. Earth Syst. Sci.* **2010**, *7*, 4411–4435. [CrossRef]

49. Sharma, A.; Tiwari, K.N. A comparative appraisal of hydrological behavior of SRTM DEM at catchment level. *J. Hydrol.* **2014**, *519*, 1394–1404. [CrossRef]

50. Petrasova, A.; Mitasova, H.; Petras, V.; Jeziorska, J. Fusion of high-resolution DEMs for water flow modeling. *Open Geospat. Data Softw. Stand.* **2017**, *2*, 6. [CrossRef]

51. McGaughey, R.J. *FUSION/LDV: Software for LIDAR Data Analysis and Visualization (Version 3.60+)*; Pacific Northwest Research Station, United States Department of Agriculture Forest Service: Seattle, WA, USA, 2009; p. 123.

52. Tarboton, D.G. A new method for the determination of flow directions and contributing areas in grid digital elevation models. *Water Resour. Res.* **1997**, *33*, 309–319. [CrossRef]

53. Horton, R.E. Erosional development of streams and their drainage basins; hydrophysical approach to quantitative morphology. *Geol. Soc. Am. Bull.* **1945**, *56*, 275–370. [CrossRef]

54. Mueller, J.E. An introduction to the hydraulic and topographic sinuosity indexes. *Ann. Assoc. Am. Geogr.* **1968**, *58*, 371–385. [CrossRef]

55. Bieger, K.; Arnold, J.G.; Rathjens, H.; White, M.J.; Bosch, D.D.; Allen, P.M. Introduction to SWAT+, a Completely Restructured Version of the Soil and Water Assessment Tool. *J. Am. Water Resour. Assoc.* **2016**, *53*, 115–130. [CrossRef]

56. Neitsch, S.L.; Arnold, J.G.; Kiniry, J.R.; Williams, J.R.; King, K.W. *Soil and Water Assessment Tool Theoretical Documentation, Version 2000*; Blackland Research Center, Texas Agricultural Experiment Station: Temple, TX, USA, 2001.

57. Rocha, J.; Roebeling, P.; Rialrivas, M.E. Assessing the impacts of sustainable agricultural practices for water quality improvements in the Vouga catchment (Portugal) using the swat model. *Sci. Total Environ.* **2015**, *536*, 48–58. [CrossRef] [PubMed]

58. Rocha, J.; Carvalho-Santos, C.; Diogo, P.; Beça, P.; Keizer, J.J.; Nunes, J.P. Impacts of climate change on reservoir water availability, quality and irrigation needs in a water scarce Mediterranean region (southern Portugal). *Sci. Total Environ.* **2020**, *736*, 139477. [CrossRef]

59. Serpa, D.; Nunes, J.P.; Santos, J.; Sampaio, E.; Jacinto, R.; Veiga, S.; Lima, J.C.; Moreira, M.; Corte-Real, J.; Keizer, J.J.; et al. Impacts of climate and land use changes on the hydrological and erosion processes of two contrasting Mediterranean catchments. *Sci. Total Environ.* **2015**, *538*, 64–77. [CrossRef]

60. Nunes, J.P.; Jacinto, R.; Keizer, J.J. Combined impacts of climate and socio-economic scenarios on irrigation water availability for a dry Mediterranean reservoir. *Sci. Total Environ.* **2017**, *584–585*, 219–233.

61. Williams, J.R.; Jones, C.A.; Dyke, P.T. A modeling approach to determining the relationship between erosion and soil productivity. *Trans. ASAE* **1984**, *27*, 129–144. [CrossRef]

62. Abbaspour, K.C. *User Manual for SWAT-CUP, SWAT Calibration and Uncertainty Analysis Programs*; Swiss Federal Institute of Aquatic Science and Technology, Eawag: Duebendorf, Switzerland, 2007.

63. Abbaspour, K.C.; Rouholahnejad, E.; Vaghefi, S.; Srinivasan, R.; Yang, H.; Kløve, B. A continental-scale hydrology and water quality model for Europe: Calibration and uncertainty of a high-resolution large-scale SWAT model. *J. Hydrol.* **2015**, *524*, 733–752. [CrossRef]

64. Gupta, H.V.; Sorooshian, S.; Yapo, P.O. Status of automatic calibration for hydrologic models: Comparison with multilevel expert calibration. *J. Hydrol. Eng.* **1999**, *4*, 135–143. [CrossRef]

65. Krause, P.; Boyle, D.; Bäse, P. Comparison of different efficiency criteria for hydrological model assessment. *Adv. Geosci.* **2005**, *5*, 89–97. [CrossRef]

66. Moriasi, D.N.; Arnold, J.G.; Van Liew, M.W.; Bingner, R.L.; Harmel, R.D.; Veith, T.L. Model evaluation guidelines for systematic quantification of accuracy in watershed simulations. *Trans. ASABE* **2007**, *50*, 885–900. [CrossRef]

67. Moriasi, D.N.; Gitau, M.W.; Pai, N.; Daggupati, P. Hydrologic and water quality models: Performance measures and evaluation criteria. *Trans. ASABE* **2015**, *58*, 1763–1785.

68. Charrier, R.; Li, Y. Assessing resolution and source effects of digital elevation models on automated floodplain delineation: A case study from the Camp Creek Watershed, Missouri. *Appl. Geogr.* **2012**, *34*, 38–46. [CrossRef]

69. Luo, Y.; Su, B.; Yuan, J.; Li, H.; Zhang, Q. GIS Techniques for Watershed Delineation of SWAT Model in Plain Polders. *Procedia Environ. Sci.* **2011**, *10*, 2050–2057. [CrossRef]

70. Strahler, A.N. Quantitative analysis of watershed geomorphology. *Trans. Am. Geophys. Union* **1957**, *38*, 913–920. [CrossRef]

71. Pang, S.J.; Wang, X.Y.; Melching, C.S.; Feger, K.H. Development and testing of a modified SWAT model based on slope condition and precipitation intensity. *J. Hydrol.* **2020**, *588*, 125098. [CrossRef]

72. Yang, P.; Ames, D.P.; Fonseca, A.; Anderson, D.; Shrestha, R.; Glenn, N.F.; Cao, Y. What is the effect of LiDAR-derived DEM resolution on large-scale watershed model results? *Environ. Model. Softw.* **2014**, *58*, 48–57. [CrossRef]
73. Malagó, A.; Vigiak, O.; Bouraoui, F.; Pagliero, L.; Franchini, M. The hillslope length impact on SWAT streamflow prediction in large basins. *J. Environ. Inform.* **2018**, *32*, 82–97. [CrossRef]

Article

Designing a Validation Protocol for Remote Sensing Based Operational Forest Masks Applications. Comparison of Products Across Europe

Angel Fernandez-Carrillo *, Antonio Franco-Nieto, Erika Pinto-Bañuls, Miguel Basarte-Mena and Beatriz Revilla-Romero

Remote Sensing and Geospatial Analytics Division, GMV, Isaac Newton 11, P.T.M. Tres Cantos, E-28760 Madrid, Spain; afranco@gmv.com (A.F.-N.); erika.pinto.b@gmv.com (E.P.-B.); mbasarte@gmv.com (M.B.-M.); brevilla@gmv.com (B.R.-R.)
* Correspondence: aafernandez@gmv.com; Tel.: +34 660-136-356

Received: 1 September 2020; Accepted: 24 September 2020; Published: 26 September 2020

Abstract: The spatial and temporal dynamics of the forest cover can be captured using remote sensing data. Forest masks are a valuable tool to monitor forest characteristics, such as biomass, deforestation, health condition and disturbances. This study was carried out under the umbrella of the EC H2020 MySustainableForest (MSF) project. A key achievement has been the development of supervised classification methods for delineating forest cover. The forest masks presented here are binary forest/non-forest classification maps obtained using Sentinel-2 data for 16 study areas across Europe with different forest types. Performance metrics can be selected to measure accuracy of forest mask. However, large-scale reference datasets are scarce and typically cannot be considered as ground truth. In this study, we implemented a stratified random sampling system and the generation of a reference dataset based on visual interpretation of satellite images. This dataset was used for validation of the forest masks, MSF and two other similar products: HRL by Copernicus and FNF by the DLR. MSF forest masks showed a good performance (OA_{MSF} = 96.3%; DC_{MSF} = 96.5), with high overall accuracy (88.7–99.5%) across all the areas, and omission and commission errors were low and balanced (OE_{MSF} = 2.4%; CE_{MSF} = 4.5%; $relB_{MSF}$ = 2%), while the other products showed on average lower accuracies (OA_{HRL} = 89.2%; OA_{FNF} = 76%). However, for all three products, the Mediterranean areas were challenging to model, where the complexity of forest structure led to relatively high omission errors (OE_{MSF} = 9.5%; OE_{HRL} = 59.5%; OE_{FNF} = 71.4%). Comparing these results with the vision from external local stakeholders highlighted the need of establishing clear large scale validation datasets and protocols for remote sensing-based forest products. Future research will be done to test the MSF mask in forest types not present in Europe and compare new outputs to available reference datasets.

Keywords: forest mask; validation; probability sampling; remote sensing; earth observations; forestry; accuracy assessment; forest classification

---

1. Introduction

One of the main data challenges forest managers face is the lack of accurate and up-to-date data to support specific activities in the silvicultural cycle, such as reporting and planning for commercial forestry or recreational activities [1,2]. Forests are dynamic environments and information related to its characterization and health condition is key [3–5]. In addition, some natural dynamics such as pest and droughts are being exacerbated by recent changes in climate, causing more frequent damages and costs [3,6–9]. Therefore, there is a need for reliable, precise and quality data which can cover large areas and with frequent updates adapted to meet forest managers' needs. Remote sensing, such as satellite [10–12] and LiDAR [13–15] sensors, offers data and added-value products to operational solutions for silviculture. In this framework, the European Union Horizon-2020 MySustainableForest (MSF) project provides a portfolio of geo-information products to support forest activities and sustainable management from the afforestation to the forest products transformation markets [16]. These products are based on satellite data, LiDAR, non-invasive sonic measurements and statistical data. The MSF project has demonstrated the advantage of incorporating remote sensing products into the daily decision making, protocols and operations of the different stakeholders across the silvicultural chain.

One of the most important variables to monitor regularly is the forest extension regularly. This can be achieved using datasets commonly called forest masks. A forest mask is a spatial binary forest/non-forest land classification which can be derived with different methods and datasets [17,18]. The information provided by forest masks helps to track changes related to forest characterization [19,20], wood quality (e.g., forest types, site index and wood strength), forest monitoring (e.g., restoration, degradation, deforestation [21] and biomass [22,23]) or vegetation stress monitoring (e.g., damage due to pest [24], drought events, etc.). One of the key products of MSF is a forest mask, which is used as baseline to produce other outputs related to forest characterization and health condition.

Maps derived from remote sensing data contain errors from different sources [25]. Validation is a key requirement and step of the production of remote sensing datasets [25,26]. Products reliability and accuracy is highly demanded by the broad range of forest stakeholders, which often claim the quality of remote sensing-derived maps to be too low for operational use [27]. Feedback from forestry end-users has been key to develop and improve MSF products during the project. Forestry experts from seven European institutions, namely the Centre Nationale de la Propriété Forestière (CNPF), the Croatian Forest Research Institute (CFRI), the Forest Owners Association of Lithuania (FOAL), the Forest Owners Association of Navarra (FORESNA), Madera Plus Calidad Forestal (MADERA+), the Instituto de Investigação da Floresta e Papel (RAIZ) and the University Forest Enterprise Masaryk Forest Křtiny of Mendel University in Brno (UFE), have carried out a validation based on a combined qualitative and quantitative analysis of the products. The qualitative analysis was based on the expert knowledge from local stakeholders, while, for the quantitative analysis, local experts used their own field measurements to compare them against MSF outputs. Nevertheless, this external validation has several problems related to the sampling strategy and performance metrics for the quantitative validation and to the subjective view of the operator in case of the qualitative validation. Sample size was often too low to yield robust accuracy estimates. Hence, very diverse results were obtained for the same product (e.g., forest mask) in areas with similar forest characteristics. This experience highlighted the need of carrying out independent validation processes with a clear and homogeneous protocol. The use of a probabilistic sampling strategy cancels out the above-mentioned problems as it produces unbiased performance metrics with known associated errors [26,28–30].

Independent ground truth datasets must be used to obtain unbiased validation metrics [26,31]. There are several large scale, global or continental, forest mask datasets which may be compared to MSF forest mask [32–34]. However, these existing datasets have associated levels of uncertainty and thus should not be considered as ground truth. Validating new products with these existing datasets would provide useful information but results would show a comparison between products (i.e., each product might have errors from different sources) rather than a validation with real ground-truth data [35]. Therefore, the performance metrics obtained would show the agreement/disagreement between different products without any precise statistical meaning about product quality. Considering the lack of accurate ground-truth data for forest/non-forest classification in Europe, another option is the generation of a validation dataset based in visual interpretation by an independent expert [30,36]

The aim of this study was to establish a validation methodology suitable for operational forest remote sensing products across different locations in Europe, assuming there is no homogeneous field dataset available. The specific objectives of this study were: (1) to design an operational validation protocol for large-scale forestry products; (2) to validate our forest masks across 16 locations and

European forest types; and (3) to assess the skill of MSF Forest Masks in comparison to other large-scale products available on those areas.

2. Study Areas and Data

2.1. Study Areas

The MSF forest mask layers used on this study were implemented on sixteen case demos in six countries and include the most representative forest types across Europe (Figure 1). These Areas of Interest (AOI) are managed by forest owners' associations and forestry research institutes which promote conservation and sustainable management values in over one million hectares in size. For simplicity, each AOI has an acronym related to the country (e.g., Spain is ESP) and the number of AOI present in each country.

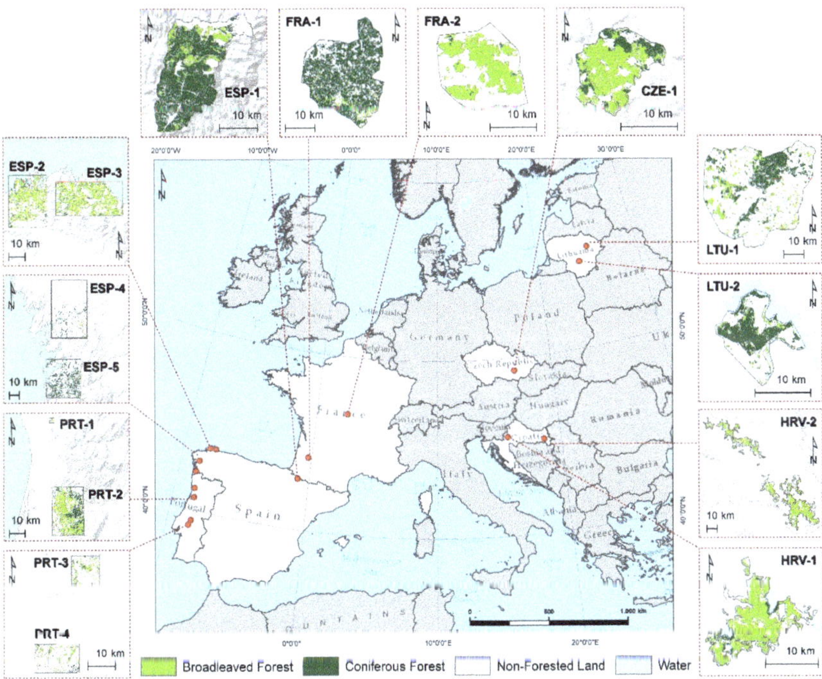

Figure 1. Location of the 16 AOIs across Europe with their abbreviations. The forest type classes showed are taken from the Joint Research Centre Forest Type Map (2006) [20,37].

Further details of each AOI can be found on the MySustainableForest demo cases website section [16], including the details of the main forest systems present:

- Croatia: Continental lowland forests by heterogeneous stands and lowland Slavonian pedunculated oak forests (HRV-1 and HRV-2).
- Czech Republic: Temperate Pannonian mixed forests: (i) conifers, namely spruces, pines and larches; and (ii) broadleaf, namely beeches, oaks and hornbeams (CZE-1).
- France: Oceanic maritime pine forest (FRA-1) and temperate continental with pedunculated oak (FRA-2).
- Lithuania: Boreal forest with forests dominated by scots pines, birches and Norway spruces (LTU-1 and LTU-2).

- Portugal: Plantations of *Eucalyptus spp.* on two climate subtypes: Atlantic-Mediterranean (PRT-1 and PRT-2) and typical Mediterranean combined with agroforestry areas (PRT-3 and PRT-4).
- Spain: Alpine forest dominated by beeches, oaks and pines (ESP-1) and Atlantic plantations of *Eucalyptus spp.* where land ownership is highly fragmented (ESP-2, ESP-3, ESP-4 and ESP-5).

2.2. High Resolution Forest/Non-Forest MSF Classification Dataset

According to Food and Agriculture Organization (FAO), forest land is any surface of more than 0.5 ha with more than 10% of tree canopy and trees higher than 5 m, excluding land predominated by agricultural or urban uses [38]. To calculate the MSF forest mask used in this study, we considered as forest the area with more than 50% of tree stratum coverage (i.e., when more that 50% of the pixel appears covered by tree canopy) [33]. We are aware that we are excluding areas with lower than 50% tree coverage which per definition may cause some disagreement with ground truth data and other forest mask datasets.

Within the MSF project, the forest mask products were derived using three different input datasets: satellite high resolution and very high resolution imagery and LiDAR data. We present here the results of developing forest masks layers based on Sentinel imagery for 2018, as well as satellite data from the autumn/winter period of 2017 for some Spanish AOIs (Table 1). A first version of this product was previously published [18], being this one an improved version based on the experiences learned from the first one. Two images (i.e., summer and winter) were used and processed to capture different phenological conditions, especially for deciduous/mixed forest discrimination. Sentinel-2 images were acquired in format Level-2A (i.e., surface reflectance). Only bands in the visible NIR and SWIR wavelengths were selected. Visible and NIR bands have a pixel size of 10 m. SWIR bands, originally at 20 m spatial resolution, were resampled to 10 m. The obtained forest masks had a spatial resolution of 10 m and Minimum Mapping Unit (MMU) of 0.1 ha (equivalent to 10 pixels/10 m pixel). A Random Forest classifier [39] was trained using as input the original Sentinel-2 bands, several vegetation and texture indices. Normalized Difference Vegetation Index (NDVI) [40], Enhanced Vegetation Index (EVI) [41], Transformed Chlorophyll Absorption Reflectance Index (TCARI) [42], Bare Soil Index [43] and Tasseled Cap (TC) Wetness [44] were added to the input data together with NDVI homogeneity and entropy [45] The model was trained using 2500 forest/non-forest random points distributed across Europe.

Table 1. Tile and acquisition date of Sentinel-2 imagery used to generate the validation dataset.

AOI Name	Sentinel-2 Tile	Date 1	Date 2
CZE-1	T33UXQ	22-03-2018	29-08-2018
ESP-1	T30TXN	26-10-2017	04-08-2018
ESP-2	T29TNJ	22-02-2018	11-08-2018
ESP-3	T29TPJ	21-12-2017	11-08-2018
ESP-4	T29TNH	22-02-2018	11-08-2018
ESP-5	T29TNG	24-02-2018	11-08-2018
FRA-1	T30TYP/T30TYQ	24-01-2018	19-08-2018
FRA-2	T31TDM	25-02-2018	19-08-2018
HRV-1	T33TWL	08-04-2018	01-08-2018
HRV-2	T33TYL/T34TCR/T34TCQ	11-03-2018	13-08-2018
LTU-1	T35ULB	18-03-2018	23-08-2018
LTU-2	T35ULA	18-03-2018	23-08-2018
PRT-1	T29TNF	26-03-2018	18-08-2018
PRT-2	T29TNF	26-03-2018	18-08-2018
PRT-3	T29SND	26-03-2018	18-08-2018
PRT-4	T29SND	26-03-2018	18-08-2018

2.3. High Resolution Forest/Non-Forest External Classification Datasets

There are many methodologies and providers of forest mask layers. However, the number of forest/non-forest products available at large scale as in this case, across Europe, is limited. Two freely available forest/non-forest classification datasets were selected to compare them against MSF forest mask. The products chosen are comparable to MSF forest mask as they were produced close in time, within 3 years of the Sentinel images used for the development of our layers, and have lower but still comparable spatial resolution (20–50 m):

- Forests High-Resolution Layer (HRL) is provided by Copernicus Land Monitoring Service (CLMS) and coordinated by the European Environment Agency (EEA) [32]. The Tree Cover Density (TCD) product provides the level of tree coverage per pixel in a range of 0–100%. It is obtained through a semi-automatic classification of multitemporal satellite images (Sentinel-2 and Landsat-8) for the year 2015 (±1 year) [32], using a combination of supervised and unsupervised classification techniques. The resulting product has 20 m spatial resolution. To make it comparable with the MSF forest masks, the TCD product was pre-processed to obtain a binary forest/non-forest classification, considering as forest all those pixels with a cover density of 50% or higher. A future update of this dataset for 2018 reference year was announced in August 2002 by Copernicus, but it was not available prior submission of this paper.
- TanDEM-X Forest/Non-Forest (FNF) is provided by Microwaves and Radar Institute of the German Aerospace Center (DLR) [33,34]. It is a global forest/non-forest map generated from interferometric synthetic aperture radar (InSAR) data from the TanDEM-X mission. The bistatic stripmap single polarization (HH) InSAR data were acquired between 2011 and 2016, being 2015 the reference year for the final product. FNF maps are produced with a final pixel size of 50 m.

2.4. Satellite Data Required to Build the Independent Validation Dataset

Two sources with different bands spatial resolutions were used to create the validation dataset: (1) High Resolution imagery: Sentinel-2 L2A provided by Copernicus (Table 1); and (2) Very High Resolution imagery: satellite data (GoogleEarth™ and Bing Aerial imagery), taking into consideration the spatial and temporal resolution of the forest/non-forest classification datasets.

3. Methods

Validation and intercomparison of the results across different areas may not be straightforward if not following a standard methodology. Considering the non-existence of homogenized forest/non-forest ground truth datasets, there are different strategies that may be implemented to perform a quantitative validation of forest masks:

- Cross validation or dataset split [46]. Commonly used for products developed using any supervised classification approach. A training dataset is needed, either provided by the user or built by the producer. The use of cross-validation techniques is widely accepted when there are not independent training and testing sets. However, the statistical distribution of the training and test samples are not independent, leading to an optimistic bias in the resulting metrics [46]. Note that the final values of the metrics will be strongly determined by the quality of the input dataset.
- Using data from national forest inventories (NFI). Many methodologies used to perform forest inventories as the selected method by each nation will depend of the purpose and the scale of the inventory, leading to significant efforts to perform data search, normalization and data engineering. In Europe, the European National Forest Inventory Network (ENFIN) [47] promotes NFIs and harmonizes forest information. However, there are still different methods and plot sizes used. Moreover, inventories contain errors from different sources, as they rely on sampling strategies and the measuring protocols are not always clear, thus adding uncertainty to the metrics generated through the validation process [4,48,49]. Hence, validation metrics might differ significantly

across areas where the product has identical quality, due to the different inventory methods used, which are in general heterogeneous.

- Building independent datasets based in visual interpretation of images [50]. This method is costly, and the validation must be carried out by independent interpreters in order to build an unbiased dataset trying to represent the variability of forest types within a determined area. The statistical metrics obtained with this method may be close to the real quality of the product if sampling and interpretation processes are carefully designed.

Considering the difficulties of using cross-validation and national forest inventories for our objectives, a new reference forest/non-forest dataset was generated by an independent operator to validate MSF forest masks and compare it with the other external products (e.g., HRL and FNF). The validation strategy was divided into four phases: (1) sampling design; (2) generation of the reference dataset through visual interpretation; (3) definition of performance metrics; and (4) intercomparison of products. The methods explained in subsequent sections are summarized in the flowchart (Figure 2).

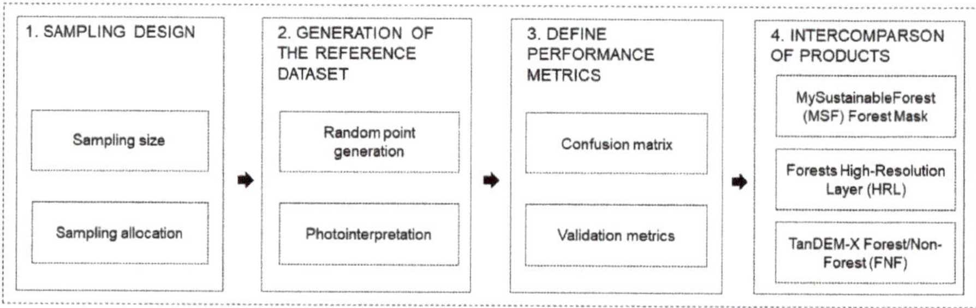

Figure 2. Methods used to validate the forest mask products.

3.1. Sampling Design

Pixels were the spatial unit selected as validation samples. Following the recommendations made by Olofsson et al. [26], a probability sampling strategy was applied to select the samples of the reference dataset. The two conditions defining a probability sample are: (1) the inclusion probability must be known for each unit selected in the sample; and (2) the inclusion probability must be greater than zero for all units in the AOI [29]. A stratified random sampling strategy was designed considering the two strata forest and non-forest. This design allows increasing the sample size in classes that occupy a small proportion of area to reduce the standard errors of the class-specific accuracy estimates for these rare classes [26].

To obtain the number of points to be included in the reference dataset, the following equation was used (Equation (1)) [51]:

$$n = \frac{(\sum W_i S_I)^2}{\left[S(\hat{O})\right]^2 + (1/N) \sum W_i S_i^2} \approx \left(\frac{\sum W_i S_i}{S(\hat{O})}\right)^2 \tag{1}$$

where N represents the number of units in the AOI, $S(\hat{O})$ is the standard error of the estimated overall accuracy, W_i is the mapped proportion of area of stratum i and S_i is the standard deviation of stratum i. $S_i = \sqrt{U_i(1 - U_i)}$, where U_i is the expected user's accuracy.

With a pessimistic user's accuracy estimate of 0.7 and a standard error of 0.01 for the overall accuracy, the total number of points for the dataset was 2100.

To choose the optimal number of points per AOI, the recommendation [26] is to use an allocation method that should meet two conditions. Firstly, the number of points per AOI should lay between equal and proportional allocation (i.e., proportional to the area of the AOI relative to the total area of all the AOIs). Secondly, the final allocation should push the proportional distribution to the equal, in order to have enough samples to robustly validate the poorer class, but without it becoming completely equal.

The specific equation (Equation (2)) used for sample allocation per AOI was:

$$n_{AOI} = \frac{p + 2e}{3} \tag{2}$$

where p is the distribution of n proportional to the area of each AOI and e is the equal distribution of n for each AOI.

Once the sample size per AOI was computed, samples need to be allocated to strata (i.e., forest/non-forest). It is important that the sample allocation allows precise estimates of accuracy and area [52]. Having a fixed number of samples per AOI to allocate in two strata, proportional allocation would increase the variances of the performance metrics for the rarer class, as it would be poorly represented. Using equal allocation would allow a more robust validation of the poorer class at the expense of increasing the variances of the bigger class. The optimal sample allocation should minimize the variance of the desired target metric. Considering the overall accuracy (OA) as the metric whose variance needs to be minimized, the variance of an estimated OA can be computed using the following equation (Equation (3)) [26]:

$$\hat{V}(\hat{O}) = \sum_{i=1}^{q} W_i^2 \hat{U}_i (1 - \hat{U}_i)/(n_i - 1) \tag{3}$$

Being n_i the number of samples per stratum i in a specific AOI, the optimal sample allocation (i.e., optimal value of n_i per stratum) was computed through an iterative process estimating the variance of a desired OA for all possible values of n_i. The values of n_i which yielded the minimum variance of OA were selected as optimal sample allocation (Table 2 and Figure 3).

Table 2. Final distribution of points per AOI and stratum.

AOIs	Total Points	Forest Points	Non-Forest Points
CZE-1	97	58	30
ESP-1	115	73	42
ESP-2	135	71	64
ESP-3	142	82	60
ESP-4	226	115	109
ESP-5	176	78	98
FRA-1	129	72	57
FRA-2	105	52	53
HRV-1	96	64	32
HRV-2	159	103	56
LTU-1	217	94	123
LTU-2	96	46	50
PRT-1	88	41	47
PRT-2	127	69	58
PRT-3	94	42	52
PRT-4	98	54	44
Total	2100	1114	984

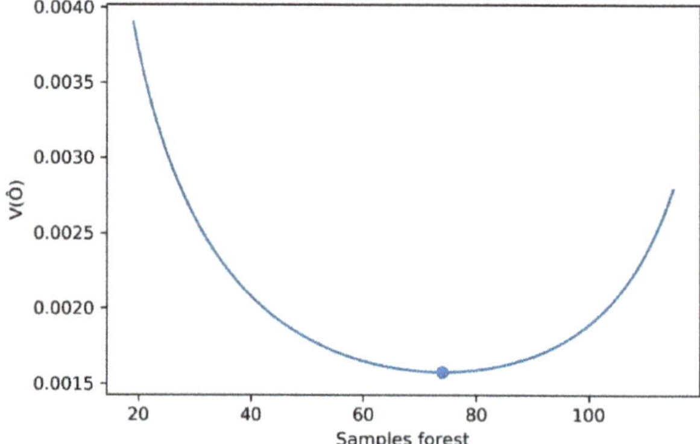

Figure 3. Changes in the variance of the overall accuracy (V(Ô)) depending on the number of samples assigned to the forest class for the AOI ESP-1. The optimal number of forest samples to minimize the variance was 73.

3.2. Dataset Generation

Once sampling size and allocation were computed, the validation dataset was created. In each AOI, the random points were generated with a minimum distance of 100 m between them. Each point was subsequently assigned to forest or non-forest via photointerpretation by an independent trained consultant. Sentinel-2 and very high resolution images (see data, Section 2.4) were used as basis for the visual interpretation.

For the Sentinel-2 images, the RGB composition used was NIR/SWIR/red, as it allows clear vegetation discrimination due to the high reflectance of the vegetation in NIR wavelengths [53]. Forests were visualized in red tones for the coniferous species and as light reds or orange for deciduous species. Non-forest areas, such as crops, roads or buildings, were visualized in green, blue or yellow. High resolution images were visualized in true color (i.e., red/green/blue).

3.3. Performance Metrics

Confusion matrices (Table 3) were built for each area using the forest mask products and the reference dataset. A total confusion matrix was computed pooling the samples of all the areas.

Table 3. Sample confusion matrix for a forest/non-forest classification.

Predicted Condition	True Condition		Total
	Forest	**Non-Forest**	**Total**
Forest	True Positive (TP)	False Positive (FP)	Predicted Condition Positive (PCP)
Non-Forest	False Negative (FN)	True Negative (TN)	Predicted Condition Negative (PCN)
Total	Condition Positive (CP)	Condition Negative (CN)	sample size (n)

Different agreement and error metrics were computed from confusion matrices (Table 4) [35]. All metrics were multiplied by 100 to be expressed as percentage.

Table 4. Confusion matrix-derived agreement and errors metrics.

	Definition	Equation	
Agreement metrics	Overall Accuracy (OA): Proportion of pixels correctly classified.	$OA = \frac{TP+TN}{n}$	(1)
	Precision (P): Proportion of correctly predicted (i.e., classified) cases from all those predicted as positive.	$P = \frac{TP}{TP+FP}$	(2)
	Recall (R): Proportion of correctly predicted cases from all the real positives.	$R = \frac{TP}{TP+FN}$	(3)
	Dice similarity Coefficient (DC) or F1-score: Harmonic mean of precision and recall.	$DC = \frac{2 \cdot TP}{2 \cdot TP+FP+FN}$	(4)
Error metrics	Commission Error (CE): Proportion of misclassified pixels from all those predicted as positive.	$CE = \frac{FP}{TP+FP}$	(5)
	Omission Error (OE): Proportion of misclassified pixels from all the real positives.	$OE = \frac{FN}{TP+FN}$	(6)
	Relative Bias (relB): It quantifies the systematic error of the classification.	$relB = \frac{(FP-FN)}{TP+FN}$	(7)

4. Results

Performance metrics for each forest/non-forest mask layer (i.e., MSF, HRL and FNF) can be found in Table 5. For the pooled set of AOIs, agreement metrics were high, the OA ranging from 88.7% to 99.5% and the DC from 91.1% to 99.5%. Highest scores were always found for MSF, followed by HRL and finally FNF. Errors and bias were low and balanced in the MSF product (OE = 2.4%; CE = 4.5; relB = 2%), while HRL (OE = 14.4%; CE = 6.3; relB = −8.6 %) and FNF (OE = 27.3%; CE = 19.7; relB = −9.5%) showed higher and imbalanced errors, which led to higher relative bias. HRL and FNF products had higher omission than commission errors (negative relB). As a summary, the relative bias of MSF product was lower and the commission error was slightly higher than the omission error, thus confirming the good calibration and balance of MSF errors.

Table 5. Performance metrics of MSF, HRL and FNF products.

	MSF	HRL	FNF
OA	96.3	89.2	76.0
DC	96.5	89.4	76.8
P	95.5	93.6	80.3
R	97.6	85.6	72.6
OE	2.4	14.4	27.3
CE	4.5	6.3	19.7
RelB	2.0	−8.6	−9.5

Analyzing the agreement metrics by AOIs (Table 6, Figure 4), we observed that results follow the same trend in overall accuracy (OA) and Dice similarity Coefficient (DC). The Interquartile Range (IQR, i.e. the difference between the first and third quartiles) was narrower in MSF (IQR$_{OA}$ = 2.93 and IQR$_{DC}$ = 2.25), followed by HRL (IQR$_{OA}$ = 13.95 and IQR$_{DC}$ = 13.28) and FNF (IQR$_{OA}$ = 16.85 and IQR$_{DC}$ = 18.83). This confirmed a good performance of MSF forest mask across different European forest types. MSF and HRL present median values higher than 90% in the OA and DC, while, in FNF, these values are 76.05% and 78.55%. The high dispersion of agreement values in FNF also pointed out a high variability of the product across different forest types.

Table 6. Performance metrics of 16 AOIs grouped by product.

		CZE-1	ESP-1	ESP-2	ESP-3	ESP-4	ESP-5	FRA-1	FRA-2	HRV-1	HRV-2	LTU-1	LTU-2	PRT-1	PRT-2	PRT-3	PRT-4
MSF	OA	96.9	98.2	97.0	88.7	97.3	97.1	97.7	97.2	93.7	98.7	99.5	97.9	96.5	95.2	92.5	90.9
	DC	97.4	98.6	97.2	91.1	97.4	96.8	97.9	97.0	95.5	99.0	99.5	97.8	96.3	95.6	91.6	92.4
	P	98.2	98.6	97.2	83.7	97.4	95.0	98.6	100	91.4	99.0	100	100	95.2	97.0	92.7	85.9
	R	96.5	98.6	97.2	100	97.4	98.7	97.2	94.2	100	99.0	98.9	95.6	97.6	94.2	90.5	100
	OE	3.4	1.3	2.8	0.0	2.6	1.2	2.8	5.8	0.0	0.9	1.0	4.3	2.4	5.8	9.5	0.0
	CE	1.7	1.3	2.8	16.3	2.6	4.9	1.4	0.9	8.6	0.9	0.0	0.0	4.7	3.0	7.3	14.0
	relB	−1.7	0.0	0.0	19.5	0.0	3.8	−1.4	−5.7	9.3	0.0	−1.0	−4.3	2.4	−3.0	−2.3	16.3
HRL	OA	96.9	96.5	92.6	88.7	89.7	98.3	82.9	99.0	88.5	96.2	97.2	94.8	77.3	81.9	68.1	56.6
	DC	97.4	97.3	92.7	89.7	89.9	98.1	83.6	99.0	91.5	97.1	96.7	94.4	69.7	84.1	53.1	35.8
	P	98.2	96.0	90.5	94.6	91.0	97.5	90.3	100	90.8	95.3	97.8	97.7	92.0	80.3	77.3	100
	R	96.5	98.6	90.1	85.4	88.7	98.7	77.8	98.0	92.2	99.0	95.7	91.3	56.0	88.4	40.5	21.8
	OE	3.4	1.3	9.8	14.6	11.3	1.2	22.2	1.9	7.81	0.9	4.2	8.7	43.9	11.6	59.5	78.2
	CE	1.7	4.0	4.5	5.4	8.9	2.6	9.7	0.0	9.23	4.7	2.1	2.3	8.0	19.7	22.7	0.0
	relB	−1.7	2.7	−5.6	−9.7	−2.6	1.3	−13.9	−1.9	1.6	3.9	−2.1	−6.5	−39.0	10.1	−47.6	−78.2
FNF	OA	81.4	63.5	77.8	79.6	73.2	81.2	60.5	95.3	76.0	76.1	91.7	91.6	64.7	72.4	56.3	57.6
	DC	85.0	65.0	77.6	81.0	72.7	79.5	59.8	95.2	82.4	82.2	90.0	90.9	63.5	75.8	36.9	44.7
	P	82.3	82.9	82.5	87.3	76.2	77.1	69.1	94.3	80.6	79.3	94.2	95.2	61.3	72.4	52.2	80.9
	R	87.9	53.4	73.2	75.6	69.6	82.0	52.8	96.1	84.4	85.4	86.2	87.0	65.8	79.7	28.6	30.9
	OE	12.1	46.6	26.7	24.9	30.4	17.9	47.2	3.8	15.6	14.5	13.8	13.0	34.1	20.3	71.4	69.1
	CE	17.7	17.0	17.4	12.7	23.8	22.9	30.9	5.6	19.4	20.7	5.8	4.7	39.6	27.6	47.8	19.0
	relB	6.9	−35.6	−11.2	−13.4	−8.7	6.4	−23.6	1.9	4.7	7.7	−8.5	−8.7	7.3	10.1	−45.2	−61.8

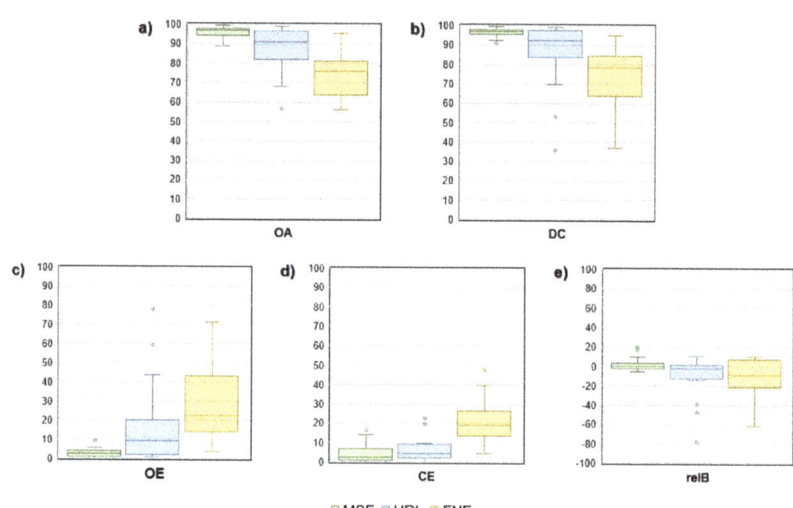

Figure 4. Box plots of performance metrics across the 16 AOIs for the three forest mask products: (**a**) overall accuracy; (**b**) Dice similarity coefficient; (**c**) omission errors; (**d**) commission errors, and (**e**) relative bias.

The IQR$_{OE}$ (Table 6 and Figure 4) was higher in FNF (22.9%), followed by HRL (13.48%) and MSF (2.65%). The IQR$_{CE}$ showed a similar behavior, being higher in FNF (8.83%) followed by HRL (6.73%) and finally MSF (4.3%). It should be noted that forest omissions were much higher than commissions in the case of HRL and FNF products. On the other hand, in MSF, the commission was greater than the omission, even though they achieved a great balance. The medians were lower than 10% for the MSF forest mask (OE = 2.5% and CE = 2.7%) and HRL (OE = 9.25% and CE = 4.6%), and it was around 20% in the case of FNF (OE = 22.6% and CE = 19.2%). The relative bias was higher for FNF and HRL than for MSF, highlighting more imbalanced errors. This fact was reinforced by the relatively high dispersion of relB in FNF and HRL. While for MSF the range of relB is 16.3, with areas with relB = 0%, the range of this metric reaches 88.3 for HRL and 71.9 for FNF. The low IQRs of MSF confirmed its consistent performance in forests with different characteristics compared to the other products analyzed.

There were some relevant differences among the AOIs (Figure 5). Independently of the forest mask used, the best metrics were obtained in Central and Northern Europe (CZE-1, FRA-2, LTU-1 and LTU-2), where continental forest have in general a continuous and dense canopy. All the forest mask products were less accurate for Portugal and Spain, especially for HRL and FNF, with OA and DC around 20% lower than MSF. Regarding the outliers of agreement metrics, most of them were located in the HRL product, mainly in Southern Portugal (PRT-4, OA; PRT-3 and PRT-4, DC). OE was up to 50% higher in areas with Mediterranean influence (PRT-3 and PRT-4). The greater omission problems were located in HRL and FNF datasets (PRT-3, OE = 71.4% and 59.5%, respectively) in areas dominated by tree–grass ecosystems, where also relative bias reached the highest values. Finally, there is an evidence that Eucalyptus plantations also presented some difficulties to be classified (CE and OE) by FNF.

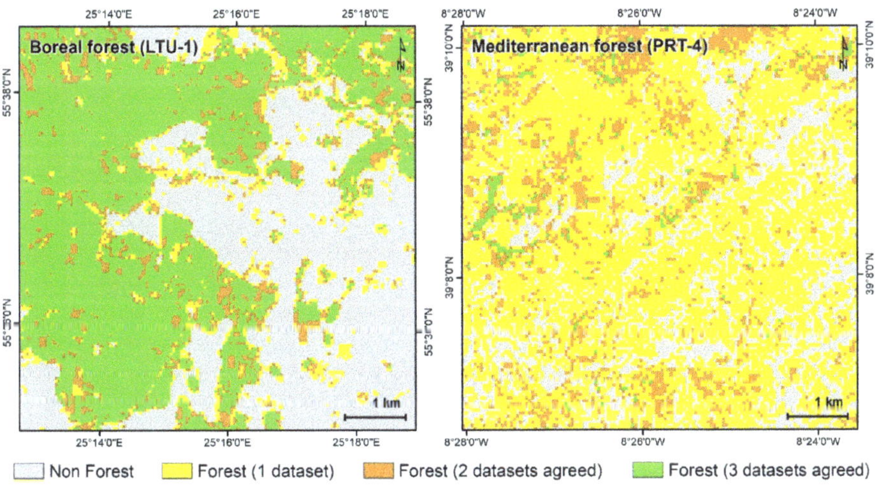

Figure 5. Comparison between two forests in the three datasets: Boreal with high accuracy (**left**); and Mediterranean with low accuracy (**right**).

5. Discussion

The validation of the MSF high resolution forest masks on average yielded high scores for the agreement metrics and low for the error metrics. Nevertheless, we found some clear performance differences between the AOIs mainly related to the dominant forest types in each area. The best results were obtained for areas dominated by conifer, broadleaf or mixed forest in Central Europe. These forests are generally characterized by homogenous tree masses and crops which are relatively easy to discriminate using remote sensing data. The shrub stratum is common in these same areas under a dense tree canopy. On the other hand, the metrics were generally worse in areas closer to the Mediterranean climate, such as Portugal and Spain. The areas with the worst results (PRT-3 and

PRT-4) are characterized by the high presence of tree–grass ecosystems, which are similar to savannas. Tree–grass ecosystems are characterized by the co-existence of a sparse tree stratum and a grassland matrix with large seasonal contrasts [54,55]. Hence, reflectance contribution to the pixel highly depends on the behavior of the seasonal grass stratum, being this contribution variable depending on the density of the sparse tree stratum [56,57]. The complexity of these ecosystems in Southern Portugal (PRT-3 and PRT-4) led to relatively high omission errors compared to the other AOIs in all the products analyzed. Nevertheless, MSF forest masks had a maximum OE of 9% in these areas, while the HRL product obtained a maximum of 78.2% and FNF a maximum of 71.4%. This highlights the challenge of accurately detecting sparse tree–grass ecosystem [58]. Determining an exact threshold to discriminate pure cover types in transitional environments may lead to misinterpretation of natural landscapes, while providing fuzzy values might be more interesting to perform further analysis [59]. In the case of MSF forest mask, using a different threshold for assigning a pixel to forest (<50%) might change significantly the binary forest mask obtained for Southern Europe. The good performance of the MSF forest mask for this type of forest in comparison to the other products result from the great effort payed on this project to adapt the algorithm to different forest types across Europe, with a strong focus in Mediterranean tree–grass areas. Transitional woody shrublands are also present in some parts of AOIs in Southern Europe, resulting in higher commission errors in Portugal and Spain. For the AOIs located in Galicia (northwest of Spain; ESP-2, ESP-3, ESP-4 and ESP-5), the commission error was relatively high due to the large areas dominated by degraded shrublands, which have a similar spectral response as the dominant *Eucalyptus* plantations of these region. This effect seemed to be attenuated in HRL, while MSF and FNF showed higher CE in these AOIs.

HRL showed more outliers than the other products when comparing the metrics across AOIs, especially for the Dice coefficient (DC = 35.8%), commission error (CE = 78.2%) and relative bias (relB = −78.2%). These three outliers were yielded by the validation of PRT-4. In this area, HRL did not classify as non-forest any real forest pixel, but this happens at the cost of classifying misclassifying 78% of the predicted forest pixels. In this area, MSF also showed its highest relB but with opposite sign. Therefore, all the pixels classified as forest were correctly predicted (CE = 0%), but many real forest pixels are classified as non-forest (OE = 16.3%), which is common when working on sparse tree–grass ecosystem. This highlights the importance of considering relative bias for a complete performance vision even where the agreement metrics show high skill. Otherwise, it might lead to an optimistic interpretation of the results while the bias indicates the possible lack of robustness of the products in some areas.

The HRL official validation report [60] yielded a R^2 = 0.84 for the Tree Cover Density (TCD) product, with significantly lower values for the Mediterranean region, especially for Portugal (R^2 = 0.66), which coincides with the challenges found in this study. The HRL document also reports low R^2 values for Iceland, the Arctic and Anatolian regions, which are not represented in MSF test areas or in this study. The best results in HRL report were found for Boreal, Pannonian and Continental bioregions (R^2 > 0.85), which also coincide with the outcomes of this study. However, the methodology and results from both studies are not comparable since in the present work a pre-processing of the TCD was carried out to obtain a binary product.

The official FNF validation [33] reported a mean agreement of 86% for Germany and Eastern Europe. This is similar to the accuracy found for FNF in the present study in similar areas (CZE-1, LTU-1 and LTU-2). Nevertheless, Martone et al. [33] validated the FNF product using as ground truth the HRL product. Hence, this value of agreement should be taken with caution as the uncertainty of this validation and the errors of HRL products are not quantified in the FNF validation report.

External validations were carried out by local stakeholders within MSF project. ESP-2, ESP-3, ESP-4 and ESP-5, located in northwestern Spain, reported accuracies ranging between 93% and 100%. However, these AOIs were validated with sets of 8, 14, 31 and 20 points, making the final accuracy not significant enough. LTU-1 reported an overall accuracy of 90.4% and a kappa coefficient (k) [61] of 0.80 comparing MSF forest mask with information from the Lithuanian National Forest Inventory. The OA given by external stakeholders in this case is 10% lower than the OA achieved in this study. For LTU-2, the difference was lower (external OA = 91.1%; external k = 0.82). FRA-1 and FRA-2 were validated by stakeholders using non-systematic visual field control, reporting accuracies of 90% for both areas, around 8% below the OA found in this work. Stakeholders from CZE-1 reported an accuracy of more than 95% through visual inspection using high resolution ortophotos and cadaster data. In this last case, the difference with our OA was under 2%. At the time of writing this article, local validation from Portuguese and Croatian stakeholders is still pending, being the former particularly interesting to compare as it was clearly the most problematic area. However, the results of these external validations are hard to interpret and compare given the variability of methods and data sources used to carry out the accuracy assessments. The differences between the independent validation here presented and the external validation carried out by local stakeholders may be explained by two main points. Firstly, local forestry institutions are not always used to deal with continuous remote sensing data (i.e., based on pixels). Hence, these institutions may not have clearly defined validation protocols for this type of products, leading to several problems such as under-sampling or not taking into account the confidence level or standard error of the metrics computed for a certain area of interest. Secondly, the main purpose of local institutions may differ. Within the MSF project consortium, some institutions were centered in public forest management at national scale (e.g., CFRI), while others were associations of private owners (e.g., FORESNA) or privately owned companies with specific objectives (e.g., RAIZ is part of the pulp industry, while MADERA+ is focused on wood technological properties). This fact, together with the different nationalities involved, led to different definition of forests depending on national forest inventories or other sources. These considerations led to significantly different external validation approaches, which highlighted the need of standard definitions and methods to validate remote sensing-based forestry products, as the protocol proposed in this study.

There are several limitations of the validation exercise presented in this study and are explained below. The characteristics of the AOIs, namely the size, locations and forest types, were defined on the basis of local stakeholders within the MSF project consortium. Hence, while some forest types and regions are well represented (e.g., *Eucalyptus* plantations in northern Spain or mixed and broadleaf forests), other forest types, such as Northern European boreal forests, are poorly represented within the AOIs. A more detailed analysis and validation could take into account a more proportional representation of all forest types present in Europe. It is also important to bear in mind that, while the MSF forest masks were specifically calibrated for the European continent, the other two datasets analyzed (i.e., Copernicus HRL and Tandem-X FNF) are global datasets. This might explain the apparent better quality of the MSF product in the selected AOIs, which is likely to decrease if the algorithm is calibrated for other environments such as tropical forests or arid regions.

We are also aware that some of the products characteristics such as the date of the imagery used and the spatial resolution may have an impact on the performance and comparison of the products (MSF, HRL and FNF). As previously explained, the three products have different spatial resolutions and were developed using satellite images from different years. The products were validated using a reference dataset built using images from 2017–2018. As there is more consistency between the satellite images used to build the MSF layers and those for the validation dataset, this might have had a penalty of the results of the other two products. However, it is unlikely that difference between the dates used on the reference dataset and product (maximum three years) would have had such a high impact. We assume that the studied forests did not experience significant changes in the last five years, as previously stated directly by local stakeholders. A multi-temporal dataset (i.e., with reference samples for different years) would be needed to identify how much is the impact of difference between products and reference datasets affecting the performance metrics. In the case of the different spatial resolutions used, even when this may have had an effect on the final results, it is not likely to be the main factor explaining such remarkable differences in performance metrics, as we confirmed by doing a visual analysis of the most problematic areas (PRT-3 and PRT-4).

6. Conclusions

In this study, a protocol was established to validate remote sensing derived forest/non-forest classification maps across Europe. Based on a stratified random sampling, a reference dataset was created using visual interpretation of satellite images in order to validate MSF forest mask. Performance metrics were compared with two similar products: HRL and FNF.

The validation protocol allowed comparing the results of the accuracy assessment for different areas across Europe and different forest mask products. Problems were found when trying to compare the results of this assessment with the validation results provided by local experts. This was caused by the high variability of techniques used by stakeholders and justified the need of clearly established operational protocols for carrying out accuracy assessments of remote sensing-based products.

Accuracies were generally high for all forest mask products (OA = 76–96.3%), with MSF showing a more precise calibration for different European forest types (MSF accuracy ranges from 89% to 99.5%, while FNF accuracy ranges from 56% to 95%), which confirms its suitability as an adequate data source for large scale forest mapping in this continent. The greatest problems to discriminate forest from other land covers were found in areas with Mediterranean influence characterized by the presence of tree–grass ecosystems. Large shrublands in degraded areas also hindered forest discrimination. Performance metrics were significantly better for MSF in these areas compared to the other datasets analyzed, while FNF performance seemed to yield worse results. Note that the reference dataset was built with satellite imagery dates similar to the MSF masks, which also have higher spatial resolution (10 m). The other products used imagery up to three years older and with slightly lower resolution (20–50 m). This difference of dates and spatial resolutions might have slightly impacted the comparison of products. The performance of MSF satellite forest mask in other continents was out of the scope in this study but should be carried out in the near future.

The design of the accuracy assessment and selection of performance metrics highlight the need of establishing clear validation protocols for remote sensing-based forestry products. Efforts should be made in creating unified reference datasets at continental or global scales in order to clearly define the advantages and limitations of incorporating remote sensing-derived forest data in operational forestry processes.

Author Contributions: Conceptualization, A.F.-C.; Methodology, A.F.-C., A.F.-N. and E.P.-B.; Software, M.B.-M.; Validation, E.P.-B., A.F.-C. and A.F.-N.; Formal Analysis, A.F.-C.; Investigation, A.F.-C. and B.R.-R.; Resources, GMV; Data Curation, M.B.-M. and A.F.-C.; Writing—Original Draft Preparation, A.F.-C., B.R.-R. and A.F.-N.; Writing—Review and Editing, B.R.-R.; Visualization, A.F.-N.; Supervision, B.R.-R.; Project Administration, MSF; and Funding Acquisition, MSF. All authors have read and agreed to the published version of the manuscript.

Funding: This research was funded by European Union's Horizon 2020 research and innovation programme, under grant agreement 776045.

Acknowledgments: The authors are grateful to MySustainableForest stakeholders from Centre Nationale de la Propriété Forestière (CNPF), Croatian Forest Research Institute (CFRI), Forest Owners Association of Lithuania (FOAL), Forest Owners Association of Navarra (FORESNA-ZURGAIA), Instituto de Investigação da Floresta e Papel (RAIZ), Madera Plus Calidad Forestal (MADERA+) and University Forest Enterprise Masaryk Forest Křtiny of Mendel University in Brno (UFE) (https://mysustainableforest.com).

Conflicts of Interest: The authors declare no conflict of interest.

References

1. Pause, M.; Schweitzer, C.; Rosenthal, M.; Keuck, V.; Bumberger, J.; Dietrich, P.; Heurich, M.; Jung, A.; Lausch, A. In situ/remote sensing integration to assess forest health-a review. *Remote Sens.* **2016**, *8*, 471. [CrossRef]

2. Masek, J.G.; Hayes, D.J.; Joseph Hughes, M.; Healey, S.P.; Turner, D.P. The role of remote sensing in process-scaling studies of managed forest ecosystems. *For. Ecol. Manag.* **2015**, *355*, 109–123. [CrossRef]

3. Trumbore, S.; Brando, P.; Hartmann, H. Forest health and global change. *Science* **2016**, *349*. [CrossRef] [PubMed]

4. White, J.C.; Coops, N.C.; Wulder, M.A.; Vastaranta, M.; Hilker, T.; Tompalski, P. Remote Sensing Technologies for Enhancing Forest Inventories: A Review. *Can. J. Remote Sens.* **2016**, *42*, 619–641. [CrossRef]

5. Forest Europe; Unece; F.A.O. *FOREST EUROPE: State of Europe's Forests 2015*; Liaison Unit: Madrid, Spain, 2015.

6. Bonan, G.B. Forests and climate change: Forcings, feedbacks, and the climate benefits of forests. *Science* **2008**, *320*, 1444–1449. [CrossRef]

7. Seidl, R.; Thom, D.; Kautz, M.; Martin-Benito, D.; Peltoniemi, M.; Vacchiano, G.; Wild, J.; Ascoli, D.; Petr, M.; Honkaniemi, J.; et al. Forest disturbances under climate change. *Nat. Clim. Chang.* **2017**, *7*, 395–402. [CrossRef]

8. Lindner, M.; Fitzgerald, J.B.; Zimmermann, N.E.; Reyer, C.; Delzon, S.; van der Maaten, E.; Schelhaas, M.-J.; Lasch, P.; Eggers, J.; van der Maaten-Theunissen, M.; et al. Climate change and European forests: What do we know, what are the uncertainties, and what are the implications for forest management? *J. Environ. Manag.* **2014**, *146*, 69–83. [CrossRef]

9. Milad, M.; Schaich, H.; Bürgi, M.; Konold, W. Climate change and nature conservation in Central European forests: A review of consequences, concepts and challenges. *For. Ecol. Manag.* **2011**, *261*, 829–843. [CrossRef]

10. Holmgren, P.; Thuresson, T. Satellite remote sensing for forestry planning—A review. *Scand. J. For. Res.* **1998**, *13*, 90–110. [CrossRef]

11. Boyd, D.S.; Danson, F.M. Satellite remote sensing of forest resources: Three decades of research development. *Prog. Phys. Geogr. Earth Environ.* **2005**, *29*, 1–26. [CrossRef]

12. Wood, J.E.; Gillis, M.D.; Goodenough, D.G.; Hall, R.J.; Leckie, D.G.; Luther, J.E.; Wulder, M.A. Earth Observation for Sustainable Development of Forests (EOSD): Project overview. In Proceedings of the IEEE International Geoscience and Remote Sensing Symposium, Toronto, ON, Canada, 24–28 June 2002; Volume 3, pp. 1299–1302.

13. Wulder, M.A.; White, J.C.; Nelson, R.F.; Næsset, E.; Ørka, H.O.; Coops, N.C.; Hilker, T.; Bater, C.W.; Gobakken, T. Lidar sampling for large-area forest characterization: A review. *Remote Sens. Environ.* **2012**, *121*, 196–209. [CrossRef]

14. Lim, K.; Treitz, P.; Wulder, M.; St-Onge, B.; Flood, M. LiDAR remote sensing of forest structure. *Prog. Phys. Geogr. Earth Environ.* **2003**, *27*, 88–106. [CrossRef]

15. Dubayah, R.O.; Drake, J.B. Lidar Remote Sensing for Forestry. *J. For.* **2000**, *98*, 44–46. [CrossRef]

16. My Sustainable Forest. Earth Observation Services for Silviculture. Available online: https://mysustainableforest.com/ (accessed on 30 August 2020).

17. Pekkarinen, A.; Reithmaier, L.; Strobl, P. Pan-European forest/non-forest mapping with Landsat ETM+ and CORINE Land Cover 2000 data. *ISPRS J. Photogramm. Remote Sens.* **2009**, *64*, 171–183. [CrossRef]

18. Fernandez-Carrillo, A.; de la Fuente, D.; Rivas-Gonzalez, F.W.; Franco-Nieto, A. A Sentinel-2 unsupervised forest mask for European sites. In Proceedings of the SPIE, Strasbourg, France, 9–12 September 2019; Volume 11156.

19. Fassnacht, F.E.; Latifi, H.; Stereńczak, K.; Modzelewska, A.; Lefsky, M.; Waser, L.T.; Straub, C.; Ghosh, A. Review of studies on tree species classification from remotely sensed data. *Remote Sens. Environ.* **2016**, *186*, 64–87. [CrossRef]

20. Kempeneers, P.; Sedano, F.; Seebach, L.; Strobl, P.; San-Miguel-Ayanz, J. Data Fusion of Different Spatial Resolution Remote Sensing Images Applied to Forest-Type Mapping. *IEEE Trans. Geosci. Remote Sens.* **2011**, *49*, 4977–4986. [CrossRef]

21. Saksa, T.; Uuttera, J.; Kolström, T.; Lehikoinen, M.; Pekkarinen, A.; Sarvi, V. Clear-cut Detection in Boreal Forest Aided by Remote Sensing. *Scand. J. For. Res.* **2003**, *18*, 537–546. [CrossRef]

22. Lu, D. The potential and challenge of remote sensing-based biomass estimation. *Int. J. Remote Sens.* **2006**, *27*, 1297–1328. [CrossRef]

23. Gleason, C.J.; Im, J. A Review of Remote Sensing of Forest Biomass and Biofuel: Options for Small-Area Applications. *GIScience Remote Sens.* **2011**, *48*, 141–170. [CrossRef]

24. Hall, R.J.; Castilla, G.; White, J.C.; Cooke, B.J.; Skakun, R.S. Remote sensing of forest pest damage: A review and lessons learned from a Canadian perspective *. *Can. Entomol.* **2016**, *148*, S296–S356. [CrossRef]

25. Foody, G.M. Assessing the accuracy of land cover change with imperfect ground reference data. *Remote Sens. Environ.* **2010**, *114*, 2271–2285. [CrossRef]

26. Olofsson, P.; Foody, G.M.; Herold, M.; Stehman, S.V.; Woodcock, C.E.; Wulder, M.A. Good practices for estimating area and assessing accuracy of land change. *Remote Sens. Environ.* **2014**, *148*, 42–57. [CrossRef]

27. Foody, G.M. Status of land cover classification accuracy assessment. *Remote Sens. Environ.* **2002**, *80*, 185–201. [CrossRef]

28. Stehman, S.V. Practical Implications of Design-Based Sampling Inference for Thematic Map Accuracy Assessment. *Remote Sens. Environ.* **2000**, *72*, 35–45. [CrossRef]

29. Stehman, S.V. Statistical rigor and practical utility in thematic map accuracy assessment. *Photogramm. Eng. Remote. Sens.* **2001**, *67*, 727–734.

30. Padilla, M.; Olofsson, P.; Stehman, S.V.; Tansey, K.; Chuvieco, E. Stratification and sample allocation for reference burned area data. *Remote Sens. Environ.* **2017**, *203*, 240–255. [CrossRef]

31. Fernandez-Carrillo, A.; Belenguer-Plomer, M.A.; Chuvieco, E.; Tanase, M.A. Effects of sample size on burned areas accuracy estimates in the Amazon Basin. In Proceedings of the SPIE—The International Society for Optical Engineering, Berlin, Germany, 10–13 September 2018; Volume 10790.

32. Langanke, T.; Herrmann, D.; Ramminger, G.; Buzzo, G.; Berndt, F. Copernicus Land Monitoring Service – High Resolution Layer Forest. Available online: https://land.copernicus.eu/user-corner/technical-library/hrl-forest (accessed on 20 July 2020).

33. Martone, M.; Rizzoli, P.; Wecklich, C.; González, C.; Bueso-Bello, J.L.; Valdo, P.; Schulze, D.; Zink, M.; Krieger, G.; Moreira, A. The global forest/non-forest map from TanDEM-X interferometric SAR data. *Remote Sens. Environ.* **2018**, *205*, 352–373. [CrossRef]

34. Esch, T.; Heldens, W.; Hirner, A.; Keil, M.; Marconcini, M.; Roth, A.; Zeidler, J.; Dech, S.; Strano, E. Breaking new ground in mapping human settlements from space—The Global Urban Footprint. *ISPRS J. Photogramm. Remote Sens.* **2017**, *134*, 30–42. [CrossRef]

35. Padilla, M.; Stehman, S.V.; Ramo, R.; Corti, D.; Hantson, S.; Oliva, P.; Alonso-Canas, I.; Bradley, A.V.; Tansey, K.; Mota, B.; et al. Comparing the accuracies of remote sensing global burned area products using stratified random sampling and estimation. *Remote Sens. Environ.* **2015**, *160*, 114–121. [CrossRef]

36. Viana-Soto, A.; Aguado, I.; Salas, J.; García, M. Identifying post-fire recovery trajectories and driving factors using landsat time series in fire-prone mediterranean pine forests. *Remote Sens.* **2020**, *12*, 1499. [CrossRef]

37. European Commission Joint Research Centre. Forest Type Map 2006. Available online: http://data.europa.eu/89h/62ec23aa-2d47-4d85-bc81-138175cdf123 (accessed on 20 July 2020).

38. FAO Global Forest Resources Assessment 2020: Terms and Definition. Available online: http://www.fao.org/forest-resources-assessment/2020 (accessed on 20 July 2020).

39. Breiman, L. Random Forests. *Mach. Learn* **2001**, *45*, 5–32. [CrossRef]

40. Tucker, C.J.; Elgin, J.H.; McMurtrey, J.E.; Fan, C.J. Monitoring corn and soybean crop development with hand-held radiometer spectral data. *Remote Sens. Environ.* **1979**, *8*, 237–248. [CrossRef]

41. Jiang, Z.; Huete, A.R.; Didan, K.; Miura, T. Development of a two-band enhanced vegetation index without a blue band. *Remote Sens. Environ.* **2008**, *112*, 3833–3845. [CrossRef]

42. Haboudane, D.; Miller, J.R.; Tremblay, N.; Zarco-Tejada, P.J.; Dextraze, L. Integrated narrow-band vegetation indices for prediction of crop chlorophyll content for application to precision agriculture. *Remote Sens. Environ.* **2002**, *81*, 416–426. [CrossRef]

43. Rikimaru, A.; Roy, P.S.; Miyatake, S. Tropical forest cover density mapping. *Trop. Ecol.* **2002**, *43*, 39–47.

44. Crist, E.P.; Cicone, R.C. A Physically-Based Transformation of Thematic Mapper Data—The TM Tasseled Cap. *IEEE Trans. Geosci. Remote Sens.* **1984**, *GE-22*, 256–263. [CrossRef]

45. Haralick, R.M. Statistical and structural approaches to texture. *Proc. IEEE* **1979**, *67*, 786–804. [CrossRef]

46. Browne, M.W. Cross-Validation Methods. *J. Math. Psychol.* **2000**, *44*, 108–132. [CrossRef]

47. Vidal, C.; Alberdi, I.; Redmond, J.; Vestman, M.; Lanz, A.; Schadauer, K. The role of European National Forest Inventories for international forestry reporting. *Ann. For. Sci.* **2016**, *73*, 793–806. [CrossRef]

48. McRoberts, R.E.; Tomppo, E.O. Remote sensing support for national forest inventories. *Remote Sens. Environ.* **2007**, *110*, 412–419. [CrossRef]

49. Tomppo, E.; Gschwantner, T.; Lawrence, M.; McRoberts, R.E.; Gabler, K.; Schadauer, K.; Vidal, C.; Lanz, A.; Ståhl, G.; Cienciala, E. National forest inventories. *Pathw. Common Rep. Eur. Sci. Found.* **2010**, *1*, 541–553.

50. Biging, G.S.; Congalton, R.G.; Murphy, E.C. *A Comparison of Photointerpretation and Ground Measurements of Forest Structure*; American Congress on Surveying and Mapping and American Soc for Photogrammetry and Remote Sensing: Baltimore, MD, USA, 25–29 March 1991.

51. Cochran, W.G. *Stratified Random Sampling*; John Willey & Sons Inc.: Toronto, Canada, 1977; ISBN 047116240X.

52. Stehman, S.V. Impact of sample size allocation when using stratified random sampling to estimate accuracy and area of land-cover change. *Remote Sens. Lett.* **2012**, *3*, 111–120. [CrossRef]

53. Guyot, G. Optical properties of vegetation canopies. In *Applications of remote sensing in agriculture.*; Steven, M.D., Clark, J.A., Eds.; Butterworth-Heinemann: London, UK, 1990; pp. 19–44. ISBN 0408047674.

54. Pacheco-Labrador, J.; El-Madany, T.S.; Martín, M.P.; Migliavacca, M.; Rossini, M.; Carrara, A.; Zarco-Tejada, P.J. Spatio-temporal relationships between optical information and carbon fluxes in a mediterranean tree-grass ecosystem. *Remote Sens.* **2017**, *9*, 608. [CrossRef]

55. Vlassova, L.; Perez-Cabello, F.; Nieto, H.; Martín, P.; Riaño, D.; Riva, J.D. La Assessment of methods for land surface temperature retrieval from landsat-5 TM images applicable to multiscale tree-grass ecosystem modeling. *Remote Sens.* **2014**, *6*, 4345–4368. [CrossRef]

56. Moore, C.E.; Beringer, J.; Evans, B.; Hutley, L.B.; Tapper, N.J. Tree–grass phenology information improves light use efficiency modelling of gross primary productivity for an Australian tropical savanna. *Biogeosciences* **2017**, *14*, 111–129. [CrossRef]

57. Liu, Y.; Hill, M.J.; Zhang, X.; Wang, Z.; Richardson, A.D.; Hufkens, K.; Filippa, G.; Baldocchi, D.D.; Ma, S.; Verfaillie, J.; et al. Using data from Landsat, MODIS, VIIRS and PhenoCams to monitor the phenology of California oak/grass savanna and open grassland across spatial scales. *Agric. For. Meteorol.* **2017**, *237–238*, 311–325. [CrossRef]

58. Whiteside, T.G.; Boggs, G.S.; Maier, S.W. Comparing object-based and pixel-based classifications for mapping savannas. *Int. J. Appl. Earth Obs. Geoinf.* **2011**, *13*, 884–893. [CrossRef]

59. Arnot, C.; Fisher, P.F.; Wadsworth, R.; Wellens, J. Landscape metrics with ecotones: Pattern under uncertainty. *Landsc. Ecol.* **2004**, *19*, 181–195. [CrossRef]

60. SIRS. GMES Initial Operations/Copernicus Land monitoring services—Validation of products: HRL Forest 2015 Validation Report. Available online: https://www.google.com.hk/url?sa=t&rct=j&q=&esrc=s&source=web&cd=&cad=rja&uact=8&ved=2ahUKEwjLuq_F1oXsAhVGXSsKHWOeBvEQFjAAegQIBxAB&url=https%3A%2F%2Fland.copernicus (accessed on 20 July 2020).

61. McHugh, M.L. Interrater reliability: The kappa statistic. *Biochem. Medica* **2012**, *22*, 276–282. [CrossRef]

Article

Detection of Longhorned Borer Attack and Assessment in Eucalyptus Plantations Using UAV Imagery

André Duarte [1],*, Luis Acevedo-Muñoz [1], Catarina I. Gonçalves [1], Luís Mota [1], Alexandre Sarmento [2], Margarida Silva [1], Sérgio Fabres [1], Nuno Borralho [1] and Carlos Valente [1]

[1] RAIZ—Forest and Paper Research Institute, Quinta de S. Francisco, Rua José Estevão (EN 230-1), 3800-783 Aveiro, Portugal; luis.munoz@thenavigatorcompany.com (L.A.-M.); catarina.goncalves@thenavigatorcompany.com (C.I.G.); bolseiro.raiz.lmota@thenavigatorcompany.com (L.M.); margarida.silva@thenavigatorcompany.com (M.S.); sergio.fabres@thenavigatorcompany.com (S.F.); nuno.borralho@thenavigatorcompany.com (N.B.); carlos.valente@thenavigatorcompany.com (C.V.)

[2] Terradrone, Av. E.U.A. Nº 97, 12 Dto. Sala 03, 1700-167 Lisboa, Portugal; alexandre@terradrone.pt

* Correspondence: andre.duarte@thenavigatorcompany.com; Tel.: +351-234-920-130

Received: 31 August 2020; Accepted: 23 September 2020; Published: 25 September 2020

Abstract: Eucalyptus Longhorned Borers (ELB) are some of the most destructive pests in regions with Mediterranean climate. Low rainfall and extended dry summers cause stress in eucalyptus trees and facilitate ELB infestation. Due to the difficulty of monitoring the stands by traditional methods, remote sensing arises as an invaluable tool. The main goal of this study was to demonstrate the accuracy of unmanned aerial vehicle (UAV) multispectral imagery for detection and quantification of ELB damages in eucalyptus stands. To detect spatial damage, Otsu thresholding analysis was conducted with five imagery-derived vegetation indices (VIs) and classification accuracy was assessed. Treetops were calculated using the local maxima filter of a sliding window algorithm. Subsequently, large-scale mean-shift segmentation was performed to extract the crowns, and these were classified with random forest (RF). Forest density maps were produced with data obtained from RF classification. The normalized difference vegetation index (NDVI) presented the highest overall accuracy at 98.2% and 0.96 Kappa value. Random forest classification resulted in 98.5% accuracy and 0.94 Kappa value. The Otsu thresholding and random forest classification can be used by forest managers to assess the infestation. The aggregation of data offered by forest density maps can be a simple tool for supporting pest management.

Keywords: *Phoracantha* spp.; unmanned aerial vehicle (UAV); multispectral imagery; vegetation index; thresholding analysis; Large Scale Mean-Shift Segmentation (LSMS); Random Forest (RF)

---

1. Introduction

Eucalyptus Longhorned Borers (ELB), *Phoracantha semipunctata* (Fabricius), and *P. recurva* Newman (Coleoptera: *Cerambycidae*), are among the most destructive eucalypt pests in regions with Mediterranean climate [1,2]. *Eucalyptus globulus* is one of the most planted eucalypt species in these regions, and it is known to have low resistance to ELB [2,3].

ELB activity generally starts in late spring when adults emerge and start laying eggs. Upon hatching, ELB larvae bore galleries along the phloem and cambium of trees, eventually preventing sap from flowing, which leads to rapid tree death during summer and fall [1,4–6]. The ability of larvae to successfully colonize the host plant depends on low bark moisture content, leaving water-stressed trees particularly susceptible to attack by ELB [1,5,7].

Under the Mediterranean climate, with low rainfall and extended dry summers periods, the attack by ELB often result in significant tree mortality, despite long-lasting efforts to select for more resistant *E. globulus* genotypes [8]. With droughts expected to increase due to climate change, ELB outbreaks will likely become more frequent and more severe [9].

ELB control methods include biological control, selection of more resistant eucalypts, and various cultural practices aimed at increasing tree adaptation and resilience, but the most effective curative measure is felling all attacked trees and removing them from the stands [6,10]. Damage caused by ELB is usually not detected using traditional surveillance techniques until significant mortality has occurred [11]. Early detection is recognized as the first step to reduce the impact of ELB [1], hence new approaches to monitoring, particularly through remote sensing, can provide an invaluable tool for forest managers in terms of planning and executing control actions.

Traditional survey techniques are restricted by small area coverage and subjectivity [12] but combined with remote-sensing technology can lead to expanded spatial coverage, minimize the response time, and reduce the costs of monitoring forested areas [13]. Following Wulder et al. [14], the appropriate sensor and resolution should be adopted according to the spatial scale which better adjusts to the situation. To date, several studies have demonstrated successful pest and diseases detection and monitoring in forests using different types of sensor and platform [15–17]. For instance, in terms of spaceborne optical sensors for large areas, Meddens et al. [18] detected multiple levels of coniferous tree mortality using multi-date and single-date Landsat imagery. Mortality in ash trees was assessed by Waser et al. [19] using multispectral WorldView-2 imagery. In the eucalypt forest context, to predict bronze bug damage Oumar and Mutanga [20] tested WorldView-2 imagery. The airborne optical sensor Compact Air-Borne Spectrographic Imager 2 (CASI-2) has been tested by Stone et al. [21] who conducted a study to assess damage caused by herbivorous insects in Australian eucalypt forests and pine plantations.

In recent years the use of unmanned aerial vehicle (UAV) platforms has become widely employed for pests and disease detection and monitoring [12,22–28]. The main attributes are very high resolution, suitability for multitemporal analysis, lower operational costs compared with airplanes and satellites, independence of cloud cover, and the ability to operate in specific phenological phases of plants or pest/disease outbreaks [12]. In addition, a large number of passive and active sensors can be assembled, such as RGB (red, green, blue), multispectral and hyperspectral cameras, LiDAR (laser imaging detection and ranging), and RADAR (radio detection and ranging) [29,30]. On the other hand, as stressed by Pádua et al. [29], the disadvantages include small area-coverage when compared with satellites, sensitivity to bad weather, increasingly strict regulations that may restrict operations, and the high volume of data generated.

The imagery classification approaches used in the more recently published UAV studies to detect pests and diseases in forests are diverse. Lehmann et al. [22] used a modified normalized difference vegetation index (NDVI) to discriminate between five classes of infestation by the oak splendor beetle through OBIA (objected-based imagery analysis) classification with an overall Kappa index of agreement of 0.81–0.77. Näsi et al. [23] investigated bark beetle damage at the tree level in South Finland using a combination of RGB, near infrared (NIR), red-edge band and NDVI. Object-based K-NN (K-nearest neighbor) classification was applied, and overall accuracy was 75%. In New Zealand, Dash et al. [12] used NDVI and red-edge NDVI to study the discoloration classes of *Pinus radiata* through the OBIA classification approach with a random forest (RF) algorithm. In Catalonia (Spain), Otsu et al. [26] detected defoliation of pine trees affected by pine processionary and distinguished pine species at a pixel level using four spectral indices and NIR band. The authors estimated the threshold values using histogram analysis. For comparisons, another classification used was the OBIA with a random forest algorithm and an overall accuracy of 93%. Finally, Iordache et al. [28] studied pine wild disease in Portugal using OBIA and machine-learning random forest algorithm for multispectral and hyperspectral imagery. The overall accuracy of both classifications was 95% and 91%, respectively. To date, no published work has been developed with UAV imagery for ELB detection.

This work arises from the necessity to find monitoring tools that could support pest management decisions regarding ELB attacks in eucalypt stands. In the past, this has been hampered since the identification of dead trees in the field is bothcostly and time-consuming. Furthermore, it is critical that ELB attacks can be identified at early stages of infestation. Current surveying methods only detect problems when large number of trees are infested.

The main objectives of this experimental work were: (1) to detect ELB attacks through UAV imagery by using selected spectral indices and Otsu thresholding; (2) to map trees crown status using large-scale mean-shift (LSMS) segmentation, as well as the machine-learning classification approach with a RF algorithm; and (3) to map tree density using the hexagonal tessellations technique to support phytosanitary interventions.

2. Materials and Methods

2.1. Study Area

A 30.56 ha *E. globulus* stand located in central Portugal close to Gavião locality (39°27.909′ N, 7°56.124′ W) was selected for this study (Figure 1). The study area is managed by The Navigator Company (NVG), a Portuguese pulp and paper company, and was selected based on reports of severe pest attacks provided by forest managers of the company.

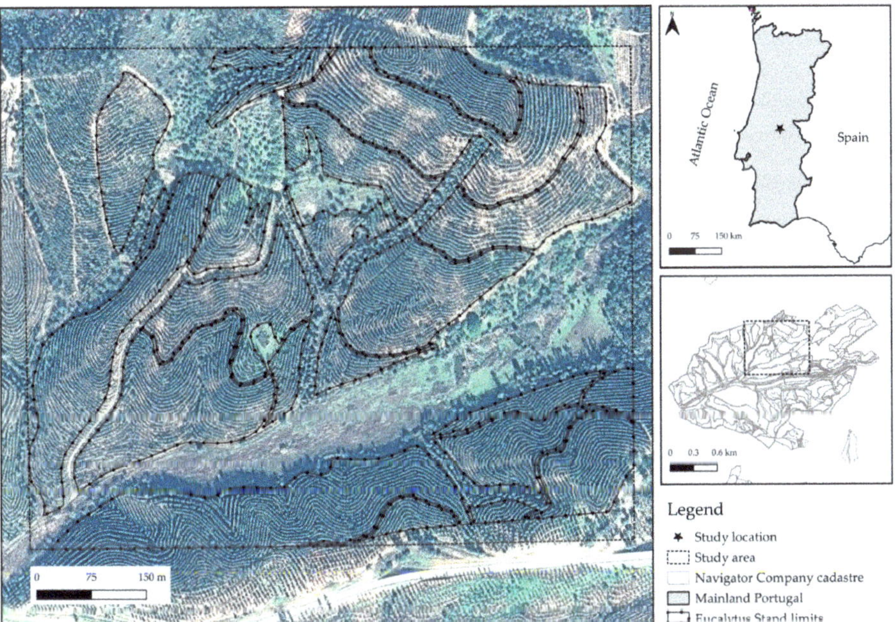

Figure 1. Location of the study site. The main element shows the stand thought Pleiades-1A imagery (image date = 11 December 2019).

The eucalypt plantation was established in 2014 with a mean stand density of 1250 plants per hectare. Initial mortality caused by ELB was detected in early 2018. Due to the severe drought that occurred in 2017, ELB attack increased over the course of a few months. This is supported by the occurrence of low rainfall in 2017, especially during the summer months, as shown in Figure 2.

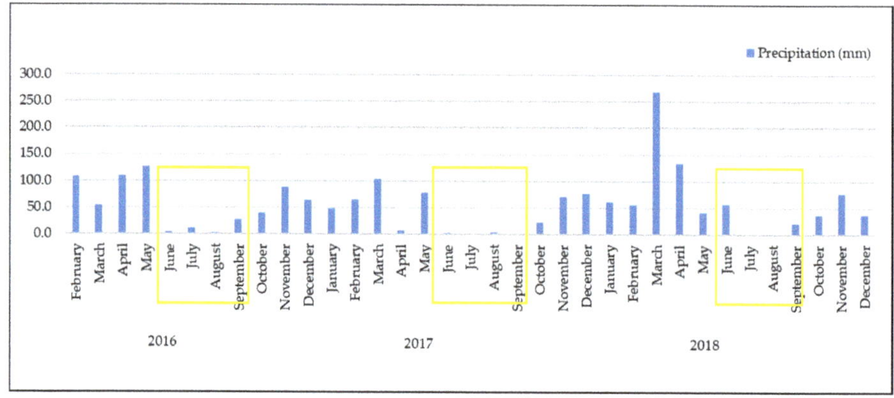

Figure 2. Precipitation distribution (mm/month) observed between 2016 and 2018. Yellow boxes indicate the summer months. Data collected from an own weather station located at 25 km of the study area.

Annual average rainfall is low in this region (700–800 mm/year), so if there is a year drier than the average, the following year's mortality is likely to increase due to water stress and ELB attacks.

This area is characterized by sandstones both in valleys and hillsides. The soils are generally poor and shallow. According to the Forest and Agriculture Organization (FAO)classification system [31], soils are Leptosols or Plinthic Arenosols. The relief is slightly inclined, and the altitude of the site varies between 160 and 250 m a.s.l. (above sea level). Southern aspects prevail in the study site.

2.2. Data Acquisition

Multispectral images were acquired through four flights on 21 January 2019 using a Parrot SEQUOIA camera (Parrot S.A., Paris, France) on a remotely piloted fixed wing eBee SenseFly drone (Parrot S.A., Paris, France). This single propeller drone allows more stable and longer flights, which in turn reduces the cost and length of missions. Considering the frequency of cloudy days and strong winds during winter season, this aerial vehicle has greater speed in the acquisition of data and efficiency in the operation. The details of flight planning parameters are presented in Table 1.

Table 1. The flight's planning parameters for the study area.

Flight Parameters	Specifications
Flight altitude (m)	190
Area (ha)	55
Start time	12:03 (p.m.)
End time	13:56 (p.m.)
Local solar noon	12:47 (p.m.)
Cloud cover (%)	10
Frontal overlap (%)	80
Side overlap (%)	80

The Parrot SEQUOIA is a compact bundle with multispectral and sunshine sensors. The camera collects four discrete and separated spectral bands with 1.2 megapixels (MP) resolution: green (530–570 nm), red (640–680 nm), red-edge (730–740 nm), and near-infrared (770–810 nm). Additionally, the RGB camera captures 16 MP of resolution [32,33].

To improve equal resolution and less likelihood of holes at higher elevation, terrain awareness was used (Figure 3a,b). The mission planning application used was SenseFly eMotion (Parrot S.A., Paris, France). To refine vertical and horizontal accuracy, nine ground control points (GCP's) were installed and measured with real-time kinematic (RTK) global navigation satellite systems (GNSS). Then, the multispectral camera was calibrated in loco, to adjust the sunshine sensors to the local lighting conditions at solar noon.

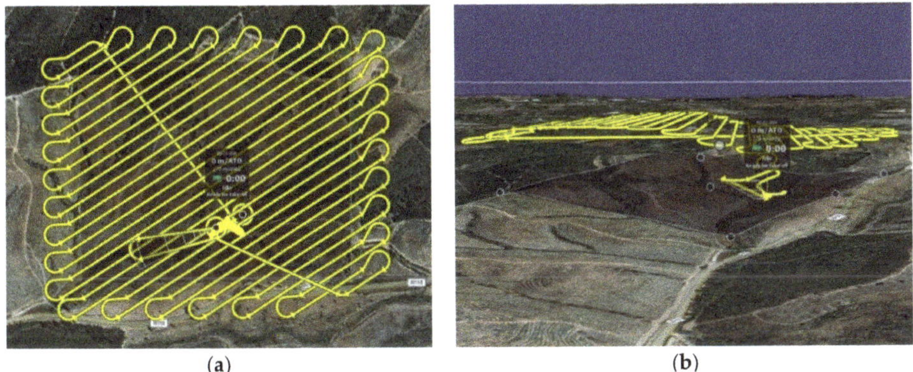

(a)   (b)

Figure 3. (a) Flight paths in the study area, (b) horizontal perspective of flight paths.

Regarding the imagery processing workflow, all discrete bands imagery and GCPs were imported to Pix4Dmapper Pro software (Version 4.2, Pix4D S.A., Prilly, Switzerland), a photogrammetry and drone mapping software broadly used for UAV imagery processing [34]. The agricultural (multispectral photogrammetry workflow was chosen, which consists in alignment, geometric calibration based on GCP's, reflectance calibration, point cloud generation, and classification, raster digital surface model (DSM) and orthomosaic generation based on the digital terrain model (DTM).

The next step was to analyze the vitality of the stand in the field. UAV multispectral imagery was used to elaborate color composites in which matching of the tree canopies pixel were compared with the real field status of trees. For this end, two different tree vitality status were assigned: healthy trees (Figure 4a) and dead trees—crown completely dead (Figure 4b). Field data collection was carried out through random sampling. Data collection was performed using Arrow Gold Antenna (Eos Positioning Systems, Inc., Terrebonne, QC, Canada) with RTK sensor. The cause of death was verified by removing the bark in order to find signs of ELB attack, namely galleries caused by larvae (Figure 4c).

Once the status of the trees was assigned in the field, spectral values were extracted for the training and validation datasets.

2.3. Selected Vegetation Indices

Among the considerable number of vegetation indices (VIs) usually used as indicators of vegetation status [35], there are a set of VIs that are generally applied to assess symptoms of stress [12,36,37]. Stressed eucalypt trees reveal chlorophyll content reduction, red discoloration produced by secondary metabolites accumulation such as anthocyanins and carotenoids, or loss of photosynthetic tissues due to defoliation or necrosis [38,39]. Commonly, NIR, red, red edge, and green bands are used to assess vegetation status due to high sensitivity to changes in moisture content, pigment indices, and vegetation health, respectively [20].

Based on their ability to reveal stress in eucalypts, a total of five vegetation indices were calculated using the available spectral reflectance bands from UAV imagery (Table 2). The difference vegetation index (DVI) and NDVI were estimated based on the evidence of correlation with plant stress [40,41]. The green normalized difference vegetation index (GNDVI) and normalized difference red-edge (NDRE) were also estimated, as both have shown high sensitivity to changes in chlorophyll concentrations [20,42–44]. In addition, some studies have shown that GNDVI has higher sensitivity to pigment changes than NDVI [45,46]. The soil-adjusted vegetation index (SAVI) was also estimated to reduce soil brightness influence [47].

(a)   (b)   (c)

Figure 4. Status of the trees according to field evaluation. (**a**) healthy tree, (**b**) dead tree, cause of death from Eucalyptus Longhorned Borers (ELB) was confirmed by the observation of galleries originated by Longhorned Borer larvae (**c**).

The values of the spectral indices were extracted for all features collected in the field using the zonal statistics tool provided by QGIS (Version 3.10) [48].

Table 2. Selected vegetation indices derived from unmanned aerial vehicle (UAV) imagery.

Vegetation Index	Equation	References
Difference Vegetation Index	DVI = Near infrared (NIR) − Red	[49]
Green Normalized Difference Vegetation Index	GNDVI = NIR − Green/NIR + Green	[45]
Normalized Difference Red-Edge	NDRE = (NIR − RedEdge)/(NIR+RedEdge)	[50]
Normalized Difference Vegetation Index	NDVI = NIR − Red/NIR + Red	[51]
Soil Adjusted Vegetation Index	SAVI = 1.5 × (NIR − Red)/(NIR + Red + 0.5)	[47]

2.4. Pixel-Based Analysis

2.4.1. Otsu Thresholding Method

The Otsu thresholding method is an unsupervised method used to select a threshold automatically from a gray level histogram [52]. This approach assumes that the image contains two classes of pixels following a bimodal histogram, and estimates a threshold value which splits the foreground and the background of an image [26,52–54]. The threshold value is calculated aiming to minimize intra-class intensity variance [52]. Although this method is usually applied to images with bimodal histograms, it might be also used for unimodal and multimodal histograms if the precision of the objects is not a requirement [53]. As stressed by Otsu [52], when the histogram valley is flat and broad, it is difficult to detect the threshold with precision. This phenomenon frequently occurs with real pictures and remote sensing imagery.

Generally, as eucalypt stands are aligned and understory vegetation is managed, tree and bare soil reflectance can be easily discriminated. Hence, by splitting soil and vegetation, gaps of eucalypt trees would be more visible. In this study, this method was applied in vegetation indices (VIs)to separate healthy and dead eucalypt. Using the Scikit-image library [55] in Python (Version 3.6.9) [56], the histogram threshold values were calculated.

2.4.2. Local Maxima of a Sliding Window

The extraction of tree location can be performed on high-resolution imagery by applying local maxima filter, as stressed by Wulder et al. [57–59] and Wang et al. [60]. Therefore, this algorithm enables the estimation of tree density, volume, basal area, and other important parameters used in forest management and planning [57]. The case of pest management is of special interest since the algorithm can be combined with other techniques in order to determine how many trees are infested or dead. To find the number of trees in the multispectral image, the QGIS Tree density plugin (Ghent University, Lieven P.C. Verbeke, Belgium) [61] was used on a brightness image calculated through the mean of the four bands. According to Crabbé et al. [61] this algorithm uses a sliding window to move over the image. The pixel is marked as a local maximum when the central pixel is the brightest of the window. From Figure 5, highlighting of local maxima filtering using a sliding window can be observed

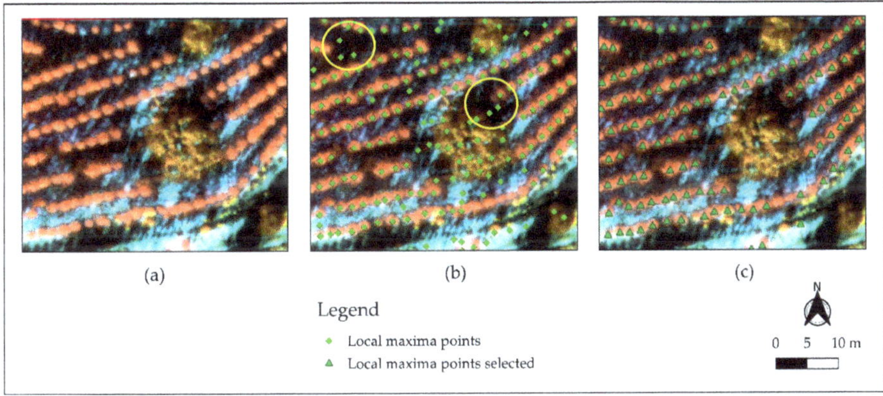

Figure 5. Highlighting local maxima filter performed: (**a**) false color composite (near infrared (NIR), red (R), green (G)), (**b**) local maxima points, and the yellow circle indicating false positives, (**c**) Local maxima points selected by visual interpretation.

In the next step, all false positives (Figure 5b) were removed through the interpretation of the false-color composite image (Figure 5c). The position of the trees will be used to estimate the number of individual dead and healthy trees. Finally, local maxima points selected will be also used to extract the crowns from the segmented image using spatial tools.

2.5 Object-Based Analysis and Classification

Geographic object-based image analysis (GEOBIA), as an extension of object-based analysis, consists on dividing imagery into different meaningful image-objects that group neighboring pixels with similar characteristics or semantic meanings such as spectral, shape (geometric), color, intensity, texture, or contextual measures [62–64]. As a result of the GEOBIA procedure, a vector file with several polygons or segments with similar properties is obtained instead of individual pixels. This approach is convenient for individual tree classification [65,66], reducing the intra-class spectral variability caused by crown textures, gaps, and shadows [67,68]. On the other hand, the main limitation of this method is related to the over- and under-segmentation, which affect the subsequent classification process.

These errors occur when the generated segments do not represent the real shape and area of the image objects [69].

Image segmentation was performed by applying the LSMS algorithm with four spectral bands stacked. The LSMS algorithm, developed by Fukunaga and Hostetler in 1975, is a non-parametric and iterative process that groups pixels with similar meanings [70] by using the radiometric mean and variance of each band [71]. The selection of the LSMS algorithm was motivated by three main reasons: (1) LSMS can be used by the ORFEO ToolBox (OTB) (CNES, Paris, France), which is an easy to use open-source project for remote-sensing imagery processing and analyses [72]; (2) LSMS has been especially developed to be applied in large very high resolution (VHR) images processing [70]; and (3) previous work has shown good results when using UAV imagery [68].

The segmentation technique is not a fully automatized process since its procedure may not create the desired "meaningful" segments. Hence, segmentation parameters are commonly "optimized" to achieve the intended entities [63]. For this aim, different segmentation parameters were tested using OTB version 7.1.0 and evaluated by visual observation. The chosen parameters were spatial radius of 5 (default value), range radius of 10, and minimum segment size of 30. Subsequently the means of bands and VI values in each segment were addressed. From Figure 6, visual comparison of the segmentation parameters tested can be ascertained.

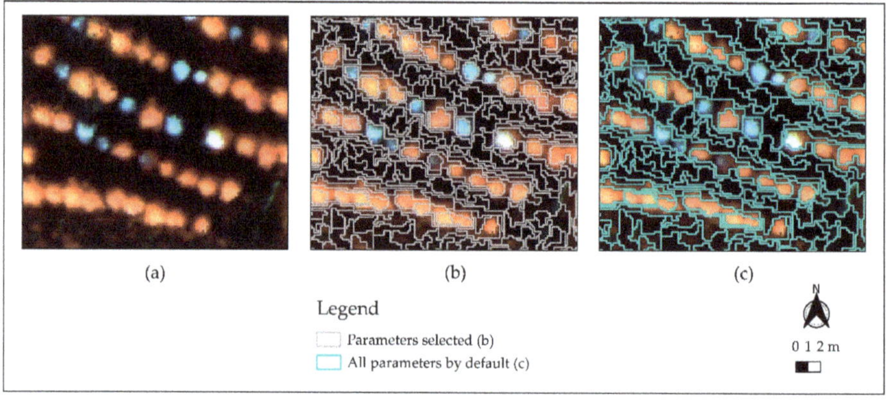

Figure 6. Highlighting some different segmentation parameters: (**a**) False color composite (near infrared (NIR), red (R), green (G), (**b**) segmentation with selected parameters, (**c**) segmentation with all parameters by default.

Given the most suitable segmentation result, one more step was undertaken to improve the segments that correspond to tree canopies. This optimization consisted of obtaining tree position by applying the local maxima algorithm included in the Tree density plugin of QGIS [61]. By using this tool, bare soil and shadows were eliminated, as they might interfere with the identification of dead trees, which had mostly already lost their leaves.

In order to classify tree canopies into two different classes, healthy and dead trees, supervised machine-learning (ML) classification was used. Among the different object-oriented ML classifiers, the RF algorithm was applied as its performance is one of the most accurate ML algorithms when supervised classification for GEOBIA is conducted [27,28,68,73]. Presented by Breiman [74], RF is an automatic ensemble method based on decision trees where each tree depends on a collection of random variables [75]. RF consists of a large number of independent individual decision trees working together and the final output is estimated based on the outputs of all decision trees involved. Thereby, the return classification is the one that has been the most recurrent. Random selection of the variables used in each decision tree is a keystone to avoid correlation among decision trees. For this purpose, the so-called bootstrap is performed and a significant percentage of samples of the training areas with replacements are used to construct each decision tree. The remaining training areas, called out-of-bag (OOB) data, are used to validate the classification accuracy of RF [74,76]. Therefore, all the decision trees are always constructed using the same parameters but on different training sets.

To supervise RF classification, 1241 out of 25,911 segments were manually selected as training areas based on field data and on-screen interpretation of different image color composites. A stratified division of the training areas was carried out according to the preset classes being divided as 80% for training and 20% for the model validation. As shown in previous studies [68,77], default values were set in OTB when RF was performed, since OTB parameters for training and classification processes worked optimally.

The summary of the performance of the machine-learning RF training model can be observed in Table 3. For each canopy class, high precision and global model performance were obtained, as shown by the Kappa value of 0.96.

Table 3. Summary of the performance of machine-learning random forest training model.

Tree Status	Precision (%)	Recall (%)	F-Score	Kappa Value
Dead	100.0	94.4	97.1	0.96
Healthy	98.3	100.0	99.1	

2.6. Accuracy Assessment

Since it is a very effective way to represent both global accuracy and individual class accuracy, an error matrix was performed to determine the agreement between the classified image and the ground truth data with regard to tree health status [78]. Field data collection through stratified random sampling was conducted for validation purposes. From the error matrix, the overall accuracy, commission error (CE), omission error (OE), producer's accuracy (PA), and user's accuracy (UA) were established. CE describes the error associated with the aerial surveyors that classify objects into a different category from the one they belong to, whereas OE represents the error associated with the aerial surveyors that place objects outside of the category that they belong to [79]. In addition, the Kappa (κ) statistic was estimated to quantitatively assess the agreement in the error matrix and the chance of an agreement, by determining the degree of association between the remotely sensed classification and the reference data [78,80].

The accuracy assessment of the classifications obtained with Otsu's method was validated with the data collected in the field. The supervised classification with the random forest was validated with 2% (517 samples) of randomly stratified selected segments using the research tools provided by QGIS [48].

2.7. Density Forest Maps

The results of the classification were joined to the tree's position layer due to the slightly touching trees phenomenon described by Wang et al. [60]. Counting the number of trees using the classification results could induce error because the tree crows are not always clearly separated.

Density maps were created by grouping the position of the trees with the hexagonal binning technique for 0.1 ha. This strategy serves to minimize the irregular shape of the stand limits and to group the trees into less than discrete groups. More importantly, hexagons are more similar to the circle than squares or diamonds [81]. This technique was first described by Carr et al. [82]. According to Carr [83] maps based on hexagon tessellations offer the opportunity to summarize point data and show the patterns for a single variable, that in this case was made into two types of the canopy.

3. Results

3.1. Spectra Details and Vegetation Indexes

Spectra details of tree canopies are shown in Figure 7a, with the mean reflectance values for each band. NIR spectral region revealed the largest differences between the two tree classes, and therefore the highest discriminating power when compared with red-edge and other bands. As shown in Figure 7b, all indices have discriminative capability, although the NDRE index showed the smallest interval difference.

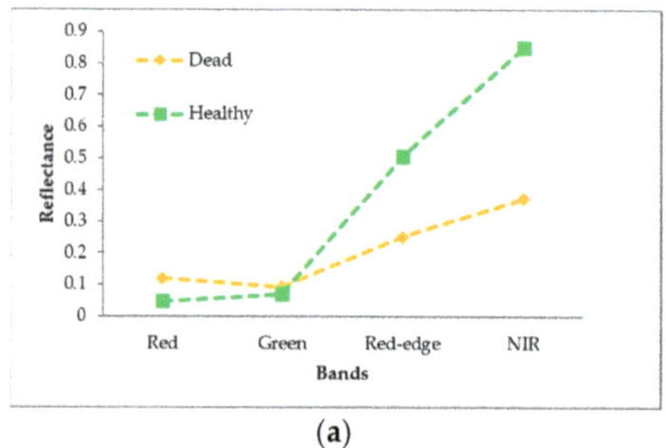
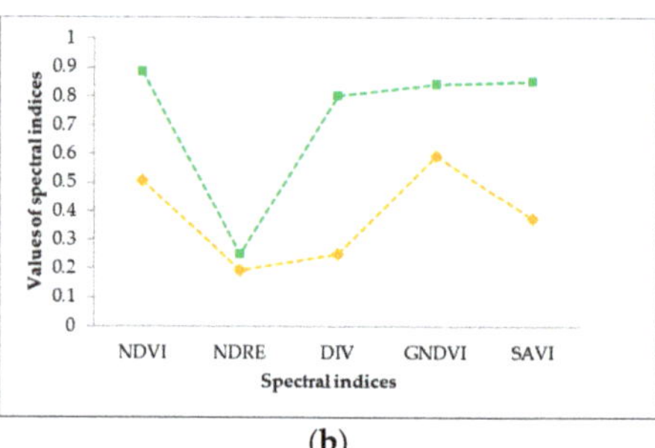

(a)   (b)

Figure 7. Spectral signatures details of all tree canopies: (**a**) extracted mean values of bands, (**b**) extracted mean values of vegetation indices (VIs) selected.

3.2. Pixel-Based Analysis

Otsu Thresholding Analysis

Imagery interval values and histogram threshold values obtained from the Otsu thresholding method, as shown in Table 4 and Figure A1 (Appendix A). The DVI index imagery interval ranged between 0.020 and 1.150, setting the threshold, value at 0.394. The GNDVI and NDVI indices showed a similar histogram threshold with global minima of 0.703 and 0.712, respectively. Their imagery intervals ranged between 0.150 and 0.930 for the NDVI index and between 0.248 and 0.890 for the GNDVI index. The other two indexes, NDRE and SAVI, had imagery intervals between −0.339 and 0.660 and between 0.057 and 0.983, respectively, whereas their threshold values were 0.215 and 0.526, respectively.

Table 4. Summary of threshold values based on the Otsu method.

Vegetation Indices (VI)	Imagery Interval Values	Histogram Threshold
DVI	0.020–1.150	0.394
NDVI	0.150–0.930	0.712
GNDVI	0.248–0.890	0.703
NDRE	−0.339–0.660	0.215
SAVI	0.057–0.983	0.526

Figure 8 highlights the threshold values of all spectral indices by applying the Otsu thresholding method. If compared with the false-color composite in Figure 8a, where red corresponds to healthy trees and blue-grey to dead trees mixed with bare soil, all indices showed good performance to discriminate between healthy and dead trees. Hence, the application of Otsu methodology seemed to improve the differentiation between classes, at least in planted stands where there is active silvicultural management.

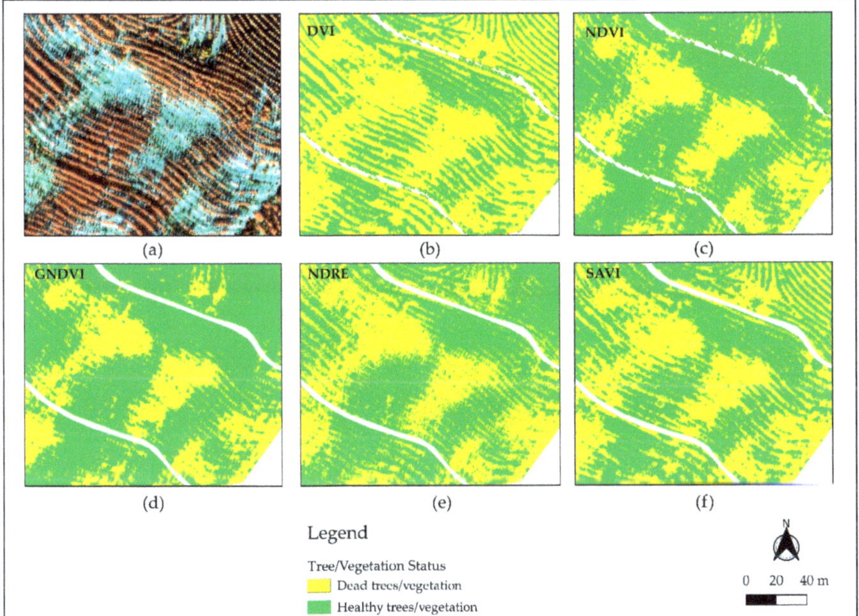

Figure 8. Contrast between healthy and dead trees for all spectral indices by applying the Otsu thresholding method: (**a**) highlighting of false-color composite, (**b**) DVI highlighting dead and healthy trees, (**c**) NDVI highlighting dead and healthy trees, (**d**) GNDVI highlighting dead and healthy trees, (**e**) NDRE highlighting dead and healthy trees, (**f**) SAVI highlighting dead and healthy trees.

3.3. Object-Based Analysis and Classification

Discrimination of tree canopy status was performed through the classification of crown segments extracted from the points generated with local maxima filtering by applying a machine-learning random forest algorithm (Figure 9). The spectral indices and band information of each segment allowed the identification of tree canopy status (Figure 9b). Large-scale mean-shift segmentation algorithm performed through OTB had good performance, improving the ability to obtain meaningful segments, which in turn resulted in improvements in RF classification capability (Figure 9c).

3.4. Accuracy Assessment

3.4.1. Otsu Thresholding

Histogram thresholding analysis and RF classifier were validated by estimating the confusion matrix and Kappa statistic (Tables 5 and 6). The confusion matrix to classify tree status with NDVI histogram analysis had the highest overall accuracy (98.2%) and Kappa value (0.96). On the other hand, NDRE histogram analysis had the lowest performance, with 86.4% overall accuracy and a Kappa value of 0.73. The other spectral indices histogram analyses had a similar performance.

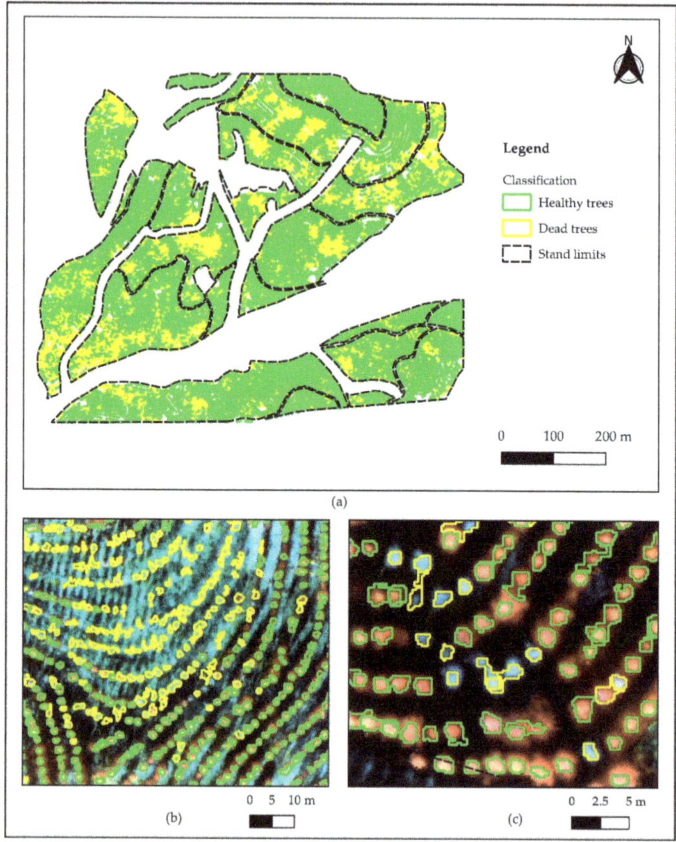

Figure 9. Discrimination of tree canopy status: (**a**) general overview of random forest (RF) classification, (**b**) highlighting dead and healthy trees, (**c**) highlighting in detail dead trees in yellow and healthy trees in green.

Table 5. Summary of the confusion matrix and Kappa statistics (K) of selected vegetation indices, validated with field data collection.

Spectral Indices	Tree Status	User's Accuracy (%)	Producer's Accuracy (%)	Overall Accuracy (%)	Kappa Value
NDVI	Dead	100.00	96.74	98.2	0.96
	Healthy	96.13	100.00		
NDRE	Dead	95.51	79.07	86.4	0.73
	Healthy	78.67	95.40		
DIV	Dead	99.51	95.35	97.2	0.94
	Healthy	94.54	99.43		
GNDVI	Dead	100.00	95.35	97.4	0.95
	Healthy	94.57	100.0		
SAVI	Dead	99.50	93.49	96.1	0.92
	Healthy	92.51	99.43		

Table 6. Confusion matrix of the RF machine learning classification model, with 517 observations selected by stratified random sampling.

Class		Predicted			
		Healthy	Dead	Total	Producer's Accuracy (%)
	Healthy	430	7	437	98.4%
Observed	Dead	1	79	79	98.8%
	Total	431	86	517	-
	User's Accuracy	99.8%	91.9%	-	98.5%

3.4.2. Object-Based Analysis and Classification

To perform the error matrix for machine-learning RF, 517 predicted observations were used, 431 for healthy trees and 86 for dead trees. Predicted observations were selected by stratified random sampling using QGIS tools (Table 6). Overall accuracy was 98.5%, which means that this model had the highest accuracy of all the models tested, while the Kappa value was 0.94.

3.5. Density Forest Maps

After RF classification, tree canopy condition data (healthy and dead) was validated and two different outputs were created by applying a hexagonal binning. Both products are representations of the classified data in order to support management decisions. Prior to the density forest map creation, the data was already validated so as no new data was generated, the validation procedures may not be required.

Hexagonal binning was created to allow the aggregation of trees in small areas and, therefore, identify hotspots and density of healthy trees (Figure 10). A total of 422 hexagons were produced with an area of 0.1 ha each, which represents one-tenth of the hectare which is the standard measure. The hexagon with the highest number of healthy trees had 130, while there was only one hexagon where all the trees were dead. Considering the total number of trees classified, 16 hexagons did not show any dead tree whereas a total of 18 hexagons had more than 40 dead trees (Figure 11). On average, each hexagon was expected to have an initial density of 125 trees, since planting density per hectare was 1250 trees.

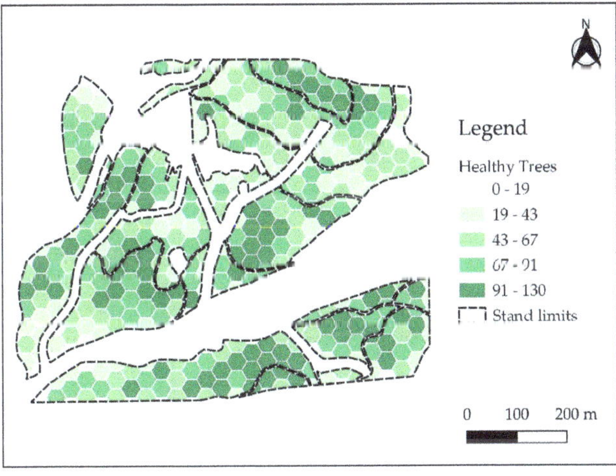

Figure 10. The number of healthy trees in each hexagon (natural breaks classification).

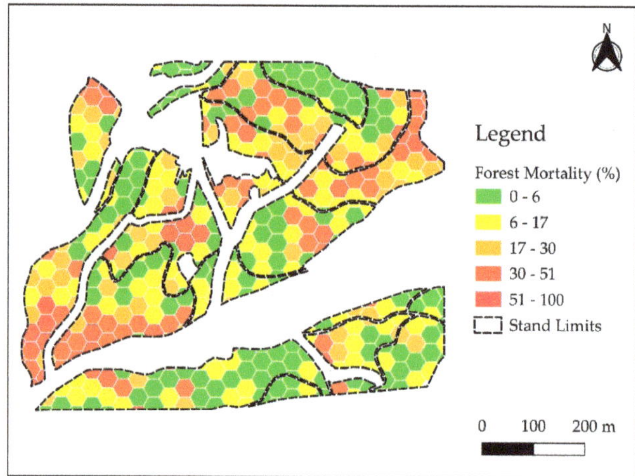

Figure 11. Percentage of mortality in each hexagon (natural breaks classification).

The total number of trees recorded in the studied stand was 28,271 of which 4507 were dead. Since total stand area is 30.56 ha, initial stocking would be about 38,000 trees at plantation, representing a 25% difference. This discrepancy may be due the leafless canopies or missing trees from early mortality events, which are common shortly after establishment.

Forest mortality maps for the plantation area (Figure 11) allow the identification of the most critical spots and are important in the planning of dedicated surveying or sanitary fellings to remove infested trees thus reducing ELB local populations and minimizing future outbreaks.

4. Discussion

This study explored the capability of multispectral images captured with a parrot sequoia camera to detect the mortality caused by ELB on *E. globulus* plantations. Before establishing a flight plan, the phenology of the trees and the life cycle of the insect must be known, in order to extract as much information as possible. The spatial resolution of the images obtained was 17 cm, which allowed the reduction of spatial scale to the individual tree level.

NIR bands had the greatest discriminating power of the all spectra analyzed and provided more information about different tree canopies. As stressed by Otsu et al. [26], NIR band was more sensitive to defoliation than the red-edge band, as observed in Figure 7b. Regarding the remaining bands, the reflectance of the green band was generally higher than that of red band because of the high absorption related to chlorophyll content [84]. Spectral indices, whose calculation includes bands such as NDVI, also presented highly discriminating behavior. This VI is particularly used to assess defoliation, and damage at high spatial resolutions using multispectral and RGB cameras mounted on UAV [12,22,23,26,85,86].

Histogram analysis allowed the discrimination between healthy and dead trees. Although the detection of the histogram valley in spectral indices was difficult to determine, this method may become a valid alternative to split the two canopy types. NDVI was the most accurate index at 98.2% and Kappa value of 0.96. The second most accurate was the GNDVI index, which had an overall accuracy of 97.4% and Kappa value of 0.95. NDRE was the less accurate index at 84.4% overall accuracy and Kappa value of 0.73. Similar overall accuracy values were obtained by Otsu et al. [26] to detect defoliation of pine needles by pine processionary in four different locations. However, in our study, the NIR band was not used to remove shadows as Miura and Midokawa et al. [87] applied. To minimize the shadow effect, the flight was undertaken at solar noon, so treetops could be easily captured, with shadows present mostly along the plantation lines.

The combination of segmentation and the maximum filtering location allowed the extraction of the tops of trees and carrying out their classification. Applying this method, shadows were removed from other vegetation and bare soil, which has a reflectance very close to that of dead trees. The RF learning machine obtained 98.5% global accuracy and the Kappa value was 0.94. This precision is explained by the great distance between reflectance values and the differences between the two classes. Iordache et al. [28] applied RF classifier on the classification of *Pinus pinaster* canopy types (infected, suspicious, and healthy) affected by pine wild and obtained an overall accuracy of 95%. Pourazar et al. [27], obtain as overall accuracy 95.58% using five spectral bands and five indices to detect dead and diseased trees.

The forest density maps produced through hexagon tessellations aimed to group the position of classified trees. The ease of reading allows the identification of the most critical areas with tree mortality and to extract important metrics for forest management such as the total number of trees, the standard deviation, and other landscape metrics. Barreto et al. [88] set a hexagonal grid to represent classes of natural Cerrado vegetation in the the Northeast of Brazil, in an area of about 25,590 km^2, to study the remaining habitats through quantitative indices. More recently, Amaral et al. [89], used a hexagonal grid subdivided into 1000 ha units to study the *restinga* (a type of coastal vegetation) in Northern Brazil.

The current strategy to control ELB relies mostly on the identification of dead trees and their removal from plantations before a new generation of adults emerges in late spring. Traditional field surveys are extremely labor-intensive, as they require visual assessment of large eucalypt plantations. As a result, outbreaks are often only detected when large numbers of trees are already dead and ELB populations are high. By allowing the identification of individual infested trees in a much more efficient way, widespread application of the methods described in this study will allow forest managers to detect ELB attacks at an early stage, thus reducing the cost and efficiency of sanitary fellings.

Future work will focus on adding additional classes to the survey in order to improve the discrimination of tree health status. The integration of a UAS-derived canopy model at a 3D tree model could be performed to automatically outline individual tree crowns. Another improvement is the ambition to implement periodical flights at different attack stages to provide a multitemporal analysis. Further research might focus on other remote-sensing platforms at different scales and considering different bioclimatic and geographical settings.

5. Conclusions

In this experimental research, ELB attack detection and assessment were proposed at pixel and object-based level using UAV multispectral imagery. The NIR band showed discriminative power for two canopy classes. In general, all selected spectral indices were effective in discriminating healthy from dead trees. Otsu thresholding analysis resulted in high overall accuracy and Kappa value for most of the vegetation indices used, with the exception of NDRE. The RF machine-learning algorithm applied also resulted in high overall accuracy (98.5%) and Kappa value (0.94). From an operational point of view, these results can have important implications for forest managers and practitioners. In particular, forest mortality and density maps using hexagonal tessellations allow a complete survey and an accurate measure of the scale of the problem and where the most critical hotspots of mortality are located. By carefully planning the time of the year when these surveys are carried out, timely quantification of mortality rates is an important metric for ensuring phytosanitary fellings are conducted as early and as thoroughly as possible.

Author Contributions: Conceptualization, A.D. and C.V.; methodology, A.D., L.A.-M. and C.I.G.; software, A.D. and A.S.; validation, A.D.; formal analysis, A.D. and L.A.-M.; data curation, A.D.; writing—original draft preparation, A.D.; writing—review and editing, A.D., L.A.-M., C.I.G., L.M., A.S., M.S., S.F., N.B. and C.V.; supervision, C.V.; project administration, M.S., S.F., N.B.; funding acquisition, M.S. All authors have read and agreed to the published version of the manuscript.

Funding: This study has been developed within the context of MySustainableForest project which has received funding from the European Union's Horizon 2020 research and innovation programme under grant agreement No 776045.

Remote Sens. **2020**, *12*, 3153

Acknowledgments: The authors would like to thank Fernando Luis Arede for his support in field data collection, João Rocha for English language check, João Gaspar for their insights, and José Vasques for providing operational data and access to the study area.

Conflicts of Interest: The authors declare no conflict of interest.

Appendix A. Histograms of Otsu Thresholding Analysis

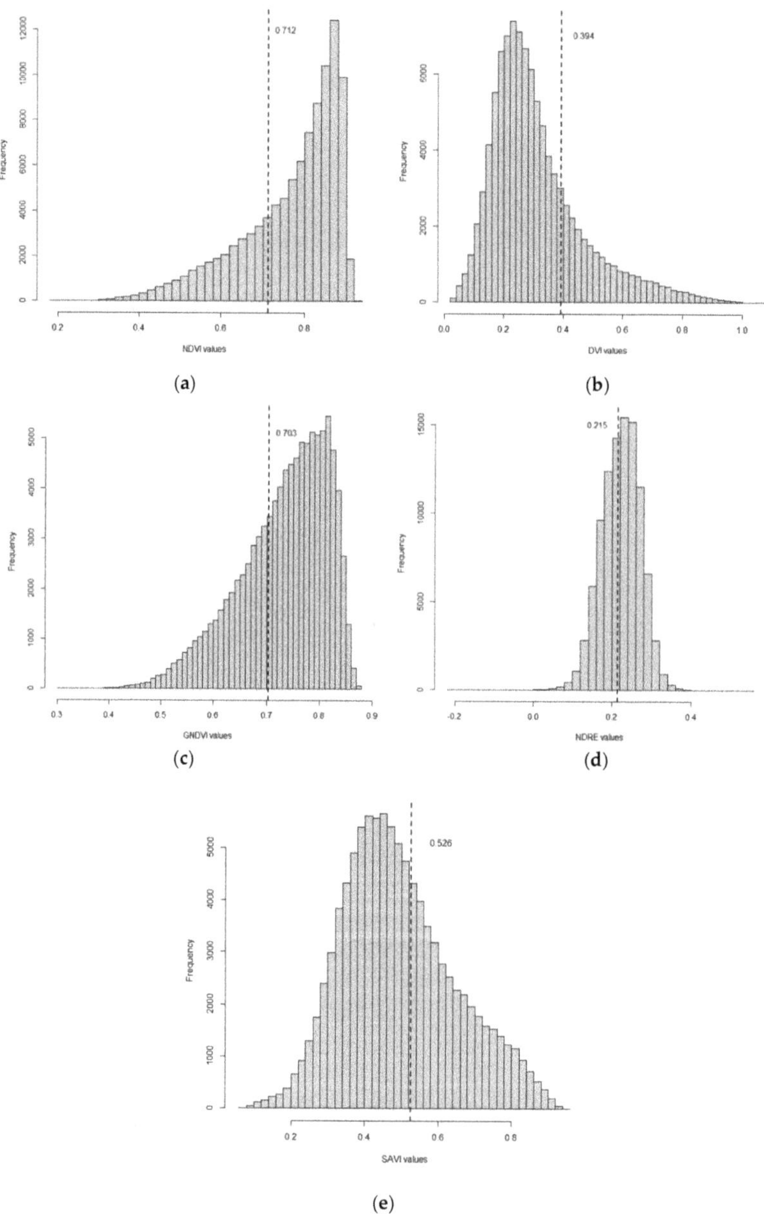

Figure A1. Histogram and Otsu thresholding values of (**a**) NDVI, (**b**) DVI, (**c**) GNDVI, (**d**) NDRE and (**e**) SAVI.

References

1. Rassati, D.; Lieutier, F.; Faccoli, M. Alien Wood-Boring Beetles in Mediterranean Regions. In *Insects and Diseases of Mediterranean Forest Systems*; Paine, T.D., Lieutier, F., Eds.; Springer International Publishing: Cham, Switzerland, 2016; pp. 293–327.
2. Estay, S. Invasive insects in the Mediterranean forests of Chile. In *Insects and Diseases of Mediterranean Forest Systems*; Paine, T.D., Lieutier, F., Eds.; Springer: Cham, Switzerland, 2016; pp. 379–396.
3. Hanks, L.M.; Paine, T.D.; Millar, J.C.; Hom, J.L. Variation among Eucalyptus species in resistance to eucalyptus longhorned borer in Southern California. *Entomol. Exp. Appl.* **1995**, *74*, 185–194. [CrossRef]
4. Mendel, Z. Seasonal development of the eucalypt borer Phoracantha semipunctata, in Israel. *Phytoparasitica* **1985**, *13*, 85–93. [CrossRef]
5. Hanks, L.M.; Paine, T.D.; Millar, J.C. Mechanisms of resistance in Eucalyptus against larvae of the Eucalyptus Longhorned Borer (Coleoptera: Cerambycidae). *Environ. Entomol.* **1991**, *20*, 1583–1588. [CrossRef]
6. Paine, T.D.; Millar, J.C. Insect pests of eucalypts in California: Implications of managing invasive species. *Bull. Entomol. Res.* **2002**, *92*, 147–151. [CrossRef] [PubMed]
7. Caldeira, M.C.; Fernandéz, V.; Tomé, J.; Pereira, J.S. Positive effect of drought on longicorn borer larval survival and growth on eucalyptus trunks. *Annu. For. Sci.* **2002**, *59*, 99–106. [CrossRef]
8. Soria, F.; Borralho, N.M.G. The genetics of resistance to Phoracantha semipunctata attack in Eucalyptus globulus in Spain. *Silvae Genet.* **1997**, *46*, 365–369.
9. Seaton, S.; Matusick, G.; Ruthrof, K.X.; Hardy, G.E.J. Outbreaks of Phoracantha semipunctata in response to severe drought in a Mediterranean Eucalyptus forest. *Forests* **2015**, *6*, 3868–3881. [CrossRef]
10. Tirado, L.G. Phoracantha semipunctata dans le Sud-Ouest espagnol: Lutte et dégâts. *Bull. EPPO* **1986**, *16*, 289–292. [CrossRef]
11. Wotherspoon, K.; Wardlaw, T.; Bashford, R.; Lawson, S. Relationships between annual rainfall, damage symptoms and insect borer populations in midrotation *Eucalyptus nitens* and *Eucalyptus globulus* plantations in Tasmania: Can static traps be used as an early warning device? *Aust. For.* **2014**, *77*, 15–24. [CrossRef]
12. Dash, J.P.; Watt, M.S.; Pearse, G.D.; Heaphy, M.; Dungey, H.S. Assessing very high resolution UAV imagery for monitoring forest health during a simulated disease outbreak. *ISPRS J. Photogramm. Remote Sens.* **2017**, *131*, 1–14. [CrossRef]
13. Ortiz, S.M.; Breidenbach, J.; Kändler, G. Early detection of bark beetle green attack using Terrasar-X and RapidEye data. *Remote Sens.* **2013**, *5*, 1912–1931. [CrossRef]
14. Wulder, M.A.; Dymond, C.C.; White, J.C.; Leckie, D.G.; Carroll, A.L. Surveying mountain pine beetle damage of forests: A review of remote sensing opportunities. *For. Ecol. Manag.* **2006**, *221*, 27–41. [CrossRef]
15. Hall, R.; Castilla, G.; White, J.C.; Cooke, B.; Skakun, R. Remote sensing of forest pest damage: A review and lessons learned from a Canadian perspective. *Can. Èntomol.* **2016**, *148*, 296–356. [CrossRef]
16. Senf, C.; Seidl, R.; Hostert, P. Remote sensing of forest insect disturbances: Current state and future directions. *Int. J. Appl. Earth Obs. Geoinf.* **2017**, *60*, 49–60. [CrossRef]
17. Gómez, C.; Alejandro, P.; Hermosilla, T.; Montes, F.; Pascual, C.; Ruiz, L.A.; Álvarez-Taboada, F.; Tanase, M.; Valbuena, R. Remote sensing for the Spanish forests in the 21st century: A review of advances, needs, and opportunities. *For. Syst.* **2019**, *28*, 1. [CrossRef]
18. Meddens, A.J.H.; Hicke, J.A.; Vierling, L.A.; Hudak, A.T. Evaluating methods to detect bark beetle-caused tree mortality using single-date and multi-date Landsat imagery. *Remote Sens. Environ.* **2013**, *132*, 49–58. [CrossRef]
19. Waser, L.T.; Küchler, M.; Jütte, K.; Stampfer, T. Evaluating the Potential of WorldView-2 Data to Classify Tree Species and Different Levels of Ash Mortality. *Remote Sens.* **2014**, *6*, 4515–4545. [CrossRef]
20. Oumar, Z.; Mutanga, O. Using WorldView-2 bands and indices to predict bronze bug (Thaumastocoris peregrinus) damage in plantation forests. *Int. J. Remote Sens.* **2013**, *34*, 2236–2249. [CrossRef]
21. Stone, C.; Coops, N.C. Assessment and monitoring of damage from insects in Australian eucalypt forests and commercial plantations. *Aust. J. Entomol.* **2004**, *43*, 283–292. [CrossRef]
22. Lehmann, J.R.K.; Nieberding, F.; Prinz, T.; Knoth, C. Analysis of Unmanned Aerial System–Based CIR Images in Forestry—A New Perspective to Monitor Pest Infestation Levels. *Forests* **2015**, *6*, 594–612. [CrossRef]

23. Näsi, R.; Honkavaara, E.; Lyytikäinen-Saarenmaa, P.; Blomqvist, M.; Litkey, P.; Hakala, T.; Viljanen, N.; Kantola, T.; Tanhuanpää, T.; Holopainen, M. Using UAV–Based Photogrammetry and Hyperspectral Imaging for Mapping Bark Beetle Damage at Tree–Level. *Remote Sens.* **2015**, *7*, 15467–15493. [CrossRef]
24. Smigaj, M.; Gaulton, R.; Barr, S.L.; Suárez, J.C. Uav–Borne Thermal Imaging for Forest Health Monitoring: Detection of Disease–Induced Canopy Temperature Increase. *ISPRS Int. Arch. Photogramm. Remote Sens. Spat. Inf. Sci.* **2015**, *40*, 349–354. [CrossRef]
25. Otsu, K.; Pla, M.; Vayreda, J.; Brotons, L. Calibrating the Severity of Forest Defoliation by Pine Processionary Moth with Landsat and UAV Imagery. *Sensors* **2018**, *18*, 3278. [CrossRef] [PubMed]
26. Otsu, K.; Pla, M.; Duane, A.; Cardil, A.; Brotons, L. Estimating the threshold of detection on tree crown defoliation using vegetation indices from UAS multispectral imagery. *Drones* **2019**, *3*, 80. [CrossRef]
27. Pourazar, H.; Samadzadegan, F.; Javan, F. Aerial multispectral imagery for plant disease detection: Radiometric calibration necessity assessment. *Eur. J. Remote Sens.* **2019**, *52* (Suppl. S3), 17–31. [CrossRef]
28. Iordache, M.D.; Mantas, V.; Baltazar, E.; Pauly, K.; Lewyckyj, N. A Machine Learning Approach to Detecting Pine Wilt Disease Using Airborne Spectral Imagery. *Remote Sens.* **2020**, *12*, 2280. [CrossRef]
29. Pádua, L.; Vanko, J.; Hruška, J.; Adão, T.; Sousa, J.J.; Peres, E.; Morais, R. UAS, sensors, and data processing in agroforestry: A review towards practical applications. *Int. J. Remote Sens.* **2017**, *38*, 2349–2391. [CrossRef]
30. Tang, L.; Shao, G. Drone remote sensing for forestry research and practices. *J. For. Res.* **2015**, *26*, 791–797. [CrossRef]
31. FAO. FAO–UNESCO Soil Map of the World. Revised Legend. In *World Soil Resources Report*; FAO: Rome, Italy, 1998; Volume 60.
32. SenseFly Parrot Group. Parrot Sequoia Multispectral Camara. Available online: https://www.sensefly.com/camera/parrot-sequoia/ (accessed on 31 August 2020).
33. Assmann, J.J.; Kerby, J.T.; Cunlie, A.M.; Myers-Smith, I.H. Vegetation monitoring using multispectral. sensors—Best practices and lessons learned from high latitudes. *J. Unmanned Veh. Syst.* **2019**, *7*, 54–75. [CrossRef]
34. Pix4D. Pix4D—Drone Mapping Software. Version 4.2. Available online: https://pix4d.com/ (accessed on 31 August 2020).
35. Nobuyuki, K.; Hiroshi, T.; Xiufeng, W.; Rei, S. Crop classification using spectral indices derived from Sentinel-2A imagery. *J. Inf. Telecommun.* **2020**, *4*, 67–90.
36. Cogato, A.; Pagay, V.; Marinello, F.; Meggio, F.; Grace, P.; De Antoni Migliorati, M. Assessing the Feasibility of Using Sentinel-2 Imagery to Quantify the Impact of Heatwaves on Irrigated Vineyards. *Remote Sens.* **2019**, *11*, 2869. [CrossRef]
37. Hawryło, P.; Bednarz, B.; Wezyk, P.; Szostak, M. Estimating defoliation of scots pine stands using machine learning methods and vegetation indices of Sentinel-2. *Eur. J. Remote Sens.* **2018**, *51*, 194–204. [CrossRef]
38. Zarco-Tejada, P.J.; Miller, J.R.; Mohammed, G.H.; Noland, T.L.; Sampson, P.H. Vegetation stress detection through Chlorophyll a + b estimation and fluorescence effects on hyperspectral imagery. *J. Environ. Qual.* **2002**, *31*, 1433–1441. [CrossRef] [PubMed]
39. Barry, K.M.; Stone, C.; Mohammed, C. Crown-scale evaluation of spectral indices for defoliated and discoloured eucalypts. *Int. J. Remote Sens.* **2008**, *29*, 47–69. [CrossRef]
40. Garcia-Ruiz, F.; Sankaran, S.; Maja, J.M.; Lee, S.; Rasmussen, J.; Ehsani, R. Comparison of two aerial imaging platforms for identification of Huanglongbing-infected citrus trees. *Comput. Electron. Agric.* **2013**, *91*, 106–115. [CrossRef]
41. Verbesselt, J.; Robinson, A.; Stone, C.; Culvenor, D. Forecasting tree mortality using change metrics from MODIS satellite data. *For. Ecol. Manag.* **2009**, *258*, 1166–1173. [CrossRef]
42. Datt, B. Remote sensing of chlorophyll a, chlorophyll b, chlorophyll a+ b, and total carotenoid content in eucalyptus leaves. *Remote Sens. Environ.* **1998**, *66*, 111–121. [CrossRef]
43. Datt, B. Visible/near infrared reflectance and chlorophyll content in Eucalyptus leaves. *Int. J. Remote Sens.* **1999**, *20*, 2741–2759. [CrossRef]
44. Deng, X.; Guo, S.; Sun, L.; Chen, J. Identification of Short-Rotation Eucalyptus Plantation at Large Scale Using Multi-Satellite Imageries and Cloud Computing Platform. *Remote Sens.* **2020**, *12*, 2153. [CrossRef]
45. Gitelson, A.A.; Merzlyak, M.N. Signature analysis of leaf reflectance spectra: Algorithm development for remote sensing of chlorophyll. *J. Plant Physiol.* **1996**, *148*, 494–500. [CrossRef]

46. Olsson, P.O.; Jönsson, A.M.; Eklundh, L. A new invasive insect in Sweden—Physokermes inopinatus: Tracing forest damage with satellite based remote sensing. *For. Ecol. Manag.* **2012**, *285*, 29–37. [CrossRef]

47. Huete, A.R. A soil adjusted vegetation index (SAVI). *Remote Sens. Environ.* **1988**, *25*, 295–309. [CrossRef]

48. QGIS Community. QGIS Geographic Information System. Open Source Geospatial Foundation Project. 2020. Available online: https:$\backslash$qgis.org (accessed on 31 August 2020).

49. Richardson, A.J.; Weigand, C.L. Distinguishing vegetation from background information. *Photogramm. Eng. Remote Sens.* **1977**, *43*, 1541–1552.

50. Barnes, E.M.; Clarke, T.R.; Richards, S.E.; Colaizzi, P.D.; Haberland, J.; Kostrzewski, M.; Waller, P.; Choi, C.; Riley, E.; Thompson, T.; et al. Coincident detection of Crop Water Stress, Nitrogen Status and Canopy Density Using Ground-Based Multispectral Data [CD Rom]. In Proceedings of the Fifth International Conference on Precision Agriculture, Bloomington, MN, USA, 16–19 July 2000.

51. Tucker, C.J. Red and photographic infrared linear combinations for monitoring vegetation. *Remote Sens. Environ.* **1979**, *150*, 127–150. [CrossRef]

52. Otsu, N. A Threshold Selection Method from Gray–Level Histograms. *IEEE Trans. Syst. Man Cybern.* **1979**, *9*, 62–66. [CrossRef]

53. Goncalves, H.; Corte-Real, L.; Goncalves, J.A. Automatic image registration through image segmentation and SIFT. *IEEE Trans. Geosci. Remote Sens.* **2011**, *49*, 2589–2600. [CrossRef]

54. Goh, T.Y.; Basah, S.N.; Yazid, H.; Safar, M.J.A.; Saad, F.S.A. Performance analysis of image thresholding: Otsu technique. *Measurement* **2018**, *114*, 298–307. [CrossRef]

55. Van der Walt, S.; Schönberger, J.L.; Nunez-Iglesias, J.; Boulogne, F.; Warner, J.D.; Yager, N.; Gouillart, E.; Yu, T.; Scikit-image Contributors. Scikit-Image: Image processing in Python. *PeerJ* **2014**, *2*, e453. [CrossRef]

56. Van Rossum, G.; De Boer, J. Interactively testing remote servers using the Python programming language. *CWI Q.* **1991**, *4*, 283–303.

57. Wulder, M.; Niemann, K.; Goodenough, D.G.; Wulder, M.A. Local maximum filtering for the extraction of tree locations and basal area from high spatial resolution imagery. *Remote Sens. Environ.* **2000**, *73*, 103–114. [CrossRef]

58. Wulder, M.A.; Niemann, K.O.; Goodenough, D.G. Error reduction methods for local maximum filtering of high spatial resolution imagery for locating trees. *Can. J. Remote Sens.* **2002**, *28*, 621–628. [CrossRef]

59. Wulder, M.A.; White, J.C.; Niemann, K.O.; Nelson, T. Comparison of airborne and satellite high spatial resolution data for the identification of individual trees with local maxima filtering. *Int. J. Remote Sens.* **2004**, *25*, 2225–2232. [CrossRef]

60. Wang, L.; Gong, P.; Biging, G.S. Individual Tree-Crown Delineation and Treetop Detection in High-Spatial-Resolution Aerial Imagery. *Photogramm. Eng. Remote Sens.* **2004**, *70*, 351–357. [CrossRef]

61. Crabbé, A.H.; Cahy, T.; Somers, B.; Verbeke, L.P.; Van Coillie, F. Tree Density Calculator Software Version 1.5.3, QGIS. 2020. Available online: https://bitbucket.org/kul-reseco/localmaxfilter (accessed on 31 August 2020).

62. Chen, G.; Weng, Q.; Hay, G.J.; He, Y. Geographic object-based image analysis (GEOBIA): Emerging trends and future opportunities. *GIScience Remote Sens.* **2018**, *55*, 159–182. [CrossRef]

63. Blaschke, T.; Hay, G.J.; Kelly, M.; Lang, S.; Hofmann, P.; Addink, E.; Queiroz Feitosa, R., Van der Meer, F., Van der Werff, H.; Van Coillie, F.; et al. Geographic Object-Based Image Analysis—Towards a new paradigm. *ISPRS J. Photogramm. Remote Sens.* **2014**, *87*, 180–191. [CrossRef]

64. Holloway, J.; Mengersen, K. Statistical machine learning methods and remote sensing for sustainable development goals: A review. *Remote Sens.* **2018**, *10*, 21. [CrossRef]

65. Murfitt, J.; He, Y.; Yang, J.; Mui, A.; De Mille, K. Ash decline assessment in emerald ash borer infested natural forests using high spatial resolution images. *Remote Sens.* **2016**, *8*, 256. [CrossRef]

66. Myint, S.W.; Gober, P.; Brazel, A.; Grossman-Clarke, S.; Weng, Q. Per-pixel vs. object-based classification of urban land cover extraction using high spatial resolution imagery. *Remote Sens. Environ.* **2011**, *115*, 1145–1161. [CrossRef]

67. Torres-Sánchez, J.; López-Granados, F.; Peña, J.M. An automatic object-based method for optimal thresholding in UAV images: Application for vegetation detection in herbaceous crops. *Comput. Electron. Agric.* **2015**, *114*, 43–52. [CrossRef]

68. De Luca, G.; Silva, J.; Cerasoli, S.; Araújo, J.; Campos, J.; Di Fazio, S.; Modica, G. Object-Based Land Cover Classification of Cork Oak Woodlands using UAV Imagery and Orfeo ToolBox. *Remote Sens.* **2019**, *11*, 1238. [CrossRef]

69. Liu, D.; Xia, F. Assessing object-based classification: Advantages and limitations. *Remote Sens. Lett.* **2010**, *1*, 187–194. [CrossRef]
70. Michel, J.; Youssefi, D.; Grizonnet, M. Stable Mean-Shift Algorithm and Its Application to the Segmentation of Arbitrarily Large Remote Sensing Images. *IEEE Trans. Geosci. Remote Sens.* **2015**, *53*, 952–964. [CrossRef]
71. OTB Development Team. *OTB CookBook Documentation*; CNES: Paris, France, 2018; p. 305.
72. Grizonnet, M.; Michel, J.; Poughon, V.; Inglada, J.; Savinaud, M.; Cresson, R. Orfeo ToolBox: Open source processing of remote sensing images. *Open Geospat. Data Softw. Stand.* **2017**, *2*, 15. [CrossRef]
73. Li, M.; Ma, L.; Blaschke, T.; Cheng, L.; Tiede, D. A systematic comparison of different object-based classification techniques using high spatial resolution imagery in agricultural environments. *Int. J. Appl. Earth Obs. Geoinf.* **2016**, *49*, 87–98. [CrossRef]
74. Breiman, L. Random forests. *Mach. Learn.* **2001**, *45*, 5–32. [CrossRef]
75. Cutler, D.R.; Edwards, T.C.; Beard, K.H.; Cutler, A.; Hess, K.T.; Gibson, J.; Lawler, J.J. Random Forests for Classification in Ecology. *Ecology* **2007**, *88*, 2783–2792. [CrossRef]
76. Feng, Q.; Liu, J.; Gong, J. UAV remote sensing for urban vegetation mapping using random forest and texture analysis. *Remote Sens.* **2015**, *7*, 1074–1094. [CrossRef]
77. Immitzer, M.; Vuolo, F.; Atzberger, C. First experience with Sentinel-2 data for crop and tree species classifications in central Europe. *Remote Sens.* **2016**, *8*, 166. [CrossRef]
78. Congalton, R.G.; Green, K. *Assessing the Accuracy of Remotely Sensed Data: Principles and Practices*; CRC Press; Taylor & Francis Group: Boca Raton, FL, USA, 2019; ISBN 1498776663.
79. Coleman, T.W.; Graves, A.D.; Heath, Z.; Flowers, R.W.; Hanavan, R.P.; Cluck, D.R.; Ryerson, D. Accuracy of aerial detection surveys for mapping insect and disease disturbances in the United States. *For. Ecol. Manag.* **2018**, *430*, 321–336. [CrossRef]
80. Landis, J.; Koch, G. The measurement of observer agreement for categorical data. *Biometrics* **1977**, *33*, 159–174. [CrossRef]
81. Birch, C.; Oom, S.P.; Beecham, J.A. Rectangular and hexagonal grids used for observation, experiment and simulation in ecology. *Ecol. Model.* **2007**, *206*, 347–359. [CrossRef]
82. Carr, D.B.; Littlefield, R.J.; Nicholson, W.L.; Littlefield, J.S. Scatterplot Matrix Techniques for Large N. *J. Am. Stat. Assoc.* **1987**, *82*, 398. [CrossRef]
83. Carr, D.B.; Olsen, A.R.; White, D. Hexagon Mosaic Maps for Display of Univariate and Bivariate Geographical Data. *Cartogr. Geogr. Inf. Sci.* **1992**, *19*, 228–236. [CrossRef]
84. Rullan-Silva, C.D.; Oltho, A.E.; Delgado de la Mata, J.A.; Pajares-Alonso, J.A. Remote Monitoring of Forest Insect Defoliation—A Review. *For. Syst.* **2013**, *22*, 377. [CrossRef]
85. Cardil, A.; Vepakomma, U.; Brotons, L. Assessing Pine Processionary Moth Defoliation Using Unmanned Aerial Systems. *Forests* **2017**, *8*, 402. [CrossRef]
86. Torresan, C.; Berton, A.; Carotenuto, F.; Di Gennaro, S.F.; Gioli, B.; Matese, A.; Miglietta, F.; Vagnoli, C.; Zaldei, A.; Wallace, L. Forestry applications of UAVs in Europe: A review. *Int. J. Remote Sens.* **2017**, *38*, 2427–2447. [CrossRef]
87. Miura, H.; Midorikawa, S. Detection of Slope Failure Areas due to the 2004 Niigata–Ken Chuetsu Earthquake Using High–Resolution Satellite Images and Digital Elevation Model. *J. Jpn. Assoc. Earthq. Eng.* **2013**, *7*, 1–14.
88. Barreto, L.; Ribeiro, M.; Veldkamp, A.; Van Eupen, M.; Kok, K.; Pontes, E. Exploring effective conservation networks based on multi-scale planning unit analysis: A case study of the Balsas sub-basin, Maranhão State, Brazil. *J. Ecol. Indic.* **2010**, *10*, 1055–1063. [CrossRef]
89. Amaral, Y.T.; Santos, E.M.D.; Ribeiro, M.C.; Barreto, L. Landscape structural analysis of the Lençóis Maranhenses national park: Implications for conservation. *J. Nat. Conserv.* **2019**, *51*, 125725. [CrossRef]

 *remote sensing*

Article

Multi-Source Remote Sensing Data Product Analysis: Investigating Anthropogenic and Naturogenic Impacts on Mangroves in Southeast Asia

Anjar Dimara Sakti [1,2,3,*], Adam Irwansyah Fauzi [4], Felia Niwan Wilwatikta [5], Yoki Sepwanto Rajagukguk [2], Sonny Adhitya Sudhana [2], Lissa Fajri Yayusman [2], Luri Nurlaila Syahid [1,2], Tanakorn Sritarapipat [6], Jeark A. Principe [7], Nguyen Thi Quynh Trang [8], Endah Sulistyawati [9], Inggita Utami [10], Candra Wirawan Arief [11] and Ketut Wikantika [1,2]

[1] Remote Sensing and Geographic Information Science Research Group, Faculty of Earth Sciences and Technology, Institut Teknologi Bandung, Bandung 40132, Indonesia; luri.nurlaila@students.itb.ac.id (L.N.S.); ketut@gd.itb.ac.id (K.W.)

[2] Center for Remote Sensing, Institut Teknologi Bandung, Bandung 40132, Indonesia; yokisepwantorg@students.itb.ac.id (Y.S.R.); Sonnyadhitya@students.itb.ac.id (S.A.S.); lissafajri@gmail.com (L.F.Y.)

[3] Center of Sustainable Development Goals, Institut Teknologi Bandung, Bandung 40132, Indonesia

[4] Department of Geomatics Engineering, Faculty of Regional and Infrastructure Technology, Institut Teknologi Sumatera, Lampung 35365, Indonesia; adam.fauzi@gt.itera.ac.id

[5] Esri Indonesia, Jakarta 12560, Indonesia; fnwilwatikta@esriindonesia.co.id

[6] School of Geoinformatics, Suranaree University of Technology, Nakhon Ratchasima 30000, Thailand; tanakorn.s@sut.ac.th

[7] Department of Geodetic Engineering, University of The Philippines Diliman, Diliman, Quezon City 1101, Philippines; japrincipe@up.edu.ph

[8] Department of Remote sensing, GIS and GPS, Vietnam Space Technology Institute, Hanoi 100000, Vietnam; ntqtrang@sti.vast.vn

[9] Forestry Technology Research Group, School of Life Sciences and Technology, Institut Teknologi Bandung, Bandung 40132, Indonesia; endah@sith.itb.ac.id

[10] Biology Department, Faculty of Applied Science and Technology, Ahmad Dahlan University, Yogyakarta 55166, Indonesia; inggitautami@bio.uad.ac.id

[11] Center for Environment and Sustainability Science, Universitas Padjadjaran, Bandung 40132, Indonesia; candra.arief@gmail.com

* Correspondence: anjards@gd.itb.ac.id

Received: 3 June 2020; Accepted: 19 August 2020; Published: 22 August 2020

Abstract: This study investigated the drivers of degradation in Southeast Asian mangroves through multi-source remote sensing data products. The degradation drivers that affect approximately half of this area are unidentified; therefore, naturogenic and anthropogenic impacts on these mangroves were studied. Various global land cover (GLC) products were harmonized and examined to identify major anthropogenic changes affecting mangrove habitats. To investigate the naturogenic factors, the impact of the water balance was evaluated using the Normalized Difference Vegetation Index (NDVI), and evapotranspiration and precipitation data. Vegetation indices' response in deforested mangrove regions depends significantly on the type of drivers. A trend analysis and break point detection of percentage of tree cover (PTC), percentage of non-tree vegetation (PNTV), and percentage of non-vegetation (PNV) datasets can aid in measuring, estimating, and tracing the drivers of change. The assimilation of GLC products suggests that agriculture and fisheries are the predominant drivers of mangrove degradation. The relationship between water balance and degradation shows that naturogenic drivers have a wider impact than anthropogenic drivers, and degradation in particular regions is likely to be a result of the accumulation of various drivers. In large-scale studies, remote sensing data products could be integrated as a remarkably powerful instrument in assisting evidence-based policy making.

Remote Sens. **2020**, *12*, 2720

Keywords: mangrove sustainability; deforestation depletion; anthropogenic; natural water balance; Southeast Asia

---

1. Introduction

Mangroves are woody plants located in the intertidal areas of tropical and subtropical regions. They typically thrive under harsh environmental conditions, such as high salinity, high temperature, extreme tides, and high sedimentation [1,2]. These productive and biologically important ecosystems provide crucial protection for coastal and marine systems, as well as humans, such as preventing abrasion [3], reducing tsunami impacts [4], providing an ecosystem for flora and fauna [5], and acting as a sink for carbon dioxide [6].

In Southeast Asia, mangrove forests cover an area of 63.2 × 105 ha, which constitutes 34.9% of the area of the world's mangrove forests (181.1 × 105 ha), with great species richness and structure [7,8]. The depletion rate of Southeast Asian mangroves from 2000 to 2012 was 4.73%, which amounts to an annual global loss of 0.39% [9]. Because of this depletion, mangrove forests are facing considerable risks [10,11]. This depletion is due to both anthropogenic and natural changes [12]. In terms of anthropogenic drivers, a significant proportion of mangrove loss is caused by the direct destruction of these forests for other land uses. These include overexploitation by coastal communities; conversion to settlements, tourist resorts, and agricultural land for rice, coconut fields, salt bed, industrial activities, and brackish-water aquaculture [13]. Natural drivers such as climate change, accelerated sea-level rise, meteorological phenomena, a changing water balance, and other aspects of global change also affect mangrove forests across the world [14]. Natural drivers of mangrove loss were frequently observed and widely distributed across the globe [15–17]. However, natural drivers constitute a substantial proportion of predicted future losses, primarily due to changes in meteorological phenomena [18,19], which can reduce precipitation levels and increase evapotranspiration [20], affecting the water balance that is vital for healthy mangrove forest growth.

Vegetation, soil, and water are interrelated factors that influence mangrove life. Soil has many functions that are very important for mangrove growth; for instance, each mangrove species requires a different type of soil texture to live [20–23]. Changes in soil characteristics can affect the capacity of mangroves to capture carbon [24,25]. In addition, soil factors can also affect vegetation. For example, changes in nutrient content in the soil can cause competition between mangroves and other vegetation, resulting in changes in vegetation zone [26,27]. Another factor that greatly influences mangrove life is water. Mangroves can only survive in tidal inundation with a range of 0.4–1.27 m [28,29]. Increases in sea level due to climate change will cause mangroves to die and the surrounding area to be no longer suitable for planting mangroves [30,31].

Changes in vegetation, water, and soil are often detected using indices such as the Normalized Difference Vegetation Index (NDVI), the Normalized Difference Water Index (NDWI), and the Soil Adjusted Vegetation Index (SAVI). A study by Gupta et al. [32] used NDVI and NDWI to distinguish mangrove and non-mangrove forests by looking at the greenness index and water content of the vegetation, which were then compared with SAVI and Simple Ratio (SR). Meanwhile, Ahmed et al. [33] detected changes in land cover from mangrove areas to water areas in the southwestern coastal areas of Bangladesh that were flooded using NDVI and NDWI. Another example is the study by Pastor-Guzman, [34], in which the phenology of mangrove forests was investigated with the Enhanced Vegetation Index (EVI), NDVI, the Green Normalized Vegetation Index (GNDVI), and NDWI.

Another developing research area, land use and land cover (LULC) change, has emphasized the generation of global land cover (GLC) products from numerous observation satellites as primary data for both global or national scale studies [35,36]. Presently, these accessible GLC products have been widely utilized through an aggregation in several research topics, including estimating agricultural enlargement into forests [37], monitoring cropland changes [38], and deriving user-specific maps [39].

In the context of mangroves, Hamilton and Casey [9] successfully integrated MFW [40] as the basis of mangrove area, world ecosystem, namely ecoregion layer, and GFC [41], namely annual forest change data, to produce global mangrove forest cover data from 2000 to 2012. This proves that the assimilation of global data products, mainly GLC products processed through robust and standardized methodology, has great potential to be addressed in investigating mangroves drivers over a large region.

Future climate models consistently increased variability in temperature and extreme precipitation [42]. According to Bathiany et al. [43], temperatures increased by 10% C^{-1} in Southeast Asia with the mechanism of soil drying and shifting of atmospheric structures. An increase in temperature will result in an increase in water loss during the evapotranspiration process [44], in turn impacting water availability for plants. According to Maslin and Austin [45], changes in the predicted total annual precipitation are very diverse and difficult to ascertain. However, according to Myhre et al. [46], precipitation will continue to increase almost twofold for any further global warming. An increase in rainfall will cause plants to die [47]. These studies indicate that an increase in temperature and extreme precipitation could be one of the reasons for the occurrence of naturogenic deforestation. Therefore, it is necessary for there to be a study related to water balance in order to determine the effect of climate change on water content in mangrove vegetation.

Remote sensing has been widely used for forest monitoring, and such multi-temporal change detection with high spatial and temporal resolution data can be used to monitor meteorological phenomena that affect water balance [48,49]. An index of mangrove forests derived from satellite imagery can be used to assess coastal impacts over large areas [50]. However, to diminish the degradation of mangrove forests, coastal areas must also be constantly monitored. However, this presents challenges to current remote sensing techniques [51]. Several studies have been conducted to monitor deforestation and degradation in mangrove forests using remote sensing and geographic information system (GIS) analysis [9–11,52]. Hamilton and Casey [9] used several global database maps to estimate the annual global mangrove forest change from 2000 to 2012. Richards and Friess [10] combined remote sensing and GIS techniques to quantify the proximate drivers of mangrove deforestation in Southeast Asia by identifying the dominant land use and land cover area that replaced the areas of deforested mangroves between 2000 and 2012, which they summarized in a one-degree spatial resolution dataset. Those data products were then improved by Fauzi et al. [52], who analyzed environmental and socio-economic data products to identify anthropogenic mangrove forest deforestation drivers in a finer resolution of 10 km grid cells. However, the results of the study show that the drivers in a high percentage of deforested mangrove areas were unidentified. This has raised the question as to whether there are contributing factors to mangrove deforestation other than land conversion, because none of the previous studies used remote sensing data products to identify both anthropogenic and naturogenic drivers of mangrove deforestation in Southeast Asia.

The main objective of this study is to identify mangrove forest areas that have been degraded and depleted due to anthropogenic and naturogenic factors. For the anthropogenic analysis, various data products of global land cover (GLC) were compared and then integrated in order to monitor mangrove deforestation in Southeast Asia. We also determined the spatiotemporal trends of remote sensing vegetation indices over the land use and land cover (LULC) conversion of mangrove forests in Southeast Asia, and characterized the relationship between vegetation indices and mangrove deforestation drivers. For the analysis of natural drivers, the effects of water balance on mangrove forests' degradation and depletion were identified. Mangrove forests' degradation and depletion must also be affected by the water balance of each species of the mangrove itself. Therefore, this research analyzed how the water balance in mangroves affected the degradation of the mangrove forests. Thus, adaptation options to avoid and minimize mangrove depletion can be identified and implemented. To the best of our knowledge, this is the first study to investigate both naturogenic and anthropogenic impacts on mangrove deforestation in Southeast Asia, by integrating the available multi-source remote sensing data products. We also attempted to develop a novel algorithm for the mangrove water balance model. Moreover, the method applied in the consistency assessment step can be adopted to evaluate the accuracy of result involving similar topics. The remainder of this article is presented in four sections detailing the materials and methods, results, discussion, and conclusion. In Section 2, we specified the type, spatial and temporal resolution, source, and characteristics of the data we used to analyze mangroves, land use, land cover, and remote sensing data products. In this section, we described the general framework of this study and detailed the methods used to process the collected data. In Section 3, we presented our findings by showing the MODIS vegetation indices' response for specific mangrove areas, the estimated land cover conversion of mangrove area based on GLC data products integration, and the mangrove coefficient growth, and identified the water balance in the mangrove areas. In Section 4, we reviewed the conformity of data products with Dominant Land Use of Deforested Mangrove Patches (DLUDMP) and The Southeast Asian Mangroves Conversion Types (SEAMCT) and the uncertainties associated with the mangrove variation data. We also present a trend analysis and breakpoint detection in particular deforested mangrove areas and future research directions. In Section 5, we summarized our findings, highlighted the strengths and limitations of this study, and elaborated on areas of further research.

2. Materials and Methods

2.1. Data Used in This Study

2.1.1. Mangrove Data Product and Change

This study used several data products produced from previous studies regarding mangrove distribution and deforestation to provide an improved spatiotemporal analysis (Table 1).

Table 1. Product specifications of mangrove data and their change.

Data Product	Data Class	Spatial Resolution	Available Year	Source
MFW-USGS	Mangrove Distribution	30 m	1997–2000	[40]
CGMFC-21	Mangrove deforestation	30 m	2001, 2005, 2009, and 2012	[9]
DLUDMP	Function of dominant land	1°	2012	[10]
SEAMCT	Mangroves Conversion Types	10 km	2000 and 2012	[52]

The Global Distribution of Mangroves (MFW) [40] is the primary data produced from the Landsat satellite image with a spatial resolution of 30 m obtained from 1997 to 2000, while the secondary data are based on the database of global, national, and local mangrove forests.

The accuracy of this product was validated by dividing the whole area into 500 × 500 grids. Each grid was qualitatively controlled in virtue of the local experts and very high-resolution images, i.e., QuickBird and IKONOS [40]. In addition to the mangrove's maps existence, although some mangroves distribution data are available, i.e., Spalding et al. [53], Saputro et al. [54], and Bunting et al. [55], we decided to only use MFW owing to the year of the existence i.e., 2000, which is more appropriate in performing long-term evaluation. Nonetheless, those mangrove datasets were processed in providing the agreement level analysis.

The Global Database of Continuous Mangrove Forest Cover for the 21st Century (CGMFC-21) [9] was produced by combining the Global Forest Change (GFC) [41], MFW [40], and the Terrestrial Ecosystem of the World (TEOW) [56] databases, as well as other data, to produce annual global mangrove forest cover maps from 2000 to 2012 with a spatial resolution of up to 30 m. The Dominant Land Use of Deforested Mangrove Patches (DLUDMP) [10] data product, compiled in 2012, presents information about the function of the dominant land in the area of deforested mangroves, with a spatial resolution of one degree. The function of dominant land is determined by estimating the largest land function range in the area of deforested mangroves, which is then indicated as the land function causing mangrove conversion. The Southeast Asian Mangroves Conversion Types (SEAMCT) by Fauzi et al. [52] was produced from environmental data products, i.e., MFW, CGMFC-21 [9], MODIS Land Cover Type Product (MCD12Q1) [57], Global MODIS Water Maps Version 6 (MOD44W) [58],

Global Human Settlement (GHS) [59], and History Database of the Global Environment Version 3.2 (HYDE 3.2) [60], and socio-economic data, i.e., Defense Meteorological Satellite Program–Operational Linescan System (DMSP–OLS) [61], Gross Domestic Product (GDP) [62], and Gridded Population of the World Version 4 (GPW) [63], to provide an enhanced mangrove forest deforestation driver map from 2000 to 2012 with a 10 km spatial resolution. This data product presents detailed information on anthropogenic deforestation drivers based on the dominant land cover changes, which were classified as conversion to agriculture, aquaculture, infrastructure, or other human activity, as well as a combination of these factors.

2.1.2. Land Use Land Cover Data Products

To identify land conversion from mangrove forest area to other land cover types, several GLC data products were used (Table 2). These databases provide adequate spatiotemporal resolution produced from various remote sensing data. The European Space Agency (ESA) CCI Land Cover [64] product was compiled from 1992 to 2015 with a spatial resolution of approximately 300 m or lower and produced through combined classification (guided and unguided) of multi-spectral data from MERIS obtained from 2003 to 2012. Such data are processed using guided and unguided classification algorithms combined automatically to obtain different land cover classes. The MODIS Land Cover Type (MCD12Q1) [65] data product was produced using guided classification of Terra and Aqua MODIS to produce annual GLC from 2001 to 2012 with a spatial resolution of 500 m. These data function as early warning indicators of GLC change. The main classes of land cover are vegetation (11 classes), land (3 classes), and non-vegetated land (3 classes). The GlobCover Global data product [66] is produced through an annual mosaic obtained from the MERIS inventory with a resolution of 300 m through the ENVISAT instrument [67]. The first product of GlobCover was produced in 2005 by the ESA in collaboration with international networks, including the Environmental Agency (EEA), United Nations Environment Program, Global Observation of Forest Cover and Land Dynamic (GOFC-GOLD), Joint Research Centre, and International Geosphere-Biosphere Programme (IGBP). The Global Land Cover by National Mapping Organizations (GLCNMO2008) [68,69] was produced through a global mapping project conducted by an international steering committee on global mapping. Version 2 of this dataset was based on 16 days of MODIS data (MCD43A4) from 2008, with a spatial resolution of 500 m due to the two directional reflectance distribution function (BDRF), with an overall accuracy of 77.9%.

Table 2. Product specification of global land cover (GLC) datasets.

Data Product	Data Class	Spatial Resolution	Data Acquisition	Source
ESA CCI Land Cover	Land cover	300 m	2001 and 2012	[64]
MCD12Q1	Land cover	500 m	2001 and 2012	[65]
GlobCover	Land cover	300 m	2005 and 2009	[66]
GLCNMO2008	Land cover	500 m	2008 and 2012	[68,69]

2.1.3. Geophysical and Vegetation Parameter Products from Remotely Sensed Data

In addition to the processed data provided by the previous studies, other remote sensing datasets were used to obtain temporal environmental information. The specifications of the remote sensing datasets used in this study are summarized below in Table 3. The MOD13Q1 Version 6 product [70] provides a vegetation index (VI) value per pixel basis at a 250 m spatial resolution. There are two primary vegetation layers (NDVI and EVI). The MODIS Vegetation Indices (MOD13A1 v06) Version 6 product [70] provides VI values at a per pixel basis at a 500 m spatial resolution. The two primary vegetation layers improved the sensitivity of the MODI3QI and the MOD13A1 datasets over high biomass regions such as the equatorial area of Central Africa, South America, and Southeast Asia [71,72]. This improvement is important because the vegetation index is prone to saturation on high biomass due to the optical signal not being able penetrate the highly dense canopy [73,74]. The MODIS

Global Terrestrial Evapotranspiration (MOD16A2 v05) [75] is a monthly composite dataset produced at a 5 km pixel resolution to estimate the global terrestrial evapotranspiration. The MOD16 global evapotranspiration product can be used to calculate the water and energy balance and the soil water status at a regional scale, providing key information for water resource management. The Climate Hazards Group InfraRed Precipitation with Stations (CHIRPS v02) dataset [76] is a blended product combining pentadal precipitation climatology and quasi-global geostationary thermal infrared satellite observations from the Climate Prediction Center (CPC), and the National Climatic Data Center (NCDC) [77] atmospheric model rainfall fields from the NOAA Climate Forecast System version 2 (CFSv2) [78] and in situ precipitation observations. These data are used because CHIRPS is an in situ station dataset used to create a gridded rainfall time series for trend analysis and seasonal drought monitoring [79]. The MOD44B Version 6 Vegetation Continuous Fields (MOD44B) [80] presents sub-pixel information about the characteristics of worldwide vegetation cover classified into three classes: percentage of tree cover (PTC), percentage of non-tree vegetation (PNTV), and percentage of non-vegetation (PNV), with a 250 m spatial resolution.

Table 3. Product specification of remote sensing datasets.

Data Product	Data Class	Spatial Resolution	Temporal Resolution	Source
MOD13Q1 v006	VI and SR	250 m	2000 and 2012	[70]
MOD13A1 v006	Vegetation Indices	500 m	2002 and 2012	[71]
MOD16A2 v005	Evapotranspiration	500 m	2002 and 2012	[75]
CHIRPS v02	Precipitation	~5.3 km	2002 and 2012	[79]
MOD44B v6	PTC, PNTV, PNV	250 m	2000–2013	[80]

2.2. Methodology

This study was divided into three main parts: the investigation of degradation and deforestation, anthropogenic drivers, and naturogenic drivers, as demonstrated in Figure 1. To investigate anthropogenic drivers, we examined the ESA CCI, GlobCover, MCD12Q1, and GLCNMO data products to estimate major LULC change in mangrove areas. Previous studies revealed LULC changes as the dominant driver of mangroves deforestation [10,11,52]. In this research, we exploited the available GLC products commonly used to identify human disturbances over Southeast Asian mangroves. Although these data products were produced based on reliable procedures, we believe that there is no perfect product. Therefore, we used all the available GLC products to improve the sensitivity of the results that were then presented through the level of agreement. The level of agreement can also help us understand how these data are related to each other. Since each product defines mangrove forests differently and refers to a different classification system, the main challenges in comparing land cover was the reconciliation of the type of land cover class [81,82]. This was done at the beginning of the process, then the obtained results were compared with those of other studies.

When investigating deforestation and degradation, we employed MOD13Q1 and MOD44B to analyze mangrove loss from 2000 to 2012. As mangrove ecosystems are inundated land that consist of vegetation, water, and soil, we explored three types of commonly used vegetation indices in mangrove studies that represent each component: NDVI, NDWI, and SAVI [32,83,84]. The derived and calculated vegetation indices from the 16 day MOD13Q1 data, annual NDVI, NDWI, and SAVI were calculated to highlight the differences of each index over the 12 years in specific regions. Other indices were identified and traced by annual PTC, PNTV, and PNV (MOD44B) datasets from 2000 to 2013, using trend analysis and breakpoint detection [85] to identify the dynamic patterns and the exact years of particular land use expansion. The trendline slope was calculated by the least-squares-based linear regression method, while the breakpoints were identified using the Bai and Perron method [86,87]. Then, we considered using high resolution imagery to validate our findings.

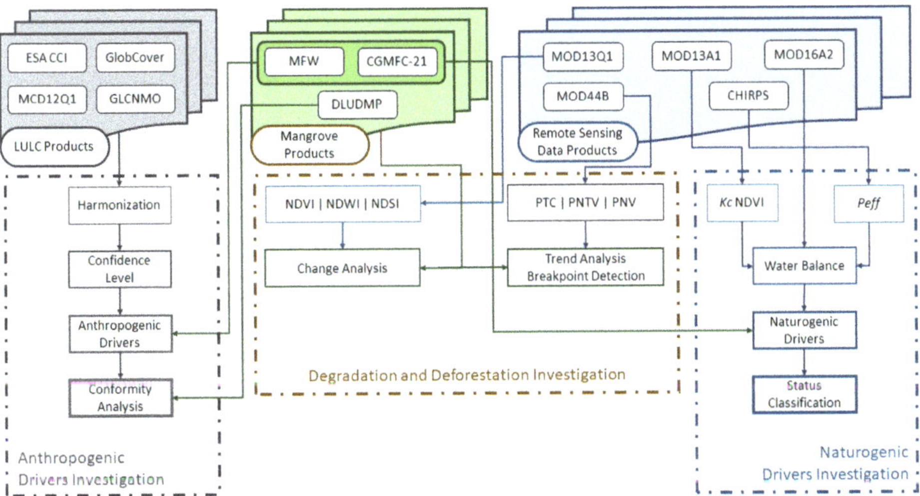

Figure 1. Research framework in this study.

To investigate the naturogenic drivers, we focused on the impact of water balance on the mangrove ecosystem. For data analysis and synthesis, linear regression models that establish the relationship between the vegetation index and mangrove coefficient growth were used as a proxy for the mangrove growth coefficient. Since mangroves are highly sensitive to changes in their water supply, the water balance model was used represent the natural drivers [88,89]. The water balance within an area can also reflect primary climatic parameters, e.g., precipitation, evapotranspiration, and temperature analysis. This model can also be used to identify the potential drought impact within mangrove ecosystems due to ongoing global climate change. The mangroves' water balance was quantified by multiplying the mangrove growth coefficient, calculated using NDVI (MOD13A1) data, by the evapotranspiration (MOD16A2) data, and then subtracting the precipitation (CHIRPS) data from it. The result of this process has a 500 m grid size. The spatial resolution was decided by considering the most common spatial resolution of the datasets and of the coverage of the study area. Specifically, we upscaled the lower resolution datasets and downscaled the higher resolution ones through resampling based on the bicubic interpolation method. To validate the results, the relationships between the mangrove's water balance and degradation and depletion were analyzed.

In this study, we tried to explore all available data products with various spatial resolution and time acquisition to reach maximal outcome. In combining multi-resolution data products, the resampling must be applied to uniform those datasets. Based on recent research articles, there are diverse approaches used to decide the basis of the grids size for the resampling result including the coarser spatial resolution among datasets [90], the modus or the most spatial resolution on the input data [91,92], the mean spatial resolution [93], and the specific resolution based on other referred data [94,95]. In this case, we assumed that there is no certain procedure to be adopted in achieving ideal spatial resolution for assorted topics. Thus, in this study, we consider the most spatial resolution and characteristic of parameters represented for each dataset (e.g., high spatial resolution is not necessary for precipitation) as the basis of the resampling result, i.e., 500-m spatial resolution. On the other hand, considering different time acquisition issues could affect the consistency of the product, we conducted agreement level analysis in the discussion as carried out by a similar research approach [96–98]. This analysis is commonly adopted by overlaying all datasets to assess the accuracy and inconsistency among these data [96,99,100].

2.2.1. Harmonization of Land Cover Data Product

In the first step, to harmonize the type of land cover data from various land cover products, the data were separated into two new classes. The first class was considered mangrove forest (Table 4), and the second class was considered the land cover causing mangrove deforestation (Table 5). The Forest and Wetland classes from ESA, GlobCover, MODIS Land Cover products, and the mangrove class from GLCNMO were all considered to be mangrove forest. In this part, we used MFW datasets to determine the mangrove extent to be overlaid with the forests and wetland classes of ESA, GlobCover, and MODIS Land Cover products. Thus, we believe that the forests and wetland classes within the MFW data are mangrove forests, as illustrated in Figure 2. Then, since fishing, farming, and urbanization are the major deforestation drivers, the urban, agriculture, wetland, and water classes in each land cover product were considered the conversion class of mangrove forest [101–103].

Table 4. Classes of each land cover product defined in the "mangrove forest" class.

Land Cover Product	Class	Description	Classification Reference
ESA CCI Land Cover	Forest	Trees with large and/or pointy greenish and yellowish leaves, open or closed alongside bush and grass, which have a canopy cover of 15% and over.	Land Cover Classification System (LCCS)
	Wetland	Trees inundated with fresh water or sea water, mixed in with the bush or grass	
GlobCover	Forest	Trees with large leaves and/or pointy greenish or yellowish leaves, open or closed with a height of 5 m, mixed in with other vegetation, such as bush and grass, with a minimum canopy cover of 15–40%.	Land Cover Classification System (LCCS) from FAO
	Wetland	Vegetation (Grass, Bush, Wood Vegetation), open and closed, inundated by fresh or sea water (>15%).	
MODIS Land Cover	Forest	Trees with large or pointy leaves, greenish or yellowish, with a height of more than 2 m and a canopy cover of more than 60%.	International Geosphere-Biosphere Programme (IGBP) Legend and Class
	Wetland	Land where 30–60% of its area is permanently inundated with fresh water or seawater, covered by at least 10% from other vegetation.	
GLCNMO	Mangrove	-	Land Cover Classification System (LCCS) from FAO

Each data GLC product was integrated to estimate land cover changes in the mangrove forest as explained in Table 4. In the early stages of data harmonization, a temporal comparison was carried out by subtracting from the CGMFC-21 data for three different time periods: between 2001 and 2012 for the ESA CCI Land Cover and MODIS datasets, between 2005 and 2009 for the GlobCover data, and between 2008 and 2012 for the GLCNMO dataset. The same process was used for each dataset of GLC to ensure that the data of land cover change obtained from both the mangrove forest class and the land cover conversion class had the same time range as the mangrove deforestation data. Furthermore, we correlated the mangrove deforestation data with the land cover change data. The amount of deforestation and land cover change of each product was estimated in a $1^0 \times 1^0$ grid cell to avoid any bias in the spatial resolution among products. In the final stage, the distribution of land cover conversion from mangrove deforestation was visualized by each land cover product, along with the unit amount of mangrove deforestation converted to the "other land cover" class with a 500 m grid size as the lowest spatial resolution of the input data.

Table 5. Estimation of each type of land cover conversion in mangrove forest areas.

Land Cover Product	Early Land Cover Class	Ultimate Land Cover Class	Type of Conversion of Land Cover from Mangrove Deforestation
ESA CCI Land Cover, MODIS Land Cover, and GlobCover	Forest	Agriculture	Mangrove to farming
		Wetlands	Mangrove to fishery
		Water	Mangrove to fishery
		Urban	Mangrove to housing
	Wetlands	Agriculture	Mangrove to farming
		Water	Mangrove to fishery
		Urban	Mangrove to housing
GLCNMO	Mangrove	Agriculture	Mangrove to farming
		Wetlands	Mangrove to fishery
		Water	Mangrove to fishery
		Urban	Mangrove to housing

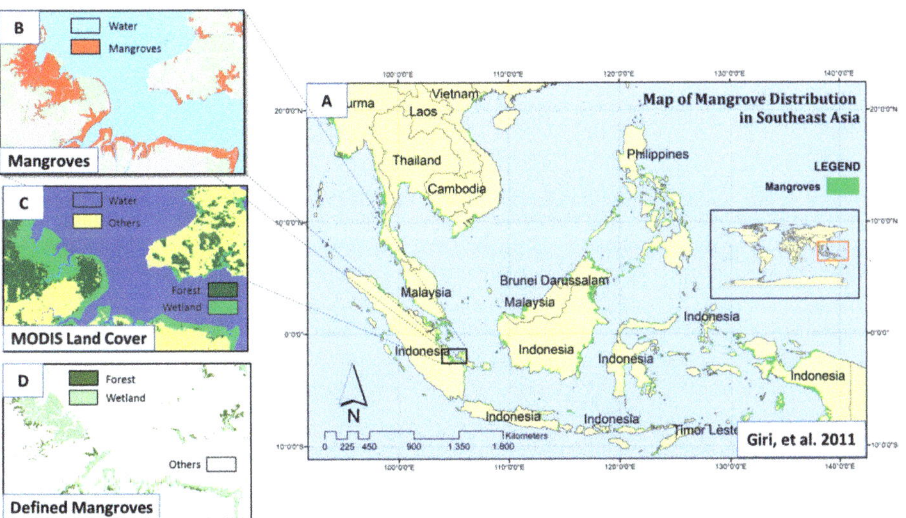

Figure 2. Illustration of how to define mangroves based on forest and wetland classes of ESA, GlobCover, and MODIS Land Cover products. (**A**) Mangroves distribution across Southeast Asia generated from MFW datasets [40]. (**B**) Mangrove patches in the northeast of Palembang, South Sumatra as sample area. (**C**) MODIS Land Cover Products that will be reclassified into mangroves and non-mangroves classes. (**D**) The result of reclassification scheme, i.e., forest and wetland classes defined as mangroves.

In this classification scheme, there are three scenarios that indicate fishery expansion, as shown in Table 5. The first indicator is the transformation of the forest class into the wetland class. The second indicator is the transformation of the forest class into the water class. The third indicator is the transformation of the wetland class into the water class. These three classification scenarios represent three levels of vegetation degradation and three levels of water increase within an area. Hence, this classification scheme could provide more sensitive signs of fishery expansion over mangrove forests.

2.2.2. MODIS Vegetation Indices

The NDVI is the most commonly used vegetation index derived from a combination of red and near-infrared bands, that indicate the existence and greenness level of vegetation [104]. The NDVI has already been widely applied in many ecological studies to observe vegetation phenology and dynamics, monitor spatial trends of forest degradation, and detect abrupt changes in ecosystems, as well as other studies. [105–107]. The SAVI is similar to the NDVI, but it suppresses the effects of soil pixels. It uses a canopy background adjustment factor as a function of vegetation density and often requires prior knowledge of vegetation amounts, as shown in Huete [108]. The NDWI is a reflectance measurement that is sensitive to changes in the water content of plant canopies. The water content is important because a higher water content indicates healthier vegetation that is likely to grow faster and be more fire-resistant [109]. The NDWI uses a normalized difference formulation instead of a simple ratio, and the index values increase with increasing water content. Applications include crop agricultural management, forest canopy monitoring, and vegetation stress detection [110–112].

2.2.3. Mangrove' Coefficient Growth

The crop coefficient (Kc) is one of the most commonly used methods for water management. Similarities between the Kc curve and a satellite-derived vegetation index showed the potential for modeling Kc as a function of the vegetation index [113]. Therefore, the possibility of directly estimating the Kc from the satellite reflectance of a plant was investigated. The Kc data used in developing the relationship with NDVI were derived from back-calculations of the FAO-56 dual Kc procedure, using field data obtained during 2007 from AmeriFlux sites that are representative of US systems in the High Plains covered by cropland area [114]. NDVI is an indicator of the density of vegetative cover that captures most of the observed variation in Kc in the absence of water stress conditions. A simple linear regression model was developed to establish a general relationship between the NDVI from moderate resolution satellite data (MODIS Vegetation Indices/MOD13A1 v06) and the Kc calculated from the flux data measured for a different plan by using AmeriFlux towers. Kc can be estimated by quantifying the fluxes of trace gases between the land and the atmosphere, which has been derived in various land cover types, e.g., cropland, mixed forests and evergreen needleleaf forests.

As reported by several previous studies, there is a flux tower that has existed since 2004 in a mangrove forest site, namely the Tower SRS-6 in Florida Everglades Shark River Slough, which can measure CO_2 and H_2O (https://ameriflux.lbl.gov/sites/siteinfo/US-Skr) [115]. Although the recorded data can be applied for either carbon or evapotranspiration studies, the preceding research mainly discusses carbon balance issues [116] Thus, the flux tower has not been utilized to assist evapotranspiration coefficient modelling. Besides In addition, according to the approach utilized by Kamble et al. [113] approach in developing Kc-NDVI relationship, there were used two flux towers used in the modelling process and the two others for validation purposes. On that account, at least two flux stations are required both for each of those two functions. Since there was only one flux tower available, the mangrove growth coefficient has not yet been determined.

The crop growth coefficient (Kc) can be applied to enhance potential evapotranspiration data based on vegetation dynamic. In this case, it can be applied for cropland phenology where at the seeding and transplanting phases, the evapotranspiration is much lower than at the ripening phase. Moreover, the phenological characteristic of cropland is not significantly different with temperate forest, where in winter the evapotranspiration is much lower than in the other seasons. Therefore, the utilization of Kc for other ecosystems has potential, especially in the mangrove ecosystem.

Therefore, we employed Kc-NDVI model developed by Kamble et al. [113] to calibrate mangroves phenology and variability of MODIS evapotranspiration data. Although the model was advanced for cropland area, we were convinced that the deviation between the crop growth coefficient and mangrove growth coefficient does not significantly affect the water balance. We tried to estimate the mangrove growth coefficient using the crop coefficient developed by Kamble et al. [113]. The mangrove growth coefficient was then multiplied by the potential evapotranspiration within mangroves area to obtain the actual mangrove evapotranspiration. The combined relationship between the NDVI and the Kc is given as

$$Kc.NDVI = 1.457\,NDVI - 0.1725 \tag{1}$$

where 1.4571 and 0.1725 are the slope and intercept coefficients, respectively. The procedure for quantifying coefficient growth from NDVI data should be useful in other regions of the globe for understanding regional water management [113].

2.2.4. Mangrove Forests' Water Balance

Many processes contribute to the ability of mangroves to maintain water uptake and limit water loss under different meteorological conditions. The water balance of mangrove forests is calculated as follows: (1) the Kc of the mangroves is calculated; (2) the effective precipitation (*Peff*) of the mangrove forest area is calculated, where *Peff* is the fraction of the total precipitation as rainfall and snowmelt available that does not run off [117]; (3) the mangrove forests' coefficient growth is multiplied with potential evapotranspiration, and the effective precipitation is subtracted; and (4) the water balance in mangrove forests is obtained by employing Equations 1, 2, and 3. Without detailed site-specific information, *Peff* is very difficult to determine; in this case, a simple approximation by following the U.S. Department of Agriculture Soil Conservation Method [117] is used.

$$Water\ Balance = (Green\ Water) - (Kc.NDVI.\ Epot) \tag{2}$$

$$\begin{aligned} Peff(Green\ Water) &= P\,(125 - 0.6P)/125 \text{ if } P \leq 250/3 \text{ mm} \\ Peff(Green\ Water) &= 1253 + 0.1P \text{ if } P > 250/3 \text{ mm} \end{aligned} \tag{3}$$

where *Peff* is the effective precipitation, *KcNDVI* is the coefficient growth of NDVI, *Epot* is the potential evapotranspiration, and *P* is the precipitation.

The relationship between the water balance of mangrove forests and mangrove forest degradation is generated by overlaying pixels and dividing them into the four classifications shown in Table 6. There are two classes based on water balance processing: surplus, where the water balance is positive, and deficit, where the water balance is negative. Meanwhile, there are also two classes as the result of mangrove deforestation processing, i.e., degraded and not degraded.

Table 6. Relationship classifications between mangrove forests' water balance and mangrove degradation.

Criteria	Classification
Water Balance Surplus and Mangroves Degraded	Anthropogenic Drivers
Water Balance Deficit and Mangroves Degraded	Naturogenic Driver
Water Balance Deficit and Mangroves Not Degraded	Mangrove at Risk
Water Balance Surplus and Mangroves Not Degraded	Sustainable Mangrove

3. Results

3.1. Spatiotemporal of MODIS Vegetation Indices in Mangrove Area

Figure 3 displays the NDVI difference images between 2000 and 2012. To achieve the most accurate visualization of changes, the largest deforested areas that corresponded to the land cover change drivers were selected. For urban conversion, the highlighted areas included Malaysia and Singapore. North Kalimantan corresponds to aquaculture conversion, Thailand corresponds to rice field conversion, South Sumatra corresponds to conversion to oil palm plantations, and mangrove regrowth was found in Papua. Positive changes imply that the index value in 2000 increased, while negative changes imply that the index value in 2000 decreased. The NDVI difference images provided more insight into the drivers of mangrove deforestation. It can be clearly observed that the vegetation index had different responses depending on the different drivers. In (**A**) Malaysia–Singapore and (**B**) North

Kalimantan, the figure shows a slight decrease in vegetation in the mangrove sites. In (**C**) Thailand, the figure shows a smaller decrease in vegetation. In (**E**) South Sumatra, the figure shows an increase in NDVI. In (**D**) Papua, the NDVI increased in some areas, but most demonstrated no signs of change. In terms of healthiness, no change in NDVI implies no vegetation damage, as the density of vegetation is the same, indicating that there was no deforestation from 2000 to 2012.

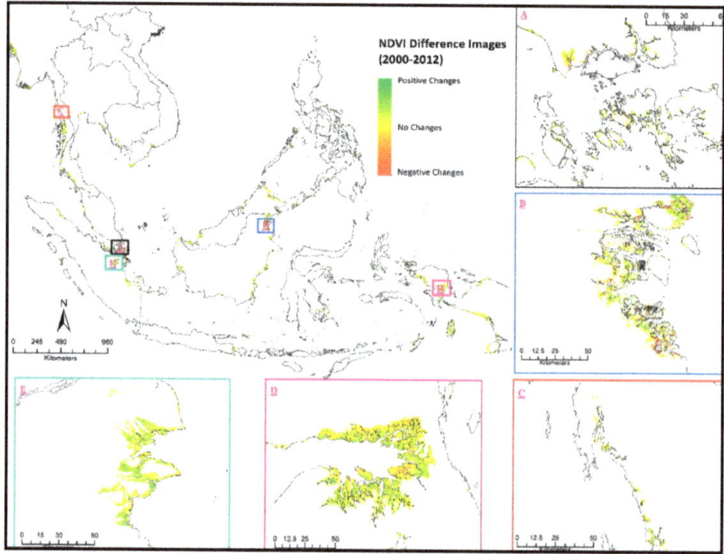

Figure 3. NDVI change between 2000 and 2012 in mangrove area. (**A**) NDVI decrease in Singapore. (**B**) NDVI decrease in North Kalimantan. (**C**) NDVI decrease in West Thailand. (**D**) NDVI increase in West Papua. (**E**) NDVI increase in South Sumatra.

The NDWI difference images provide more insight into the mangrove moisture condition. A decrease in the NDWI implies that the water content of mangroves is decreasing [109]. Figure 4 displays the NDWI difference between 2000 and 2012 with the highlighted areas representing the NDVI difference images. It can be clearly observed that the water index had a different response to each driver. In (**A**) Malaysia–Singapore and (**B**) North Kalimantan, which were deforested for urban and aquaculture conversion, respectively, the figure shows a slight decrease in water content for the mangrove sites. In (**C**) Thailand, the NDWI shows a smaller decrease. In (**E**) South Sumatra, there was an increase in the NDWI. This trend is also displayed in (**D**) Papua, where the NDWI increased in some areas, but overall, there was no relative change. As mangroves are considered an evergreen vegetation, a higher or decreasing NDWI indicates that the mangrove's condition is damaged.

The SAVI difference images provide more insight into the canopy height of the mangrove forests. The higher the SAVI value, the higher the vegetation canopy [108]. In Figure 5, it can be clearly observed that the SAVI has different responses to each driver. In (**A**) Malaysia–Singapore and (**B**) North Kalimantan, the figure shows a severe decrease in canopy height. In (**C**) Thailand, there was deforestation due to rice plantations; hence, the figure shows a lower decrease in canopy height. In (**E**) South Sumatra, there was deforestation due to oil palm plantations, and the figure shows an increase in SAVI. The same trend was observed in (**D**) Papua, where the SAVI increased for some areas but most had no signs of change. From these results, the SAVI could be a potential index for detecting oil palm expansion over mangrove area, as reported by previous research [118]. However, it could be misinterpreted if no other information is available.

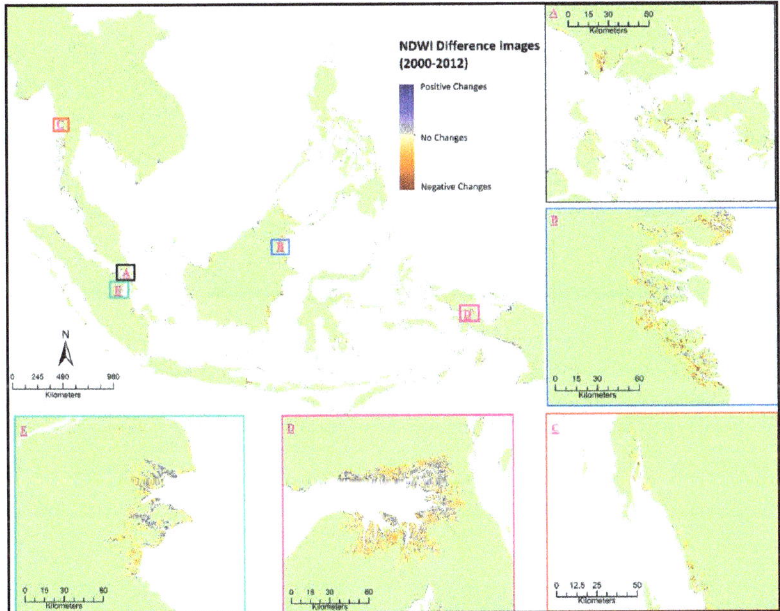

Figure 4. NDWI change between 2000 and 2012 in mangrove area. (**A**) NDWI decrease in Singapore. (**B**) NDWI decrease in North Kalimantan. (**C**) NDWI decrease in West Thailand. (**D**) NDWI increase in West Papua. (**E**) NDWI increase in South Sumatra.

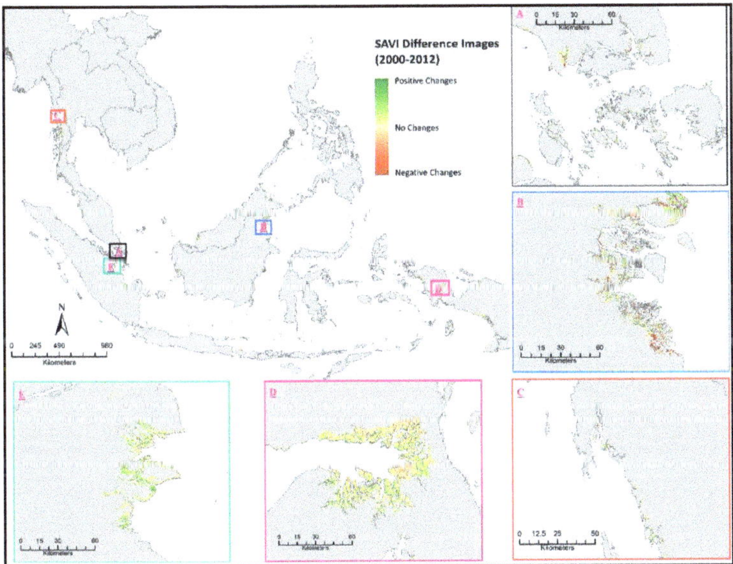

Figure 5. SAVI changes between 2000 and 2012 in mangrove areas. (**A**) SAVI decrease in Singapore. (**B**) SAVI decrease in North Kalimantan. (**C**) SAVI decrease in West Thailand. (**D**) SAVI increase in West Papua. (**E**) SAVI increase in South Sumatra.

3.2. Results of Land Cover Conversion from Deforested Mangroves in Southeast Asia

From the ESA CCI Land Cover product, Figure 6a shows that the rate of deforestation in Myanmar is large, where, averaged over one grid, more than 50 ha of mangroves were converted into the farming class. This was followed by Indonesia, where fishery and farming in Kalimantan and Sulawesi are the major contributors to mangrove deforestation, with the average over one grid cell being 5–25 ha. The product of MODIS Land Cover (Figure 6b) shows that Indonesia has the highest amount of mangrove deforestation owing to the fishery class that has spread across Sumatra, Kalimantan, Sulawesi, and Papua, averaging at 10–25 ha of deforested mangrove. Myanmar also has a high deforestation level; approximately 25–50 ha of deforested mangroves averaged over one grid cell were converted to the farming and fishery classes. From the GLCNMO results, Figure 6c shows that Indonesia has the highest number of deforestation points due to the spread of the farming classes from Sumatra to Papua, with an average deforested area of 50–100 ha. Meanwhile, Malaysia and the Philippines have mangrove deforestation points due to changes in land cover, with an average of less than 50 ha. As observed from the GlobCover product results (Figure 6d), Indonesia has the highest deforestation point distribution as a consequence of the farming class spreading across Sumatra and Kalimantan, with an average deforestation of less than 10 ha. Myanmar has a significant deforestation point where one point of mangrove deforestation of more than 7.5 ha has been converted into the farming class.

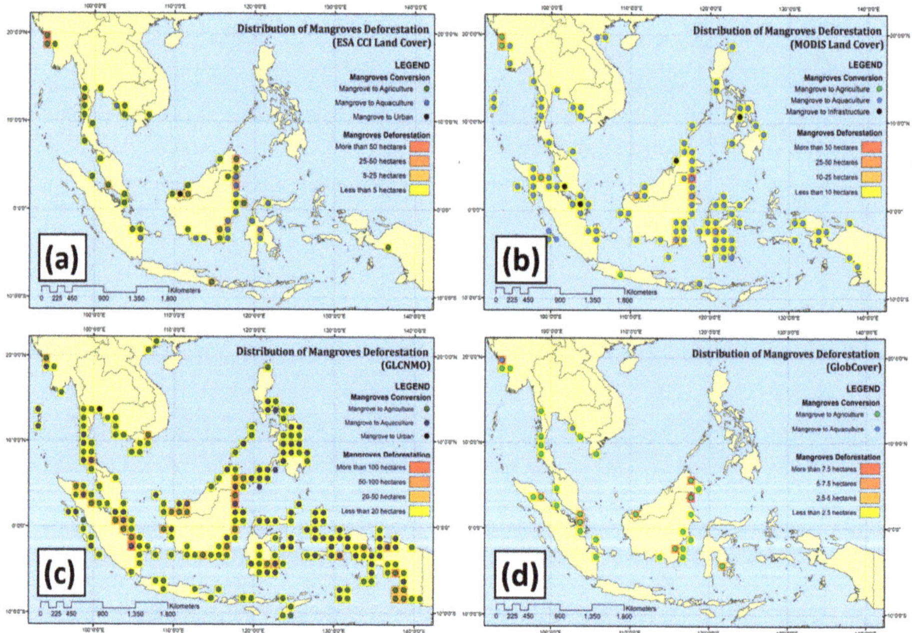

Figure 6. Distribution of land as the result of mangrove deforestation conversion and its quantity based on (**a**) ESA CCI Land Cover 2001 and 2012, (**b**) MODIS Land Cover 2001 and 2012, (**c**) GLCNMO 2008 dan 2012 and (**d**) GlobCover 2005 and 2009.

Figure 7 is a map displaying the confidence levels in mangrove forest conversion obtained by combining all GLC products with the same land cover conversion classes from mangrove forests (farming, fishery, and housing) within a 1° grid cell. The confidence level is ranked from lowest to highest; level 1 means that the land cover conversion class is recognized by one GLC product, while level 4 means that the land cover conversion class in that location is recognized by all GLC products. This agreement level map (Figure 7) does not show the accuracy of each GLC product while detecting deforestation classes or the mangrove land cover changes. However, the confidence level illustrates the occurrence of spatial conversion from mangrove forests to land cover classes, as recognized by the GLC products in this study. Figure 7 shows that the highest confidence levels are in degraded mangrove areas, i.e., in Myanmar caused by farming, Indonesia caused by fisheries, and Malaysia caused by housing.

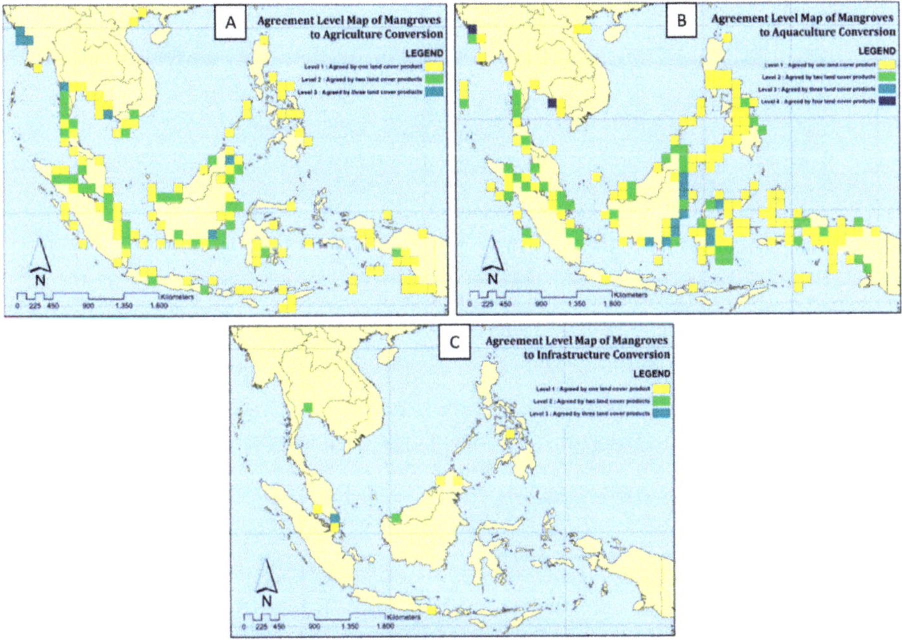

Figure 7. Confidence levels for land cover class conversion of mangrove forest. (**A**) Agreement level of mangroves to agriculture. (**B**) Agreement level of mangroves to aquaculture. (**C**) Agreement level of mangroves to infrastructure.

3.3. Mangrove Coefficient Growth

The value of Kc in each country in Southeast Asia ranges from 0.01 to approximately 1.3. This value indicates the growth rate of mangroves; a value close to 0 implies that the mangrove is not growing, and a value close to 1.3 implies the opposite. Some mangrove forests in Southeast Asia follow the dry rainy seasons, whereas others do not. The dry season usually occurs from April to September, and the rainy season from October to March. As shown in Figure 8, the countries that follow the seasonal pattern are Cambodia, Myanmar, and Thailand; other countries, such as Indonesia, Brunei Darussalam, Malaysia, and the Philippines, do not follow this seasonal pattern.

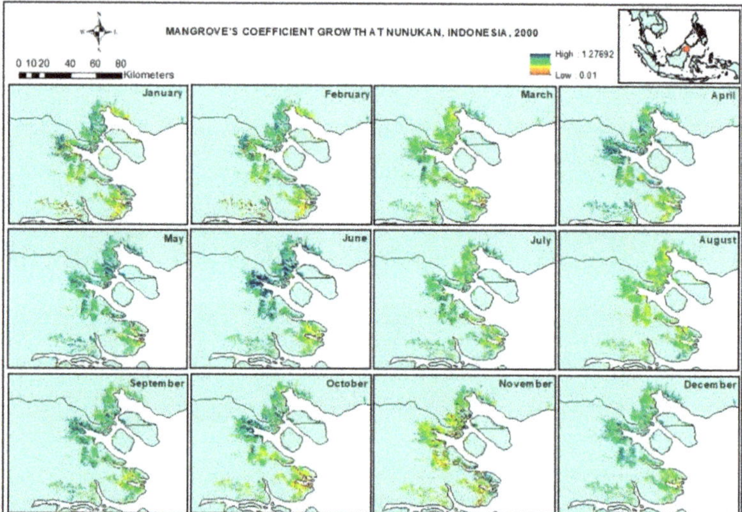

Figure 8. Mangrove coefficient growth (Kc) at Nunukan region, Indonesia, January to December 2000.

3.4. Mangrove Forests' Water Balance for Degradation and Depletion Identification

The resulting values in the form of water balance deficits and surpluses are shown in Figure 9. In Cambodia, Myanmar, and Thailand, the values of deficit and surplus followed the pattern of dry and rainy seasons, respectively, while in Indonesia, Brunei Darussalam, Malaysia, and the Philippines, they did not follow the seasonal patterns.

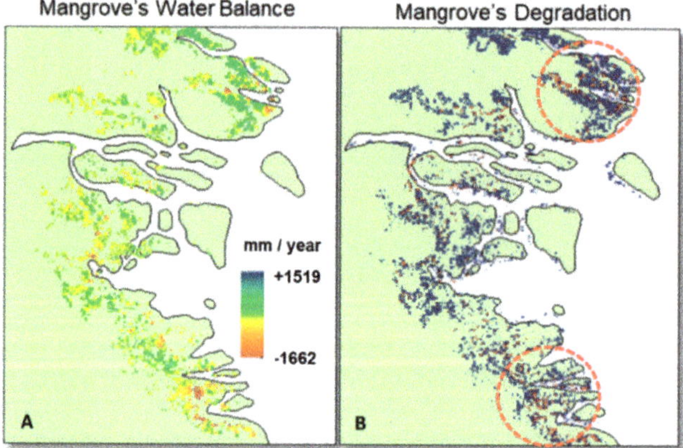

Figure 9. (**A**) Mangrove's net water balance (in this study) and (**B**) mangrove's net degradation between 2000 and 2012 in Southeast Asia.

As shown in Figure 10, this research shows that natural drivers have a larger effect on mangrove forests' degradation and depletion compared with anthropogenic drivers. In this study, the determination of the effect of natural drivers focuses on evapotranspiration and precipitation data, as well as other meteorological phenomena such as whether or not regional patterns follow seasonal variation. The classification "mangroves at risk" refers to regions where the water balance is in deficit but not yet degraded. This should encourage policy makers to enforce the necessary steps to prevent mangrove degradation, recognizing the many advantages mangroves bring to society and the natural environment.

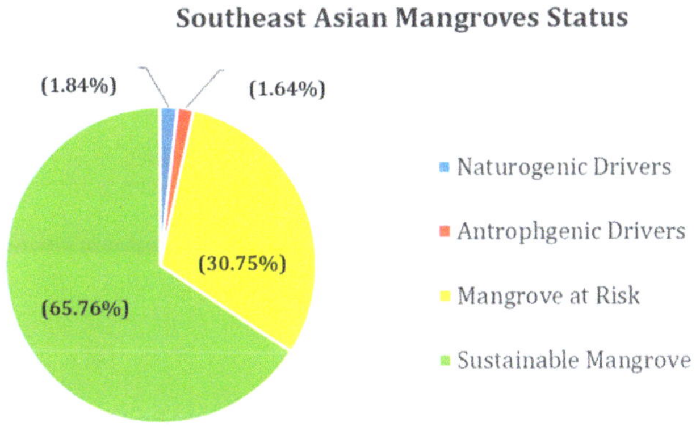

Southeast Asian Mangroves Status

Figure 10. Relationship classifications between land use land cover and mangrove forests' water balance with the degradation and depletion condition of them.

4. Discussion

4.1. Mangrove Agreement Level

Agreement analysis is used to compare the similarity of various data with the same product level [119]. In this study, agreement analysis was used to compare four mangrove distribution datasets by making four levels of agreement. The first level indicates low agreement, level 2 indicates moderate agreement, level 3 indicates high agreement, and level 4 indicates very high agreement. Mangrove distribution data that will be compared are World Atlas of Mangroves v 1.1 for the period 1999–2003 [53], Global Distribution of Mangrove USGS v 1.3 for the period 1997–2000 [40], Indonesia Mangrove Map for the period 2006–2009 [54], and Global Mangrove Watch for the period 2010 [55]. Figure 11 shows that the majority of the level 4 mangrove agreements are in the Bintuny Bay (Papua) area. In the area of Sumatra, especially South Sumatra, the majority of mangrove agreements are level 3. Meanwhile, the majority of mangrove agreements for level 1 and 2 are on the island of Kalimantan.

Figure 12 shows the percentage level of agreement for each mangrove distribution. The USGS Global Distribution of Mangrove (v 1.3) has the highest percentage agreement among other products at level 4 (47%), while level 3, level 2 and level 1 have a percentage of 31%, 13%, and 9%, respectively. The mangrove distribution dataset that has the second highest value at level 4 is Global Mangrove Watch at 45%, while level 3, level 2 and level 1 are 33%, 14%, and 8%, respectively. World Atlas of Mangroves has a value of 43% for level 4, 16% for level 3, 14% for level 2, and 27% for level 1, while Indonesia Mangrove Map has a value of 25% for level 4, 21% for level 3, 16% for level 2, and 38% for level 1. It can be seen that the USGS Global Distribution of Mangroves (v 1.3) has the highest agreement value, so this study uses the USGS Global Distribution of Mangroves (v 1.3) as the distribution of mangroves.

4.2. Conformity of Data Products with DLUDMP and SEAMCT

The consistency of each land cover product was evaluated with reference to DLUDMP [10] and SEAMCT [52] in terms of the rate and the trigger of mangrove deforestation in Southeast Asia from 2000 to 2012. For this purpose, research related to the function of the dominant land in deforested mangrove areas from 2001 to 2012 as well as Southeast Asian mangrove conversion types were used for comparative data to assess the accuracy of the land cover conversion results obtained from each product of GLC. The calculation of the percentage of data products existing in one single land cover class was carried out in a 1° × 1° grid.

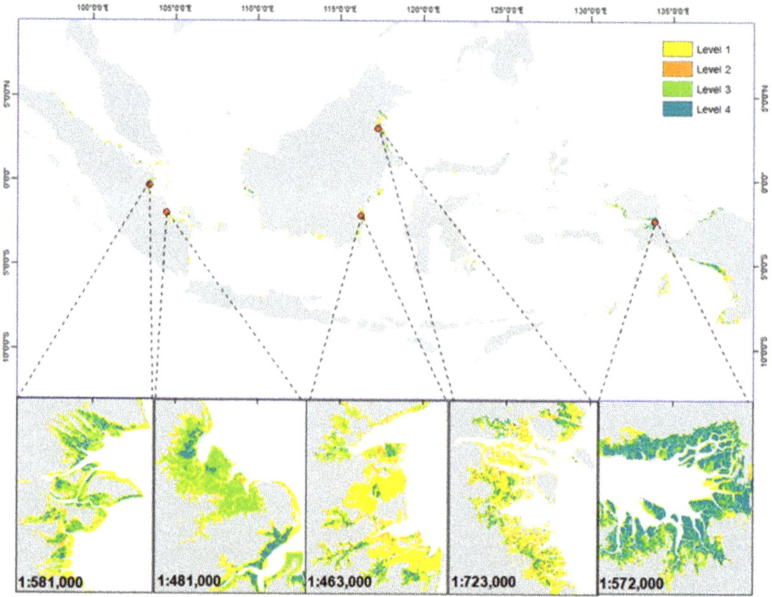

Figure 11. Agreement level of mangrove distribution in Indonesia.

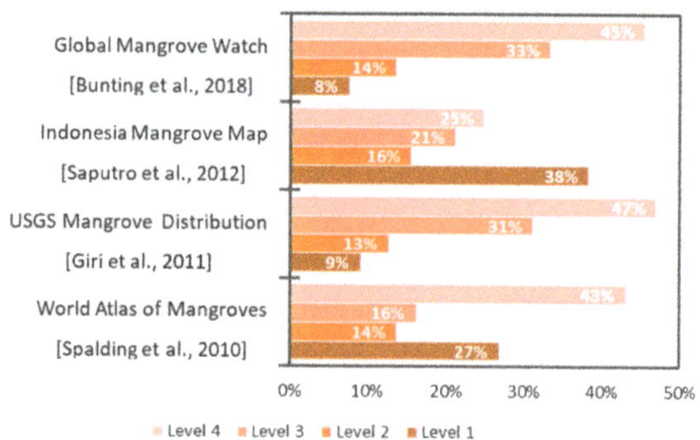

Figure 12. The percentage level of agreement for each mangrove distribution.

In the spatial context, this study illustrates that the points of mangrove forest conversion are similar to those in DLUDMP and SEAMCT. Figure 13 shows the locations of mangrove forest conversion, which is recognized by three to four GLC products, as mentioned previously. The results show that the conversion of mangrove forests into the farming class mostly occurred in the Rakhine region, Myanmar. The Myanmar government has strong policies that support national food security through increased rice production. As a result, many mangrove forests have been converted into paddy fields [120]. The conversion of mangrove forests into the fishery class mostly occurred in East Kalimantan and Sulawesi, Indonesia. The expansion of fisheries was driven by Indonesia's government policies, which were aimed at making the country the largest fish producer in the world by 2015 [121].

Meanwhile, mangrove forests were converted to land for the housing class in the peninsulas of Malaysia and Singapore.

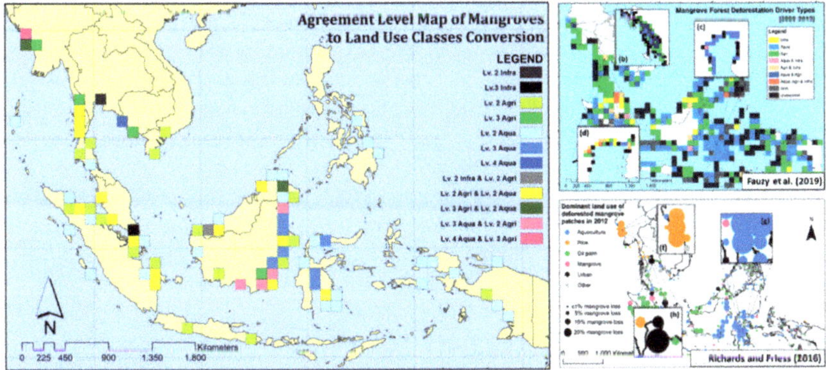

Figure 13. Comparison of mangrove forests converted to other classes at confidence levels of 3 and 4 with Dominant Land Use of Deforested Mangrove Patches (DLUDMP)-The Southeast Asian Mangroves Conversion Types (SEAMCT) from a spatial context.

To the best of our knowledge, this research is the first to compare GLC products for monitoring mangrove deforestation over large areas. The approach used in this study is accompanied by certain assumptions to make the comparison of each GLC product easier (owing to the fact that these products have different characteristics). These assumptions, therefore, limit the quantitative comparison among different GLC products and increase the potential for bias in the results. In addition, the results cannot represent an assessment of the overall quality of each GLC product. However, the study focuses on drawing conclusions and finding policy recommendations for monitoring mangrove deforestation.

In general, based on the areas of mangrove forests converted into other land cover classes as shown in Table 7, the similarity of each GLC product to DLUDMP-SEAMCT varies by different proportions, from 40% to 60% to less than 2%. Among the global land cover products, GLCNMO shows the highest conformity to DLUDMP, with an average difference of 3.81%. ESA CCI LC has the highest conformity to SEAMCT, with a mean difference of approximately 1.21%.

Table 7. Conformity percentage between each data product with DLUDMP land use classes and SEAMCT based on the area of deforested mangrove

Mangrove Forest Conversion Area Based on Global Land Cover Data Products (GLCM)		Width Percentage (GLCM)	Width Percentage (DLUDMP)	Width Percentage (SEAMCT)	Difference between GLCM and DLUDMP	Difference between GLCM and SEAMCT
MODIS (2001 dan 2012)	Farming	8.19%	52.70%	77.59%	44.52%	69.40%
	Fishery	88.12%	41.47%	20.05%	46.65%	68.08%
	Housing	3.69%	5.83%	2.36%	2.14%	1.33%
GlobCover (2005 and 2009)	Farming	92.42%	52.70%	77.59%	39.71%	14.83%
	Fishery	5.76%	41.47%	20.05%	35.71%	14.29%
	Housing	1.83%	5.83%	2.36%	4.00%	0.54%
ESA CCI LC (2001 and 2012)	Farming	77.07%	52.70%	77.59%	24.36%	0.52%
	Fishery	21.85%	41.47%	20.05%	19.62%	1.81%
	Housing	1.08%	5.83%	2.36%	4.75%	1.29%
GLCNMO (2008 and 2012)	Farming	55.63%	52.70%	77.59%	2.92%	21.96%
	Fishery	44.26%	41.47%	20.05%	2.79%	24.21%
	Housing	0.12%	5.83%	2.36%	5.71%	2.25%

4.3. Uncertainties in Mangrove Change Data

As shown in Figure 14, the number of pixels located in the mangrove pixel class produced by MODIS Land Cover (2001 and 2012) reaches 4000 km^2, with 1000 km^2 of deforestation. In ESA CCI LC (2001 and 2012), the total area of the mangrove forest class reached 37,500 km^2, with a deforested area of 6500 km^2. In GlobCover (2005 and 2009), the total mangrove forest reached 28,000 km^2 and showed no signs of significant deforestation. Lastly, the area of mangrove forest class produced by GLCNMO (2008 and 2012) from the global land cover products was very similar to the area of mangrove owned by CGMFC-21. Hence, the assumptions made for the harmonization of the mangrove forest classes can be considered correct. Furthermore, the area of the mangrove forest measured by all products of GLC declined for several years, indicating widespread mangrove deforestation in that time period. The variations in the total area of mangrove forest could be caused by different methods of data acquisition, classification techniques, class definitions, and the production year of each land cover product.

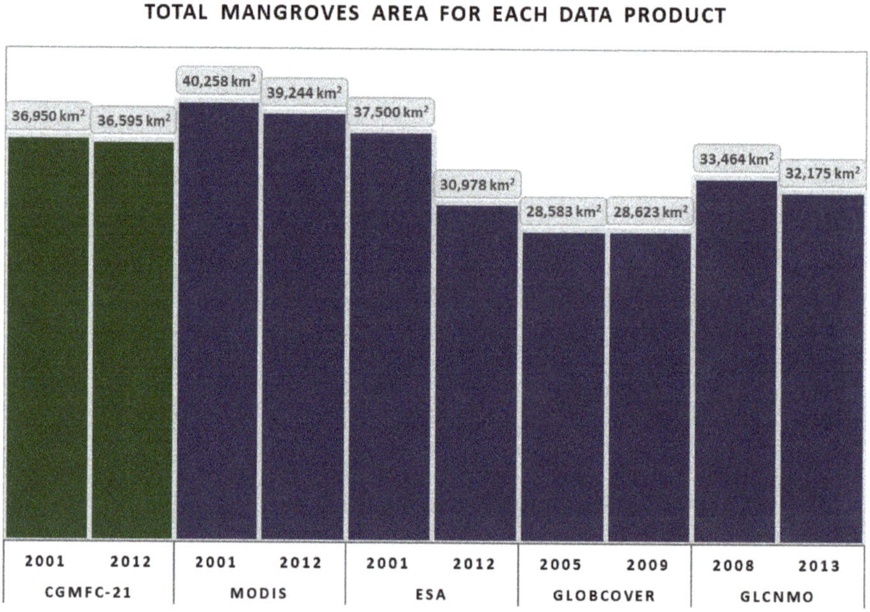

Figure 14. Total area of mangrove forest class of each GLC product.

4.4. Trend Analysis and Breakpoint Detection on Deforested and Degraded Mangrove Area

PTC, PNTV, and PNV data were applied to identify extreme deforestation events (e.g., agriculture expansion in Myanmar, infrastructure expansion in Thailand, and aquaculture expansion in Indonesia) using trend analysis and breakpoint detection as explained below. Figure 15 shows a sample of a mangrove deforestation area driven by agriculture expansion in Rakhine State, Myanmar, as explained in the PTC trend. Positive values indicate an increase in PTC, while negative values indicate a decrease.

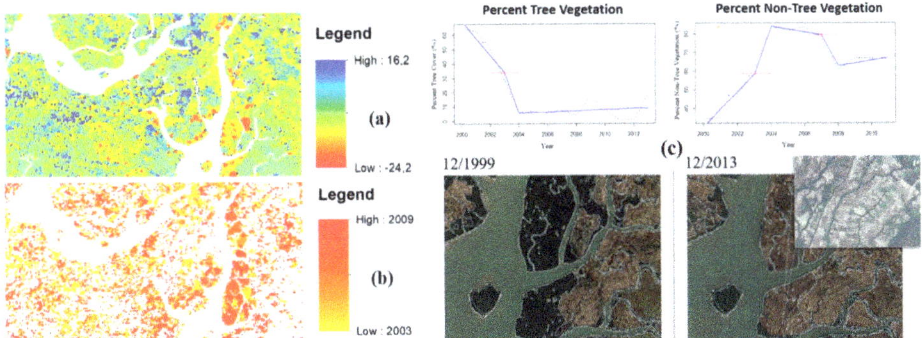

Figure 15. Mangrove deforestation driven by agriculture expansion in Rakhine State, Myanmar. The (**a**) slope and (**b**) breakpoint maps are produced by annual percentage of tree cover (PTC) and percentage of non-tree vegetation (PNTV) (MOD44B) data from 2000 to 2013 using the green-brown and bfast package in R [85,87]. (**c**) shows the breakpoint detection processed by annual PTC and PNTV (MOD44B) data from 2000 to 2013 using the bfast package in R [85] and a land cover comparison before deforestation in 1999 (left) and after deforestation in 2013 (right) using the Google Earth archives.

We detected 1269 grids with a 250 m spatial resolution that show the area converted for agriculture use in Rakhine State, Myanmar. Using trend analysis, at least 911 or 71.79% of those grids showed there was a negative trend in PTC and a positive trend in PNTV from 2000 to 2013. Using breakpoint detection in a particular grid, we revealed that the expansion occurred in 2003 when the PTC dropped and the PNTV jumped (Figure 15c). In Myanmar, especially in Rakhine State, nearly 70% of mangrove forests have been degraded since 1983 due to the vast expansion of rice farming as a consequence of the low productivity levels and competition [122,123].

Figure 16a portrays infrastructure expansion in Samut Sakhon State, Thailand. Trend analysis of PTC and PNV data indicated that 18 out of the 21 grids (75%) showed a negative trend in PTC and a positive trend in PNV. Meanwhile, the breakpoint detection discovered that the expansion happened in 2007, when the PTC dropped and PNV jumped in the same year (Figure 16a). According to the earlier studies, mangrove forests around the coastal areas of the Thailand Peninsula, which has a dense urban population, e.g., Phuket, Songkhla, and Phang Nga, were converted to residential areas and tourism facilities [124].

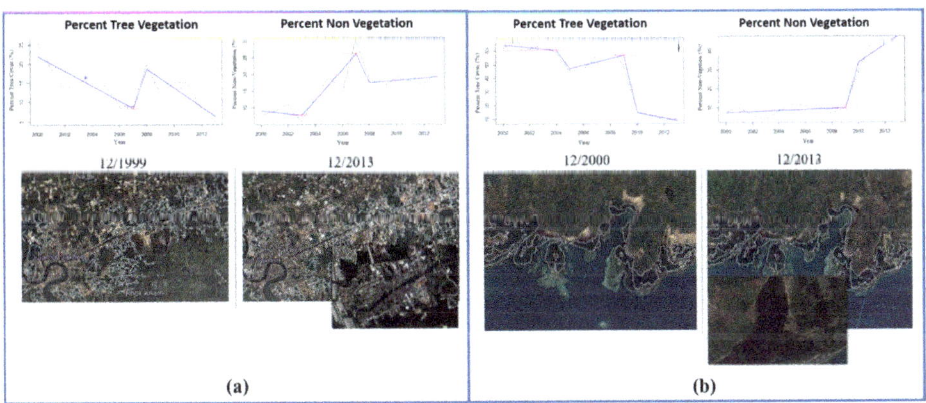

Figure 16. (**a**) Mangrove deforestation driven by infrastructure expansion in Samut Sakhon State, Thailand; (**b**) mangrove deforestation driven by aquaculture expansion in Northern Sulawesi, Indonesia.

In Northern Sulawesi, Indonesia, trend analysis of PTC and PNV data revealed that 291 out of 309 grids (94.14%) showed a negative trend in PTC and positive trend in PNV. Meanwhile, the breakpoint detection identified that most of the mangrove forest conversion in this region occurred in 2009 (Figure 16b). As reported in previous research, the very rapid decline of mangrove forest in this area was due to major brackish-water pond developments [121].

4.5. Future Possible Directions

To detect other drivers of mangrove deforestation, other satellite-derived indices should be studied. For more information on the indices, higher spatial resolution datasets, such as Landsat and Sentinel, are needed to better establish their utility beyond the initial findings of this study. Implementing trend analysis and a break point detection algorithm to NDVI, NDWI, and SAVI for a time series analysis could form the basis of future studies. Another potential direction is a deeper phenology analysis based on water balance and remote sensing indices using long-term data. The continuous use of remote sensing indices will help to detect spatial and temporal changes in mangrove deforestation; thus, necessary steps toward mitigation can be planned and regulated.

The addition of other anthropogenic and naturogenic factors is crucial to the validation and completion of this research. Moreover, the spatial resolution used in this study should be improved and synchronized to increase the minimum level of detail and make it easily comparable. Notably, the approach used in this study can be used as an alternative for conducting a spatiotemporal analysis of a phenomenon on a regional or global scale. In addition, the method used in the consistency test stage can also be adopted to test the accuracy of information on various remote sensing data products involving similar topics. The combination of all available multi-temporal GLC products with a robust methodology for investigating global forest change would be the next stage of this research. Although this study has revealed that mangroves have been converted for various land uses, mangrove rehabilitation is still being conducted. Many concerned players have already made collaborative efforts in designing pilot projects for mangrove protection. Rehabilitation activities are abundant in Southeast Asian countries [125]; however, these have been deployed in relatively small areas, and the results are not noticeable because of the coarse spatial resolution of the data products. This research could be useful for designing mangrove forest rehabilitation strategies by combining the results of this study with supporting data [125].

Future applications that address environmental and socio-economic implications arising from these findings could be explored. The impact of this mangrove forest change on the environment, including on climate change [126], biodiversity [127], sea level rise, and coastal economies [128], could also be explored. Further investigations are required to elucidate the reason for this degradation change, such as the impact of long-term anthropogenic factors, that include cropland expansion and intensification change [129], urban change models [130], and water surface change [131] would enhance our ability to measure and improve future mangrove forest management [132,133].

5. Conclusions

The summary presented herein offers the primary points that will serve as the basis of future investigations. First, the relationships between remote sensing indices and deforestation drivers highly depend on the type of drivers. In future studies, exploring more satellite-derived indices could allow us to expose more deforestation drivers. Second, by adopting a trend analysis and break point detection in three difference sites, Rakhine State, Samut Sakhon, and North Sulawesi, we confirmed that mangrove forests were converted for agriculture, infrastructure, and aquaculture in 2003, 2007, and 2009, respectively. Moreover, the aggregation of PTV, PNV, and PNTV datasets is a highly recommended method of measuring the rate of changes, examining the degradation drivers, and tracing the exact year of expansion. Thirdly, the assimilation of GLC products revealed that agricultural and fishery classes are the predominant drivers in Southeast Asia, notably from 2001 to 2012. Although the study could not accurately describe the amount of mangrove conversion, the data processing and accuracy assessment method can be applied as an alternative method for conducting spatiotemporal analysis of either regional or global scale studies. Fourth, by analyzing the connection between the water balance and degradation of mangrove forests, we discovered that the natural drivers have a greater effect than the anthropogenic drivers. This finding could be investigated further by increasing the number of natural driver variables in the same spatial resolution, making the output more robust and comparable with the data of anthropogenic drivers. We hope that longer-term studies will be undertaken to determine how the water balance in Southeast Asian mangrove forests affect their level of degradation and depletion. Fifth, this extensive investigation leads us to deduce that we cannot easily determine a single factor as a sole driver of degradation in a particular mangrove patch. The combination of both anthropogenic and naturogenic drivers greatly affects and perhaps accelerates mangrove degradation. Finally, the abundance of remote sensing data and products recorded from extraterrestrial sensors could reveal valuable information required to understand and protect our terrestrial ecosystem, and it could be the foundation of important recommendations for creating impactful policy making.

Author Contributions: A.D.S., A.I.F., L.F.Y., T.S., J.A.P., N.T.Q.T. and K.W. conceived and designed the experiments; A.D.S., A.I.F., F.N.W., Y.S.R., L.N.S. and S.A.S. performed the experiments; A.D.S., E.S., I.U. and C.W.A. analyzed the data; A.D.S., T.S., J.A.P. and N.T.Q.T. preprocessed the base datasets; A.D.S., A.I.F., L.F.Y., F.N.W., Y.S.R. and S.A.S. wrote the paper; and all authors read the paper and provided revision suggestions. All authors have read and agreed to the published version of the manuscript.

Funding: This project was funded by Kurita Asia Research Grant (19Pid017) provided by Kurita Water and Environment Foundation, and PMDSU scholarship from the Ministry of Research, Technology and Higher Education Indonesia (RisetDikti).

Acknowledgments: The authors are grateful to acknowledge the support from Kurita Water and Environment Foundation and the Ministry of Research, Technology and Higher Education Indonesia. We also thank the anonymous reviewers whose valuable comments greatly helped us to prepare an improved and clearer version of this paper. All persons and institutes who kindly made their data available for this analysis are acknowledged.

Conflicts of Interest: The authors declare no conflict of interest. The funders had no role in the design of the study; in the collection, analysis, or interpretation of data; in the writing of the manuscript, or in the decision to publish the results.

References

1. Tomlinson, P.B. *The Botany of Mangroves*; Cambridge University Press: New York, NY, USA, 1986; ISBN 0-521-25567-8.
2. Alongi, D.M. *Introduction in the Energetics of Mangrove Forests*; Springer Science and Business Media BV: New York, NY, USA, 2009; pp. 47–64.
3. Kathiresan, K.; Bingham, B.L. Biology of mangroves and mangrove ecosystems. *Adv. Mar. Biol.* **2001**, *40*, 81–251. [CrossRef]
4. Brown, D. *Mangrove: Nature's Defences against Tsunamis*; Environmental Justice Foundation: London, UK, 2004; ISBN 1-904523-05-7.
5. Feller, I.C.; Sitnik, M. *Mangrove Ecology Workshop Manual*; Smithsonian Institution: Washington, DC, USA, 1996.
6. Kirui, B.; Kairo, J.G.; Karachi, M. Allometric Equations for Estimating Above Ground Biomass of Rhizophora Mangroves at Gazi Bay, Kenya. *West. Indian Ocean J. Mar. Sci.* **2006**, *5*, 27–34. [CrossRef]
7. Spalding, M.; Blasco, F.; Field, C. *World Atlas of Mangroves*, Routledge. London, UK, 1997; p. 336.
8. Giesen, W. Indonesian mangroves part I: Plant diversity and vegetation. *Trop. Biodivers.* **1998**, *5*, 99–111.
9. Hamilton, S.E.; Casey, D. Creation of a high spatio-temporal resolution global database of continuous mangrove forest cover for the 21st century (CGMFC-21). *Glob. Ecol. Biogeogr.* **2016**, *25*, 729–738. [CrossRef]
10. Richards, D.R.; Friess, D.A. Rates and drivers of mangrove deforestation in Southeast Asia, 2000–2012. *Proc. Natl. Acad. Sci. USA* **2016**, *113*, 344–349. [CrossRef] [PubMed]

11. Fauzi., A.I.; Sakti, A.D.; Yayusman, L.F.; Harto, A.B.; Prasetyo, L.B.; Irawan, B.; Wikantika, K. Evaluating Mangrove Forest Deforestation Causes in Southeast Asia by Analyzing Recent Environment and Socio-Economic Data Products. In Proceedings of the 39th Asian Conference on Remote Sensing, Kuala Lumpur, Malaysia, 15–19 October 2018; pp. 880–889.

12. Foti, R.; Jesus, M.; Rinaldo, A.; Rodriguez-Iturbe, I. Signs of Critical Transition in The Everglades Wetlands in Response to Climate and Anthropogenic Changes. *Proc. Natl. Acad. Sci. USA* **2013**, *10*, 6296–6300. [CrossRef]

13. FAO. *The World's Mangroves 1980–2005*; Forestry Paper FAO 153; FAO: Rome, Italy, 2007.

14. Woodroffe, C.D. The Impact of Sea-Level Rise on Mangrove Shorelines. *Prog. Phys. Geogr. Earth Environ.* **1990**, *14*, 483–520. [CrossRef]

15. Thomas, N.; Lucas, R.; Bunting, P.; Hardy, A.; Rosenqvist, A.; Simard, M. Distribution and drivers of global mangrove forest change 1996–2010. *PLoS ONE* **2017**, *12*, e0179302. [CrossRef]

16. Rioja-Nieto, R.; Barrera-Falcón, E.; Torres-Irineo, E.; Mendoza-González, G.; Cuervo-Robayo, A.P. Environmental drivers of decadal change of a mangrove forest in the North coast of the Yucatan peninsula, Mexico. *J. Coast Conserv.* **2017**, *21*, 167–175. [CrossRef]

17. Songsom, V.; Koedsin, W.; Ritchie, R.J.; Huete, A. Mangrove Phenology and Environmental Drivers Derived from Remote Sensing in Southern Thailand. *Remote Sens.* **2019**, *11*, 955. [CrossRef]

18. Valiela, I.; Bowen, J.; York, J. Mangrove Forests: One of The World's Threatened Major Tropical Environments. *Bioscience* **2001**, *51*, 807–815. [CrossRef]

19. Cahoon, D.R.; Hensel, P.F.; Spencer, T.; Reed, D.J.; McKee, K.L.; Saintilan, N. Coastal Wetland Vulnerability to Relative Sea-Level Rise: Wetland Elevation Trends and Process Controls. In *Wetlands and Natural Resource Management*; Varhoeven, J.T.A., Beltman, B., Bobbink, R., Whigham, D.F., Eds.; Springer: Berlin/Heidelberg, Germany, 2006; Volume 190, pp. 271–292. ISBN 978-3-540-33187-2. [CrossRef]

20. MacKay, F.; Cyrus, D.; Russell, K.L. Macrobenthic invertebrate responses to prolonged drought in South Africa's largest estuarine lake complex. *Estuar. Coast. Shelf S.* **2010**, *86*, 553–567. [CrossRef]

21. Bhat, N.R.; Suleiman, M.K. Classification of soils supporting mangrove plantation in Kuwait. *Arch. Agron. Soil Sci.* **2004**, *50*, 535–551. [CrossRef]

22. Almulla, L. Soil site suitability evaluation for mangrove plantation in Kuwait. *World Appl. Sci. J.* **2013**, *22*, 1644–1651.

23. Nusantara, M.A.; Hutomo, M.; Purnama, H. Evaluation and planning of mangrove restoration programs in Sedari Village of Kerawang District, West Java: Contribution of PHE-ONWJ coastal development programs. *Procedia Environ. Sci.* **2015**, *23*, 207–214. [CrossRef]

24. Boto, K.G.; Wellington, J.T. Soil characteristics and nutrient status in a Nothern Australian mangrove forest. *Estuaries* **1984**, *7*, 61–69. [CrossRef]

25. Chimner, R.A. Soil respiration rates of tropical peatlands in Micronesia and Hawaii. *Wetlands* **2004**, *24*, 51–56. [CrossRef]

26. Crain, C.M.; Silliman, B.R.; Bertness, S.L.; Bertness, M.D. Physical and biotic drivers of plaant distribution across estuarine salinity gradients. *Ecology* **2004**, *85*, 2539–2549. [CrossRef]

27. Ribeiro, J.P.N.; Tiberio, F.C.S.; de Oliveira, A.A. The role of soil nutrients in boundaries between mangrove and herbaceous assemblages in a tropical estuary. *Biotropica* **2015**, *47*, 517–520. [CrossRef]

28. Clarke, L.D.; Hannon, N.J. The mangrove swamp and salt marsh communities of the Sydney District: II. the holocoenotic complex with particular reference to physiography. *J. Ecol.* **1969**, *57*, 213–234. [CrossRef]

29. Primavera, J.H.; Savaris, J.D.; Bajoyo, B.; Coching, J.D.; Curnick, D.J.; Golbeque, R.; Guzman, A.T.; Henderin, J.Q.; Joven, R.V.; Loma, R.A.; et al. *Manual on Community-Based Mangrove Rehabilitation - Mangrove Manual Series No.1*; Zoological Society of London: London, UK, 2012; 240p, ISBN 978-971-95370-1-4.

30. Cohen, M.C.L.; Lara, R.J.; Cuevas, E.; Oliveras, E.M.; Sternberg, L.D.S. Effects of sea-level rise and climatic changes on mangroves from southwestern littoral of Puerto Rico during the middle and late holocene. *Catena* **2016**, *143*, 187–200. [CrossRef]

31. Asbridge, E.F.; Bartolo, R.; Finlayson, C.M.; Lucas, R.M.; Rogers, K.; Woodroffe, C.D. Assessing the distribution and drivers of maangrove dieback in Kakadu National Paark, nothern Australia. *Estuar. Coast. Shelf Sci.* **2019**, *228*, 1–12. [CrossRef]

32. Gupta, K.; Mukhopadhyay, A.; Giri, S.; Chanda, A.; Majumdar, S.D.; Samanta, S.; Mitra, D.; Samal, R.N.; Pattnaik, A.K.; Hazra, S. An index for discrimination of mangroves from non-mangroves using LANDSAT 8 OLI imagery. *MethodsX* **2018**, *5*, 1129–1139. [CrossRef] [PubMed]

33. Ahmed, K.R.; Akter, S. Analysis of landcover change in southwest Bengal delta due to floods by NDVI, NDWI and K-means cluster with landsat multi-spectral surface reflectance satellite data. *Remote Sens. Appl. Soc. Environ.* **2017**, *8*, 168–181. [CrossRef]

34. Pastor-Guzman, J.; Dash, J.; Atkinson, P.M. Remote sensing of mangrove forest phenology and its environmental drivers. *Remote Sens. Environ.* **2018**, *205*, 71–84. [CrossRef]

35. Yu, L.; Wang, J.; Li, X.; Li, C.; Zhao, Y.; Gong, P. A multi-resolution global land cover dataset through multisource data aggregation. *Sci. China Earth Sci.* **2014**, *57*, 2317–2329. [CrossRef]

36. Hailu, B.T.; Fekadu, M.; Nauss, T. Availability of global and national scale land cover products and their accuracy in mountainous areas of Ethiopia: A review. *J. Appl. Remote Sens.* **2018**, *12*, 041502. [CrossRef]

37. Pendrill, F.; Persson, U.M. Combining global land cover datasets to quantify agricultural expansion into forests in Latin America: Limitations and challenges. *PLoS ONE* **2017**, *12*, e0181202. [CrossRef]

38. Pérez-Hoyos, A.; Rembold, F.; Kerdiles, H.; Gallego, J. Comparison of Global Land Cover Datasets for Cropland Monitoring. *Remote Sens.* **2017**, *9*, 1118. [CrossRef]

39. Tsendbazar, N.; Bruin, S.D.; Herold, M. Integrating global land cover datasets for deriving user-specific maps. *Int. J. Digit. Earth* **2017**, *10*, 219–237. [CrossRef]

40. Giri, C.; Ochieng, E.; Tieszen, L.L.; Zhu, Z.; Singh, A.; Loveland, T.; Duke, N. Status and distribution of mangrove forests of the world using earth observation satellite data. *Glob. Ecol. Biogeogr.* **2011**, *20*, 154–159. [CrossRef]

41. Hansen, M.C. High-Resolution Global Maps of 21st-Century Forest Cover Change. *Science* **2013**, *342*, 850–853. [CrossRef] [PubMed]

42. Coumou, D.; Rahmstorf, S. A decade of seather extremes. *Nat. Clim. Chang.* **2012**, *2*, 491–496. [CrossRef]

43. Bathiany, S.; Dakos, V.; Scheffer, M.; Lenton, T.M. Climate models predict increasing temperaature variability in poor countries. *Sci. Adv.* **2018**, *4*, 1–10. [CrossRef]

44. Arnell, N.W. Climate change and global water resources: SRES emissions and socio-economic scenarios. *Glob. Environ. Chang.* **2004**, *14*, 31–51. [CrossRef]

45. Maslin, M.; Austin, P. Climate models at their limit? *Nature* **2012**, *486*, 183–184. [CrossRef] [PubMed]

46. Myhre, G.; Alterskjaer, K.; Stjern, C.W.; Hodnebrog, Ø.; Marelle, L.; Samset, B.H.; Sillmann, J.; Schaller, N.; Fischer, E.; Schulz, M.; et al. Frequency of extreme precipitation increases extensively with event rareness under global warming. *Sci. Rep.* **2019**, *9*, 16063. [CrossRef] [PubMed]

47. Fischer, E.M.; Knutti, R. Detection of spatially aggregated changes in temperature and precipitation extremes. *Geophys. Res. Lett.* **2014**, *41*, 547–554. [CrossRef]

48. Duke, N.C.; Kovacs, J.M.; Griffths, A.D.; Preece, L.; Hill, D.J.E.; Oosterzee, P.; Mackenzie, J.; Morning, H.S.; Burrows, D. Large-Scale Dieback of Mangroves in Australia's Gulf of Carpentaria: A Severe Ecosystem Response, Coincidental with an Unusually Extreme Weather Event. *Mar. Freshw. Res.* **2017**, *68*, 1816–1829. [CrossRef]

49. Jones, R. Quantifying Extreme Weather Event Impacts on the Northern Gulf Coast Using Landsat Imagery. *J. Coast. Res.* **2015**, *31*, 1229–1240. [CrossRef]

50. Pettorelli, N. *The Normalized Difference Vegetation Index*; Oxford University Press: Oxford, UK, 2013; ISBN 978-0199693160.

51. Ibharim, N.A.; Mustapha, M.A.; Lihan, T.; Mazlan, A.G. Mapping Mangrove Changes in The Matang Mangrove Forest Using Multi Temporal Satellite Imageries. *Ocean Coast. Manag.* **2015**, *114*, 64–76. [CrossRef]

52. Fauzi, A.; Sakti, A.; Yayusman, L.; Harto, A.; Prasetyo, L.; Irawan, B.; Kamal, M.; Wikantika, K. Contextualizing Mangrove Forest Deforestation in Southeast Asia Using Environmental and Socio-Economic Data Products. *Forests* **2019**, *10*, 952. [CrossRef]

53. Spalding, M.; Kainuma, M.; Collins, L. *World Atlas of Mangrove, Version 3.0*; A collaborative project of ITTO, ISME, FAO, UNEP-WCMC, UNESCO-MAB, UNU-INWEH and TNC; Earthscan: London, UK, 2010; 319p.

54. Saputro, G.B.; Hartini, S.; Sukardjo, S.; Susanto, A.; Poniman, A. *Peta Mangroves Indonesia*; Bakosurtanal: Bogor, Indonesia, 2012.

55. Bunting, P.; Rosenqvist, A.; Lucas, R.M.; Rebelo, L.-M.; Hilarides, J.; Thomas, N.; Hardy, A.; Itoh, T.; Shimada, M.; Finlayson, C.M. The global mangrove watch–a new 2010 global baseline of mangrove extent. *Remote Sens.* **2018**, *10*, 1669. [CrossRef]

56. Olson, D.M.; Dinerstein, E.; Wikramanayake, E.D.; Burgess, N.D.; Powell, G.V.N.; Underwood, E.C.; D'amico, J.A.; Itoua, I.; Strand, H.E.; Morrison, J.C.; et al. Terrestrial Ecoregions of the World: A New Map of Life on Earth: A new global map of terrestrial ecoregions provides an innovative tool for conserving biodiversity. *BioScience* **2001**, *51*, 933–938. [CrossRef]

57. Friedl, M.; Sulla-Menashe, D. MCD12Q1 MODIS/Terra+Aqua Land Cover Type Yearly L3 Global 500m SIN Grid V006 [Data set]. *NASA EOSDIS Land Process. DAAC* **2015**. [CrossRef]

58. Carroll, M.L.; DiMiceli, C.M.; Wooten, M.R.; Hubbard, A.B.; Sohlberg, R.A.; Townshend, J.R.G. MOD44W MODIS/Terra Land Water Mask Derived from MODIS and SRTM L3 Global 250m SIN Grid V006 [Data set]. *NASA EOSDIS Land Process. DAAC* **2017**. [CrossRef]

59. Pesaresi, M.; Ehrlich, D.; Ferri, S.; Florczyk, A.J.; Freire, S.; Halkia, S.; Julea, A.M.; Kemper, T.; Soille, P.; Syrris, V. *Operating Procedure for the Production of the Global Human Settlement Layer from Landsat Data of the Epochs 1975, 1990, 2000, and 2014*; Publications Oce of the European Union EUR 27741 EN: Ispra, Italy, 2016. [CrossRef]

60. Goldewijk, K.K.; Beusen, A.; Doelman, J.; Stehfest, E. Anthropogenic land-use estimates for the Holocene—HYDE 3.2. *Earth Syst. Sci. Data* **2017**, *9*, 927–953. [CrossRef]

61. NGDC NOAA, Version 4 DMSP-OLS Nighttime Lights Time Series. Available online: https://ngdc.noaa.gov/eog/dmsp/downloadV4composites.html (accessed on 8 August 2019).

62. Kummu, M.; Taka, M.; Guillaume, J.H.A. Gridded global datasets for Gross Domestic Product and Human Development Index over 1990–2015. *Sci. Data* **2018**, *5*, 180004. [CrossRef]

63. Doxsey-Whitfield, E.; MacManus, K.; Adamo, S.B.; Pistolesi, L.; Squires, J.; Borkovska, O.; Baptista, S.R. Taking Advantage of the Improved Availability of Census Data: A First Look at the Gridded Population of the World, Version 4. *Pap. Appl. Geogr.* **2015**, *1*, 226–234. [CrossRef]

64. ESA-European Space Agency. *Land Cover CCI Product User Guide, Version 2*; ESA CCI LC Project; European Space Agency: Paris, France, 2017.

65. Friedl, M.A.; Menashe, D.S.; Tan, B.; Schneider, A.; Ramankutty, N.; Sibley, A.; Huang, X. MODIS Collection 5 global land cover: Algorithm refinements and characterization of new datasets. *Remote. Sens. Environ.* **2010**, *114*, 168–182. [CrossRef]

66. Bontemps, S.; Defourny, P.; Van Bogaert, E.; Arino, O.; Kalogirou, V.; Ramos Pérez, J. GlobCover 2009. Products Description and Validation Report. Available online: http://due.esrin.esa.int/files/GLOBCOVER2009_Validation_Report_2.2.pdf (accessed on 12 June 2019).

67. Bicheron, P.; Huc, M.; Henry, C.; Bontemps, S.; Lacaux, J.P. *GlobCover Products Description Manual*; European Space Agency: Paris, France, 2008; p. 25.

68. Tateishi, R.; Uriyangqai, B.; Al-Bilbisi, H.; Aboel Ghar, M.; Tsend-Ayush, J.; Kobayashi, T.; Kasimu, A.; Thanh Hoan, N.; Shalaby, A.; Alsaaideh, B.; et al. Production of global land cover data—GLCMO. *Int. J. Digit. Earth* **2011**, *4*, 22–49. [CrossRef]

69. Tateishi, R.; Hoan, N.; Kobayashi, T.; Alsaaideh, B.; Tana, G.; Pong, D. Production of Global Land Cover Data—GLCNMO2008. *J. Geogr. Geol.* **2014**, *6*, 99–122. [CrossRef]

70. Didan, K. MOD13Q1 MODIS/Terra Vegetation Indices 16-Day L3 Global 250 m SIN Grid V006 [Data set]. *NASA EOSDIS Land Process. DAAC* **2015**. [CrossRef]

71. Didan, K. MOD13A1 MODIS/Terra Vegetation Indices 16-Day L3 Global 500m SIN Grid V006 [Data set]. *NASA EOSDIS Land Process. DAAC* **2015**. [CrossRef]

72. Hu, T.; Su, Y.; Xue, B.; Liu, J.; Zhao, X.; Fang, J.; Guo, Q. Mapping Global Forest Aboveground Biomass with Spaceborne LiDAR, Optical Imagery, and Forest Inventory Data. *Remote Sens.* **2016**, *8*, 565. [CrossRef]

73. Mutanga, O.; Skidmore A., K. Narrow band vegetation indices overcome the saturation problem in biomass estimation. *Int. J. Remote Sens.* **2004**, *25*, 3999–4014. [CrossRef]

74. Spawn, S.A.; Sullivan, C.C.; Lark, T.J.; Gibbs, H.K. Harmonized global maps of above and belowground biomass carbon density in the year 2010. *Sci. Data* **2020**, *7*, 112. [CrossRef]

75. Running, S.; Mu, Q. MOD16A2 MODIS/Terra Evapotranspiration 8-day L4 Global 500m SIN Grid. *NASA LP DAAC* **2015**. [CrossRef]

76. Funk, C.; Peterson, P.; Landsfeld, M.; Pedreros, D.; Verdin, J.; Shukla, S.; Husak, G.; Rowland, J.; Harrison, L.; Hoell, A.; et al. The climate hazards infrared precipitation with stations—A new environmental record for monitoring extremes. *Sci. Data* **2015**, *2*, 150066. [CrossRef]

77. Janowiak, J.E.; Joyce, R.J.; Yarosh, Y. A Real-Time Global Half-Hourly Pixel-Resolution Infrared Dataset and Its Applications. *Bull. Amer. Meteor. Soc.* **2001**, *82*, 205–218. [CrossRef]

78. Saha, S.; Moorthi, S.; Pan, H.; Wu, X.; Wang, J.; Nadiga, S.; Tripp, P.; Kistler, R.; Woollen, J.; Behringer, D.; et al. The NCEP Climate Forecast System Reanalysis. *Bull. Amer. Meteor. Soc.* **2010**, *91*, 1–146. [CrossRef]

79. Peterson, P.; Funk, C.C.; Landsfeld, M.F.; Husak, G.J.; Pedreros, D.H.; Verdin, J.P.; Rowland, J.; Shukla, S.; McNally, A.; Michaelsen, J.; et al. The Climate Hazards Group Infrared Precipitation with Stations (CHIRPS) Dataset: Quasi-Global Precipitation Estimates for Drought Monitoring and Trend Analysis. In Proceedings of the American Geophysical Union Fall Meeting, San Francisco, CA, USA, 15–19 December 2014.

80. Dimiceli, C.; Carroll, M.; Sohlberg, R.; Kim, D.H.; Kelly, M.; Townshend, J.R.G. MOD44B MODIS/Terra Vegetation Continuous Fields Yearly L3 Global 250m SIN Grid V006 [Data set]. *NASA EOSDIS Land Process. DAAC* **2015**. [CrossRef]

81. Herold, M.; Schmullius, C. *Report on the Harmonization of Global and Regional Land Cover Products Meeting*; FAO: Rome, Italy, 2004; Available online: https://gofcgold.org/sites/default/files/docs/GOLD_20.pdf (accessed on 14 April 2020).

82. Herold, M.; Woodcock, C.; Gregorio, A.D.; Mayaux, P.; Latham, A.B.; Schmullius, C.C. A joint initiative for harmonization and validation of land cover datasets. *IEEE Trans. Geosci. Remote Sens.* **2006**, *44*, 1719–1727. [CrossRef]

83. Roy, S.; Mahapatra, M.; Chakraborty, A. Mapping and monitoring of mangrove along the Odisha coast based on remote sensing and GIS techniques. *Model. Earth Syst. Environ.* **2019**, *5*, 217–226. [CrossRef]

84. Rhyma, P.; Norizah, K.; Hamdan, O.; Faridah-Hanum, I.; Zulfa, A. Integration of normalised different vegetation index and Soil-Adjusted Vegetation Index for mangrove vegetation delineation. *Remote Sens. Appl.* **2020**, *17*, 100280. [CrossRef]

85. Verbesselt, J.; Hyndman, R.; Zeileis, A.; Culvenor, D. Phenological change detection while accounting for abrupt and gradual trends in satellite image time series. *Remote Sens. Environ.* **2010**, *114*, 2970–2980. [CrossRef]

86. Bai, J.; Perron, P. Computation and analysis of multiple structural change models. *J. Appl. Econom.* **2003**, *18*, 1–22. [CrossRef]

87. Forkel, M.; Carvalhais, N.; Verbesselt, J.; Mahecha, M.D.; Neigh, C.S.R.; Reichstein, M. Trend change detection in NDVI time series: Effects of inter-annual variability and methodology. *Remote Sens.* **2013**, *5*, 2113–2144. [CrossRef]

88. Reef, R.; Lovelock, C.E. Regulation of water balance in mangroves. *Ann. Bot.* **2015**, *115*, 385–395. [CrossRef]

89. Santini, N.S.; Reef, R.; Lockington, D.A.; Lovelock, C. The use of fresh and saline water sources by the mangrove Avicennia marina. *Hydrobiologia* **2015**, *745*, 59–68. [CrossRef]

90. Feng, M.; Bai, Y. A global land cover map produced through integrating multi-source datasets. *Big Earth Data* **2019**, *3*, 191–219. [CrossRef]

91. Jaafar, H.H.; Ahmad, F.A.; El Beyrouthy, N. GCN250, new global gridded curve numbers for hydrologic modeling and design. *Sci. Data* **2019**, *6*, 145. [CrossRef] [PubMed]

92. Buchhorn, M.; Lesiv, M.; Tsendbazar, N.-E.; Herold, M.; Bertels, L.; Smets, B. Copernicus Global Land Cover Layers—Collection 2. *Remote Sens.* **2020**, *12*, 1044. [CrossRef]

93. Lloyd, C.T.; Sorichetta, A.; Tatem, A.J. High resolution global gridded data for use in population studies. *Sci. Data* **2017**, *4*, 170001. [CrossRef]

94. Vancutsem, C.; Marinho, E.; Kayitakire, F.; See, L.; Fritz, S. Harmonizing and Combining Existing Land Cover/Land Use Datasets for Cropland Area Monitoring at the African Continental Scale. *Remote Sens.* **2013**, *5*, 19–41. [CrossRef]

95. Song, X.; Huang, C.; Townshend, J.R. Improving global land cover characterization through data fusion. *Geo-Spat. Inf. Sci.* **2017**, *20*, 141–150. [CrossRef]

96. Herold, M.; Mayaux, P.; Woodcock, C.E.; Baccini, A.; Schmullius, C. Some challenges in global land cover mapping: An assessment of agreement and accuracy in existing 1 km datasets. *Remote Sens. Environ.* **2008**, *112*, 2538–2556. [CrossRef]

97. Lu, M.; Wu, W.; Zhang, L.; Liao, A.; Peng, S.; Tang, H. A comparative analysis of five global cropland datasets in China. *Sci. China Earth Sci.* **2016**, *59*, 2307–2317. [CrossRef]

98. Fritz, S.; See, L.; Perger, C.; McCallum, I.; Schill, C.; Schepaschenko, D.; Duerauer, M.; Karner, M.; Dresel, C.; Laso-Bayas, J.-C.; et al. A global dataset of crowdsourced land cover and land use reference data. *Sci. Data* **2017**, *4*, 170075. [CrossRef]

99. Yang, Y.K.; Xiao, P.F.; Feng, X.Z.; Li, H.X. Accuracy assessment of seven global land cover datasets Journal Pre-proof Journal Pre-proof over China. *ISPRS-J. Photogramm. Remote Sens.* **2017**, *125*, 156–173. [CrossRef]

100. Islam, S.; Zhang, M.; Yang, H.; Ma, M. Assessing inconsistency in global land cover products and synthesis of studies on land use and land cover dynamics during 2001 to 2017 in the southeastern region of Bangladesh. *J. Appl. Remote Sens.* **2019**, *13*, 048501. [CrossRef]

101. Barnabe, G. *Aquaculture*; Ellis Horwood: New York, NY, USA, 1990; Volume I, ISBN 0-203-16883-6.

102. Webb, E.L.; Jachowski, N.R.A.; Phelps, J.; Friess, D.A.; Than, M.M.; Ziegler, A.D. Deforestation in the Ayeyarwady Delta and the conservation implications of an internationally-engaged Myanmar. *Glob. Environ. Chang.* **2014**, *24*, 321–333. [CrossRef]

103. Lai, S.; Loke, L.H.L.; Hilton, M.J.; Bouma, T.J.; Todd, P.A. The effects of urbanization on coastal habitats and the potential for ecological engineering: A Singapore case study. *Ocean Coast. Manag.* **2015**, *103*, 78–85. [CrossRef]

104. Weier, J.; Herring, D. *Measuring Vegetation (NDVI & EVI)*; NASA Earth Observatory: Washington, DC, USA, 2000. Available online: https://earthobservatory.nasa.gov/features/MeasuringVegetation (accessed on 24 April 2020).

105. Cai, Z.; Jonsson, P.; Jin, H.; Eklundh, L. Performance of Smoothing Methods for Reconstructing NDVI Time-Series and Estimating Vegetation Phenology from MODIS Data. *Remote Sens.* **2017**, *9*, 1271. [CrossRef]

106. Meneses-Tovar, C. NDVI as indicator of degradation. In *Measuring Forest Degradation*; Unasylva No. 238; Obstler, R., Ed.; Food and Agriculture Organization of the United Nations: Rome, Italy, 2011; Volume 62.

107. Jamali, S.; Jonsson, P.; Eklundh, L.; Ardo, J.; Seaquist, J. Detecting changes in vegetation trends using time series segmentation. *Remote. Sens. Environ.* **2015**, *156*, 182–195. [CrossRef]

108. Huete, A.R. A soil-adjusted vegetation index (SAVI). *Remote Sens. Environ.* **1988**, *25*, 295–309. [CrossRef]

109. Gao, B. NDWI–A normalized difference water index for remote sensing of vegetation liquid water from space. *Remote Sens. Environ.* **1996**, *58*, 257–266. [CrossRef]

110. Cancela, J.J.; Gonzalez, X.P.; Vilanova, M.; Miras-Avalos, J.M. Water Management Using Drones and Satellites in Agriculture. *Water* **2019**, *11*, 874. [CrossRef]

111. Marusig, D.; Petruzzellis, F.; Tomasella, M.; Napolitano, R.; Altobelli, A.; Nardini, A. Correlation of Field-Measured and Remotely Sensed Plant Water Status as a Tool to Monitor the Risk of Drought-Induced Forest Decline. *Forests* **2020**, *11*, 77. [CrossRef]

112. Kim, D.M.; Zhang, H.; Zhou, H.; Du, T.; Wu, Q.; Mockler, T.C.; Berezin, M.Y. Highly sensitive image-derived indices of water-stressed plants using hyperspectral imaging in SWIR and histogram analysis. *Sci. Rep.* **2015**, *5*, 15919. [CrossRef]

113. Kamble, B.; Kilic, A.; Hubbard, K. Estimating Crop Coefficients Using Remote Sensing-Based Vegetation Index. *Remote Sens.* **2013**, *5*, 1588–1602. [CrossRef]

114. Kumar, J.; Hoffman, F.M.; Hargrove, W.W.; Collier, N. Understanding the representativeness of FLUXNET for upscaling carbon flux from eddy covariance measurements. *J. Earth Syst. Sci. Data* **2016**. [CrossRef]

115. Rodda, S.R.; Thumaty, K.C.; Jha, C.S.; Dadhwal, V.K. Seasonal Variations of Carbon Dioxide, Water Vapor and Energy Fluxes in Tropical Indian Mangroves. *Forests* **2016**, *7*, 35. [CrossRef]

116. Wang, L.; Jia, M.; Yin, D.; Tian, J. A Review of Remote Sensing for Mangrove Forests: 1956–2018. *Remote Sens. Environ.* **2019**, *231*, 111223. [CrossRef]

117. Smith, M. CROPWAT—A Computer Program for Irrigation Planning and Management. In *Irrigation and Drainage Paper*; Food and Agriculture Organization of the UN: Rome, Italy, 1994; Volume 46, ISBN 978-9251031063.

118. Oon, A.; Shafri, H.Z.M.; Lechner, A.M.; Azhar, B. Discriminating between large-scale oil palm plantations and smallholdings on tropical peatlands using vegetation indices and supervised classification of LANDSAT-8. *Int. J. Remote Sens.* **2019**, *40*, 7287–7296. [CrossRef]

119. Sakti, A.D.; Takeuchi, W.; Wikantika, K. Development of Global Cropland Agreement Level Analysis by Integrating Pixel Similarity of Recent Global Land Cover Datasets. *J. Environ. Prot.* **2017**, *8*, 1509–1529. [CrossRef]

120. Matsuda, M. Dynamics of rice production in Myanmar: Growth centers, technological changes, and driving forces. *Trop. Agric. Dev.* **2009**, *53*, 14–27. [CrossRef]

121. Ilman, M.; Dargusch, P.; Dart, P.; Onrizal. A historical analysis of the drivers of loss and degradation of Indonesia's mangroves. *Land Use Policy* **2016**, *54*, 448–459. [CrossRef]

122. Blasco, F.; Aizpuru, M.; Gers, C. Depletion of the mangroves of Continental Asia. *Wetl. Ecol. Manag.* **2001**, *9*, 245–256. [CrossRef]

123. Storey, D. *A Socio-Ecological Assessment of Mangrove Areas in Sittwe, Pauktaw, Minbya, and Myebon Townships, North Rakhine State*; REACH: Geneva, Switzerland, 2015.

124. Plathong, S.; Plathong, J. Past and Present Threats on mangrove ecosystem in peninsular Thailand. In *Coastal Biodiversity in Mangrove Ecosystems: Paper Presented in UNU-INWEH-UNESCO International Training Course, Held at Centre of Advanced Studies*; Annamalai University: Chidambaram, India, 2004; pp. 1–13.

125. Kusmana, C. Lesson Learned from Mangrove Rehabilitation Program in Indonesia. *J. Pengelolaan Sumberd. Alam Dan Lingkung.* **2017**, *7*, 89–97. [CrossRef]

126. Popkin, G. How Much Can Forests Fight Climate Change? *Nature* **2019**, *565*, 280–282. [CrossRef] [PubMed]

127. Betts, M.G.; Wolf, C.; Ripple, W.J.; Phalan, B.; Millers, K.A.; Duarte, A.; Butchart, S.H.M.; Levi, T. Global forest loss disproportionately erodes biodiversity in intact landscapes. *Nature* **2017**, *547*, 441–444. [CrossRef] [PubMed]

128. Kulp, S.A.; Strauss, B.H. New elevation data triple estimates of global vulnerability to sea-level rise and coastal flooding. *Nat. Commun.* **2019**, *10*, 4844. [CrossRef] [PubMed]

129. Sakti, A.D.; Takeuchi, W. A Data-Intensive Approach to Address Food Sustainability: Integrating Optic and Microwave Satellite Imagery for Developing Long-Term Global Cropping Intensity and Sowing Month from 2001 to 2015. *Sustainability* **2020**, *12*, 3227. [CrossRef]

130. Zhou, Y.; Varquez, A.C.G.; Kanda, M. High-resolution global urban growth projection based on multiple applications of the SLEUTH urban growth model. *Sci. Data* **2019**, *6*, 34. [CrossRef]

131. Pekel, J.-F.; Cottam, A.; Gorelick, N.; Belward, A.S. High-resolution mapping of global surface water and its long-term changes. *Nature* **2016**, *540*, 418–422. [CrossRef]

132. Bryan-Brown, D.N.; Connolly, R.M.; Richards, D.; Adame, F.; Friess, D.A.; Brown, C.J. Global trends in mangrove forest fragmentation. *Sci. Rep.* **2020**, *10*, 7117. [CrossRef]

133. Friess, D.A.; Rogers, K.; Lovelock, C.E.; Krauss, K.W.; Hamilton, S.E.; Lee, S.Y.; Lucas, R.; Primavera, J.; Rajkaran, A.; Shi, S. The State of the World's Mangrove Forests: Past, Present, and Future. *Annu. Rev. Environ. Resour.* **2019**, *44*, 89–115. [CrossRef]

Letter

Assessing the Applicability of Mobile Laser Scanning for Mapping Forest Roads in the Republic of Korea

Hyeongkeun Kweon [1], Jung Il Seo [2,*] and Joon-Woo Lee [3]

[1] Institute of Ecological Restoration, Kongju National University, 54 Daehak-ro, Yesan-eup, Yesan-gun, Chungcheongnam-do 32439, Korea; hkkweon00@gmail.com

[2] Department of Forest Resources, Kongju National University, 54 Daehak-ro, Yesan-eup, Yesan-gun, Chungcheongnam-do 32439, Korea

[3] Department of Environment and Forest Resources, Chungnam National University, 99 Daehak-ro, Yuseong-gu, Daejeon 34134, Korea; jwlee@cnu.ac.kr

* Correspondence: jungil.seo@kongju.ac.kr; Tel.: +82-41-330-1302

Received: 8 April 2020; Accepted: 6 May 2020; Published: 8 May 2020

Abstract: Forest roads are an essential facility for sustainable forest management and protection. With advances in survey technology, such as Light Detection and Ranging, forest road maps with greater accuracy and resolution can be produced. This study produced a 3D map for establishment of a forest road inventory using a Mobile Laser Scanning (MLS) device mounted on a vehicle in four study forest roads in Korea, in order to review its precision, accuracy and efficiency based on comparisons with mapping using Total Station (TS) and Global Navigation Satellite System (GNSS). We counted the points that consist of the cloud data of the maps to determine the degree of precision density, and then compared this with 50 points at 20-m intervals on the centerlines bisecting the widths of the study forest roads. Then, we evaluated the relative positional accuracy of the MLS data based on three criteria: the total length of each forest road; the Root Mean Square Error (RMSE) obtained from coordinate values of the MLS and TS surveys compared to the GNSS survey; and the ratios of the centerlines extracted by the MLS and TS surveys overlaid to the buffer zone by the GNSS survey. Finally, we estimated the time and cost per unit length for producing the map to examine the efficiency of MLS mapping compared to the other two surveys. The results showed that the point cloud data acquired by the MLS survey on the study forest roads had very high precision and so is sufficient to produce a 3D forest road map with high-precision density and a low RMSE value. Although the equipment rental cost is somewhat high, the fact that information targeting on all spatial elements of forest roads can be obtained with a low cost of labor is a benefit when evaluating the efficiency of MLS survey and mapping. Our findings are expected to provide a quantitative assessment of both maintaining sustainable effectiveness and preventing potential environmental damage of forest roads.

Keywords: forest road inventory; total station; global navigation satellite system; point cloud; precision density; positional accuracy; efficiency

---

1. Introduction

Worldwide, forest roads are essential facilities for the sustainable management and conservation of forests with economic and/or public value [1–4]. Forest roads, however, are low-volume roads built not only on lowland areas but also on rough terrain in the forest, so they may have both steep longitudinal lines and complex planar features [5]. They therefore need to be well-managed to maintain their sustainable effectiveness and to prevent potential environmental damage. For the systematic management of forest roads, the acquisition and inventory of spatial information of the roads should be prioritized. In addition, monitoring changes in the spatial information and follow-up treatments based on these changes are required.

The acquisition of spatial information related to forest roads is drawing keen attention due to the recent development of survey technology [6]. In particular, the development of technology for processing and analyzing data obtained through laser and photographic measurements has enabled efficient surveys and analysis of forest resources. Light Detection and Ranging (LiDAR) surveys using Aerial Laser Scanning (ALS) and Terrestrial Laser Scanning (TLS) technologies can efficiently acquire and visualize high-resolution forest spatial information, and thus are widely used for characterizing forest resources and topography. Conceptually, aircraft-based ALS should be suitable for measurements targeting a relatively wide range of forest spaces [3,7–9], and ground point-based TLS should be suitable for measurements targeting a relatively narrow range of forest spaces [10–13]. However, LiDAR surveys are limited to obtaining spatial information on forest roads in the form of lines. For example, it is challenging if not impossible to use ALS for surveying cut- and fill-slopes that are shielded by stand crowns; and it is difficult to survey long stretches of forest roads within a short timeframe by using TLS. Unmanned Aerial Vehicles (UAVs) can be suggested as an alternative, but their low battery charge greatly limits survey time [14]. As one of the ways to overcome these limitations, Mobile Laser Scanning (MLS) is a highly advanced mapping technology that can acquire and visualize spatial information on and around forest roads [15]. In particular, MLS equipped vehicles is a technology that is accurate and efficient, and can help rational decision making for systematic forest road management by obtaining the best spatial information for given spatiotemporal conditions.

Research on MLS dates back to the early 1970s when research on measurement technology using film cameras was conducted. Later, the use of Global Positioning Systems (GPS) expanded and the production and analysis techniques of images were developed, evolving into an image-based mapping technique by the late 1980s [16]. In the 1990s, MLS techniques went through a technological sophistication phase, and in the 2000s, high-performance MLS laser scanning sensors began to be used [17]. As a representative study, Grejner-Brzeinska et al. [18] suggested ways to overcome the effects of gravity, eliminate signal noise, and mitigate Inertial Measurement Unit (IMU) errors to increase the positioning accuracy of GPS and Inertial Navigation Systems (INS), which are the basic elements of MLS. During the 2010s, practical application and industrialization of MLS technology became well-established. For example, De Agostino [13] obtained spatial information related to rock walls adjacent to roads using vehicle MLS and then corrected that information with a Global Navigation Satellite System (GNSS), thereby providing a reliable technique to manage areas vulnerable to rock collapse and falls. In addition, Qin et al. [19] proposed a line-based model to improve accuracy by combining Point Cloud Data acquired using MLS with images taken on the ground. Cui et al. [20] set a line-based model for point cloud data and panoramic images acquired through MLS, applying a multisensor in urban areas, and evaluated the accuracy of the converted model by Root Mean Square Error (RMSE) analysis. Recently, MLS surveys are used to acquire and analyze on- and off-road information inventory, including the detection and extraction of on-road objects (e.g., road surface, road crack, and driving line) and off-road objects (e.g., traffic lights, signs, and utility poles), to use as the basis of Advanced Driving Assistant Systems [21,22]. Nevertheless, few studies have dealt with cases in which MLS has been applied to the field of forestry because that there are severe limitations on using laser scanning systems due to vegetation in forested areas, unlike urban areas [15,21]. Exception only includes several studies have mainly investigated the accuracy of measuring tree diameter by handheld MLS system [23–25]. Upon considering the spatial peculiarity of forests, it may be very labor-intensive to collect large quantities of high-quality spatial information. However, it is expected that it will be relatively easy to obtain large quantities of high-quality spatial information on and around the forest roads where the crown layer of forests is opened due to the tree cutting, which is considered a great advantage in building an inventory for forest road management.

From this background, this study was aimed at using a MLS system to accurately and conveniently acquire rich spatial information on and around forest roads, and thereby to provide precise data for efficient forest road management. The objectives of this study were: (i) to produce 3D forest road maps using MLS system, including characterization of both on-road surface and cut- and fill-slopes, and (ii) to evaluate MLS precision density, relative positional accuracy, and mapping efficiency by comparing it to data obtained by GNSS and Total Station (TS) techniques. To achieve these objectives, we conducted the study in the order shown in Figure 1.

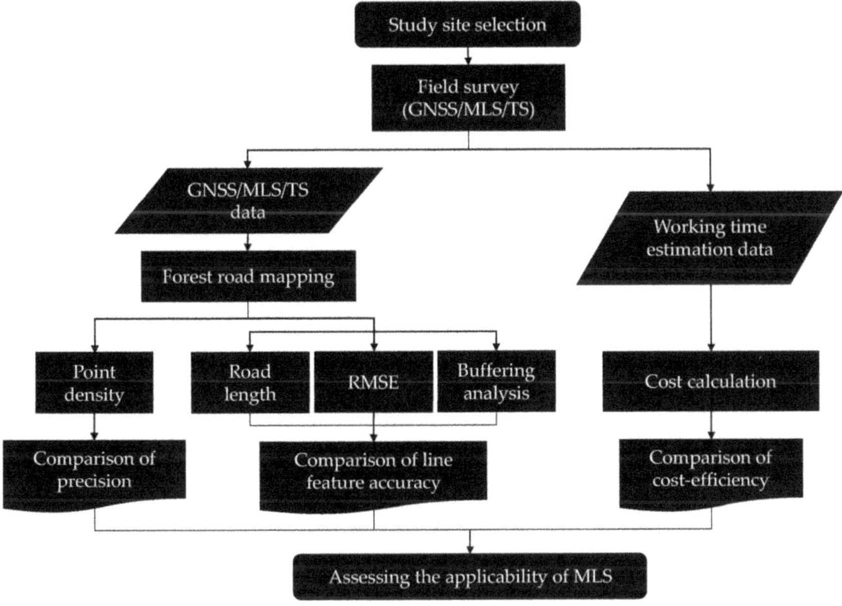

Figure 1. Flowchart displaying an approach of the study.

2. Materials and Methods

2.1. Study Site Descriptions

In the study, four forest roads, one each from four forested areas in Gongju-si, Nonsan-si, Hongseong-gun, and Geumsan-gun in Chungcheongnam-do, Republic of Korea, were selected as the study sites for the 3D forest road mapping using MLS. These forests, covered with mixed pines and oaks of approximately 40 years in age, are managed by the local government and on steep and rugged terrain at an altitude of 100–500 m above sea level.

The four road routes selected for the study were unpaved with lengths approximately equal to each other. Based on the GNSS survey, the Gongju route was 1022 m, the Nonsan route was 1015 m, and the Hongseong and Geumsan routes were 1019 m. The difference in altitude above sea level within each road route was the largest in the Nonsan route with 62.7 m, the smallest in the Gongju route with 11.7 m, and the Geumsan route (34.7 m) and Hongseong route (19.3 m) were intermediate. The forest roads were newly built within the last two years on hillslopes with steep cross-sectional gradient, and in a condition that both the cut-and-fill slope surfaces of the road routes were not yet fully covered with even understory vegetation.

The locations and general characteristics of the sites subject to field investigation are shown in Figure 2 and Table 1.

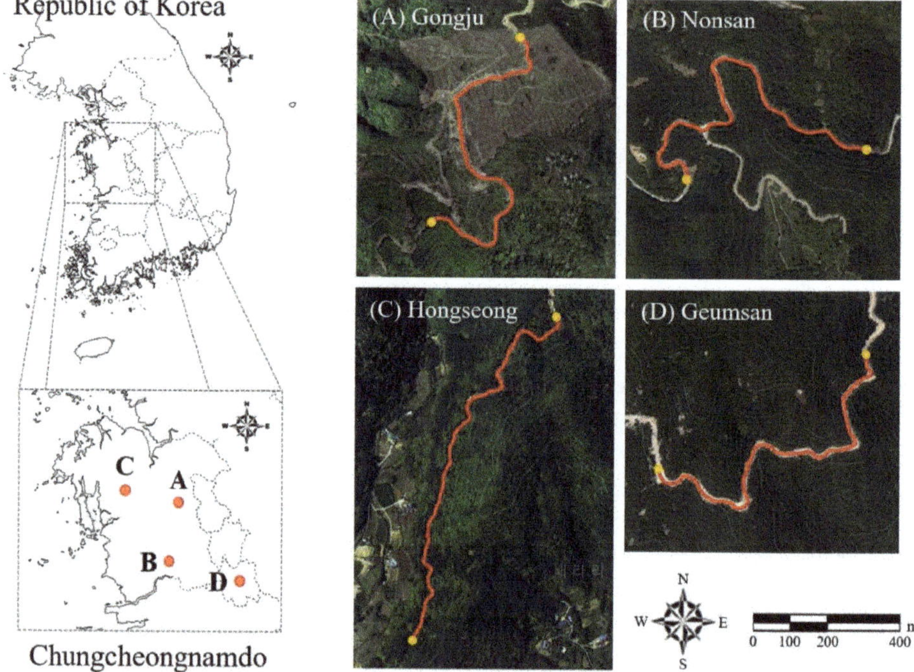

Figure 2. Locations of the study sites.

Table 1. General characteristics of the study sites.

Index	Study Area	Road Length Surveyed (m)	Altitude above Sea Level (m)		Construction Year	Forest Type
			Minimum	Maximum		
A	Gongju	1022	439.1	450.8	2017	Mixed forest
B	Nonsan	1015	127.9	190.6	2017	Mixed forest
C	Hongseoung	1019	135.4	154.7	2018	Mixed forest
D	Geumsan	1019	424.5	459.2	2017	Mixed forest

2.2. Collection and Mapping of MLS Data

Field collection of point cloud data was conducted using a vehicle equipped with a model MX2 MLS instrument (Trimble Inc., Sunnyvale, CA, USA, Figure 3). The performance of the Trimble MX2 model is very accurate and efficient (Table 2). Each on the left and right sides of the model's body has one laser scanner, and therefore two data sets were collected as the vehicle moved forward, providing an accurate and complete point cloud of the terrain along the road.

To secure the accuracy of the point cloud data acquired from not only the on-road and cut-slope surfaces (with relatively large viewing angles while driving) but also the fill-slope surfaces (with relatively small viewing angles while driving) of forest roads, we drove the MLS vehicle back and forth. Here, we drove it close to the right edge of the road based on the direction of driving, especially to survey the fill-slopes even at a sharp angle. We also maintained a speed of less than 20 km/h because the MLS survey was conducted on rough unpaved roads, although the maximum design speed of vehicles on forest roads in Korea is 40 km/h [26]. Consequently, we therefore acquired four sets of point cloud data from two laser sensors for each study road.

Figure 3. The vehicle equipped with Mobile Laser Scanner (MLS, Trimble MX2 model).

Table 2. Specifications of Trimble MX2 [27].

Mobile Laser Scanning (MLS) Trimble MX2	
Type	Single or dual SLM-250 Class 1 lasers
Range	Up to 250 m
Accuracy	±1 cm at 50 m
Scanner FOV	360 degrees
Scan rate	Dual laser head: 2 × 20 Hz (1200 rpm)
Maximum effective measurement rate	Single laser head: 36,000 points per second Dual laser head: 72,000 points per second
Pulse rate	Dual laser head: 2 × 36 kHz

The coordinates of the acquired point cloud data were calibrated based on the national reference points, and the 3D maps of the studied road routes were produced using the calibrated data in the Trimble Realworks 10.4 software (Trimble Inc.). However, these maps include spatial information relevant to not only topographic and structural features belonging to cut-slope, on-road, and fill-slope surfaces, but also vegetation on those surfaces and outside the cut-and-fill slopes. We therefore removed only the information relevant to vegetation from the maps, and finally extracted a 3D map that visualized the topographic and structural features that actually constituted the forest roads. To examine the change in the degree of precision density during these 3D mapping processes as well as to evaluate quantitatively MLS survey with relatively high-precision, we counted the number of points that consisted of the cloud data for both the initial and final 3D forest road maps.

On each final 3D map produced previously, we digitized the centerline bisecting the width of forest road and extracted it as a polyline-type shape file. We then extracted coordinate values (i.e., x, y, z) of 50 points with 20-m intervals on the centerline shape file of each forest road, and estimated the distances between two consecutive points.

2.3. Collection and Mapping of TS and GNSS Data

TS and GNSS are instruments that are commonly used to survey the geographic position of any given point. Using these instruments (i.e., NPL 352, Nikon, Tokyo, Japan, and Trimble R8, respectively; Table 3), we surveyed the 50 points at 20 m intervals on the centerline that bisects the width of each forest road route. The data obtained by the GNSS, which have absolute coordinates of the 50 points, were used to make the reference map displaying the centerline, and those by TS, which have relative coordinates of the exact same 50 points, were used to make the comparative one to the reference map (Figure 4).

Table 3. Specifications of Total Station (TS) [28] and Global Navigation Satellite System (GNSS) [27].

	TS (Nikon NPL 352)	GNSS (Trimble R8)
Operating temperature	−40 to +60 °C	−40 to +65 °C
Distance measurement	1.6 to 200 m	-
Weight	5.5 kg	1.52 kg
Accuracy	x : 3 mm y : 3 mm z : 3 mm	x : 8 mm y : 8 mm z : 15 mm

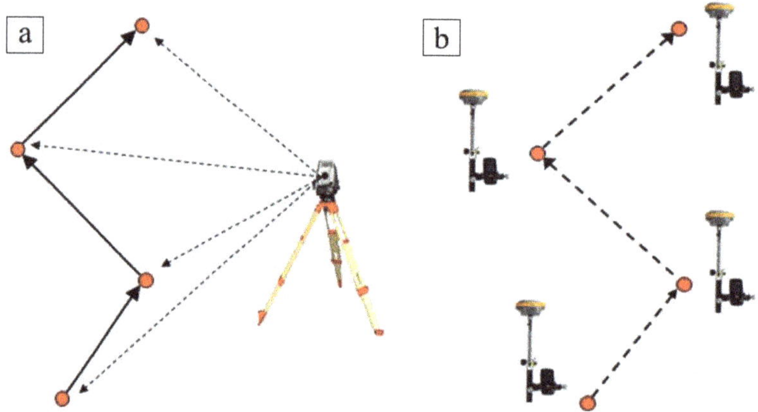

Figure 4. Schematic diagram displaying the difference between (**a**) Total Station (TS) survey with relative coordinates) and (**b**) Global Navigation Satellite System (GNSS) survey with absolute coordinates.

2.4. Comparison of the Relative Positional Accuracy on MLS, TS, and GNSS Data

The evaluation of the relative positional accuracy of the three types of survey data above was based on three criteria. First, using the length of each 20 m section extracted previously on the centerline, the length of the road route on the comparative map surveyed by the MLS or TS survey was compared to the GNSS survey reference map. Second, the 50 coordinate values (i.e., x, y, z) determined from the MLS or TS survey for each road route were applied to calculate the Root Mean Square Error (RMSE), Equation (1) [29], based on a comparison to the same points established by the GNSS survey:

$$RMSE = \sqrt{\frac{\sum (C_i - C_t)^2}{(n-1)}} \tag{1}$$

where C_i are the coordinates from the GNSS survey, C_t are the coordinates from the MLS or TS survey results, and n represents the total number of stations. Third, the ratio of the comparative map (polyline) by the MLS or TS survey overlaid to the buffer zone (polygon) of the GNSS survey reference map for each road route was calculated. Here, the radius of the buffer zone was increased by 0.1 m, and the buffer diameters were compared with each other when 95% and 100% of the comparative maps by the MLS and TS surveys were included within the buffer zone of the GNSS survey reference map. All spatial analysis was performed using QGIS 2.18 (QGIS development team, Boston, MA, USA) to compare the relative positional accuracy of the studied roads.

2.5. Comparison of the Mapping Time and Cost Using MLS, TS, and GNSS

The time and cost of mapping the survey results are typically expressed as the value per unit area [30,31]. However, because this study targeted the linear object, i.e., forest road, the time per unit length (unit time, min/km) and the cost per unit length (unit cost, USD/km) required to produce the map were estimated. Here, the mapping by the MLS survey generated a 3D spatial map displaying on-road surfaces and cut-and-fill slopes of the forest road as the final output, while the mapping by TS or GNSS surveys provided a 3D line map representing only the centerline of the forest road as the final output. Although these outputs do not contain the same level of spatial information, the time and cost spent obtaining the outputs were compared.

3. Results

3.1. 3D Forest Road Maps Created Using MLS and The Degree of Precision Density

The 3D forest road maps produced initially by the MLS survey in the four study roads are shown in Figures 5a and 6a. By removing vegetation information from these maps, the 3D maps show only the topography and structures (e.g., ditches and drainage facilities exposed to the ground) of the forest roads (Figures 5b and 6b).

The results examining the number of point data comprising these two types of 3D maps (Table 4) showed that the number of points lost in removing the vegetation information ranged from 54.4% to 75.1%, with a mean of 65.6%, based on the initial number of point data. In addition, the numbers of point data per unit road surface area (unit point numbers), which were estimated in the final 3D maps, ranged from 725.8 point/m^2 to 1249.3 point/m^2, with a mean of 878.2 point/m^2.

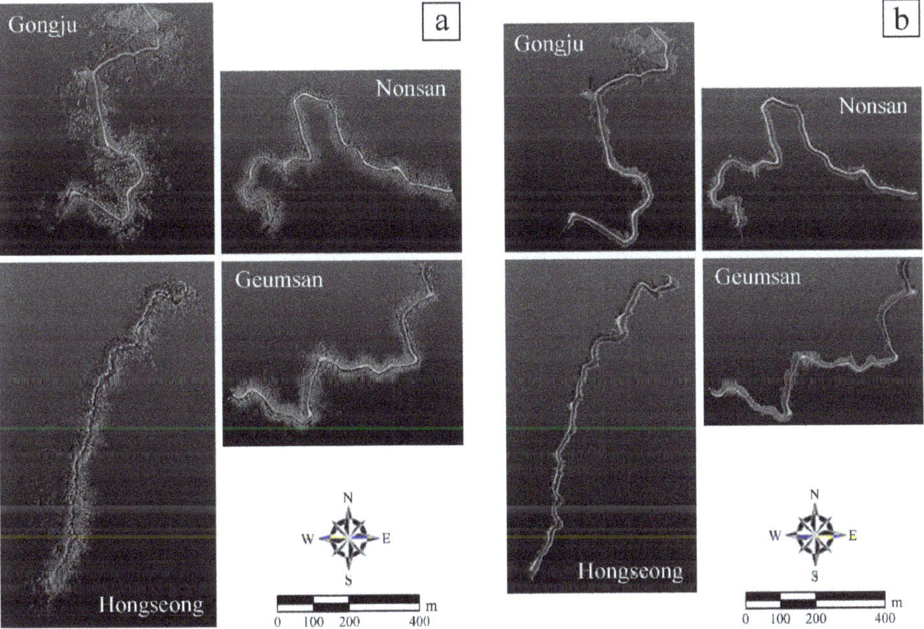

Figure 5. The 3D maps created using point cloud data obtained from the Mobile Laser Scanner (MLS) survey on the study forest roads. The map (**a**) before and (**b**) after removing point cloud data displaying vegetation on the outsides of cut-and-fill slopes.

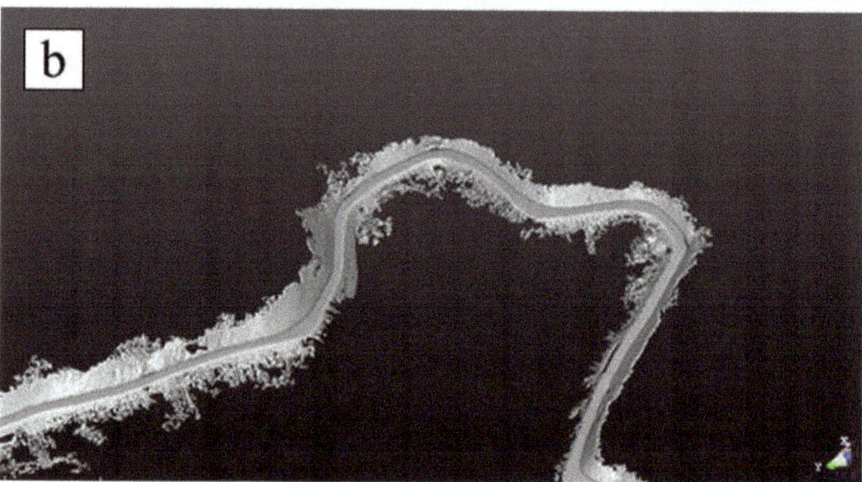

Figure 6. An example of the close range view of the 3D map displaying forest roads using point cloud data obtained from the Mobile Laser Scanner (MLS) survey. The view (**a**) before and (**b**) after removing point cloud data displaying vegetation on the outsides of cut-and-fill slopes.

Table 4. Change in the number of point data by excluding vegetation information from the initial 3D maps of the four forest roads studied.

Study Road	Number of Point Data (point/km)		Point Data Loss by Excluding Vegetation Information (point/km, %)	Forest Road Area in the Final 3D Map (m²/km)	Unit Point Number in the Final 3D Map (point/m²)
	Initial 3D Map	Final 3D Map			
Gongju	26,310,704	9,729,385	16,581,319 (63.0)	12,427.4	782.9
Nonsan	33,920,193	10,174,684	23,745,509 (70.0)	13,480.0	754.8
Hongseong	36,070,736	8,971,478	27,099,258 (75.1)	12,360.8	725.8
Geumsan	37,813,718	17,237,514	20,576,204 (54.4)	13,797.7	1249.3
Mean	33,528,838	11,528,265	22,000,573 (65.6)	13,016.5	878.2

3.2. Relative Positional Accuracy Estimated from MLS, TS, and GNSS Data

The length of the road route surveyed was determined as the first criterion to compare the relative positional accuracy of the MLS survey with other survey results. As noted in the site descriptions, the length of each route was approximately 1 km. However, the lengths showed slight differences according to the survey methods (Table 5); that is, difference in the road lengths (i.e., lengths of the centerlines bisecting the widths of forest road routes) between GNSS and MLS surveys only ranged from 0.1 to 0.7 m, while the difference between GNSS and TS surveys revealed a larger range, from 2.1 to 4.7 m.

Table 5. Lengths of road routes surveyed by Mobile Laser Scanner (MLS), Total Station (TS), and Global Navigation Satellite System (GNSS) in the forest roads studied.

Study Road	Road Length (m)				
	GNSS	MLS	MLS–GNSS	TS	TS–GNSS
Gongju	1022.2	1022.3	0.1	1025.1	2.9
Nonsan	1015.1	1015.8	0.7	1019.8	4.7
Hongseong	1019.2	1019.7	0.5	1021.3	2.1
Geumsan	1019.4	1019.6	0.2	1023.2	3.8
Mean			0.4		3.4

Calculating the RMSEs of the MLS or TS survey to the GNSS survey for 50 corresponding points for each road route, which is the second criterion to examine the relative positional accuracy of the MLS survey, showed that there is no significant differences in the RMSEs between the MLS and the TS at all study road routes (Table 6).

Table 6. RMSEs of the Mobile Laser Scanner (MLS) or Total Station (TS survey) to the Global Navigation Satellite System (GNSS) survey for 50 corresponding points for each road route.

Survey Method	Total ($n = 200$)		Gongju ($n = 50$)		Nonsan ($n = 50$)		Hongseong ($n = 50$)		Geumsan ($n = 50$)	
	Mean	SD	Mean	SD	Mean	SD	Mean	SD	Mean	SD
MLS	0.802	0.473	0.685	0.328	0.745	0.364	0.961	0.481	0.817	0.631
TS	0.716	0.593	0.553	0.401	0.813	0.642	0.868	0.724	0.625	0.501
t	1.778		1.801		−0.833		0.828		1.796	
p	0.076		0.074		0.406		0.409		0.076	

The SD denotes standard deviation.

Calculating the ratio of the comparative map by MLS or TS survey overlaid to the buffer zone of the GNSS survey reference map for each road route, which is the third criterion to examine the relative positional accuracy of the MLS survey, gave the results shown in Figure 7. The centerline comparative maps produced with the MLS and TS surveys were nested by 95% within diameters of 1.88 m (0.94 m radius) and 1.90 m (0.95 m radius), respectively, and also were nested by 100% within diameters of 2.6 m (1.3 m radius) and 4.4 m (2.2 m radius), respectively, from the centerline produced by the GNSS survey.

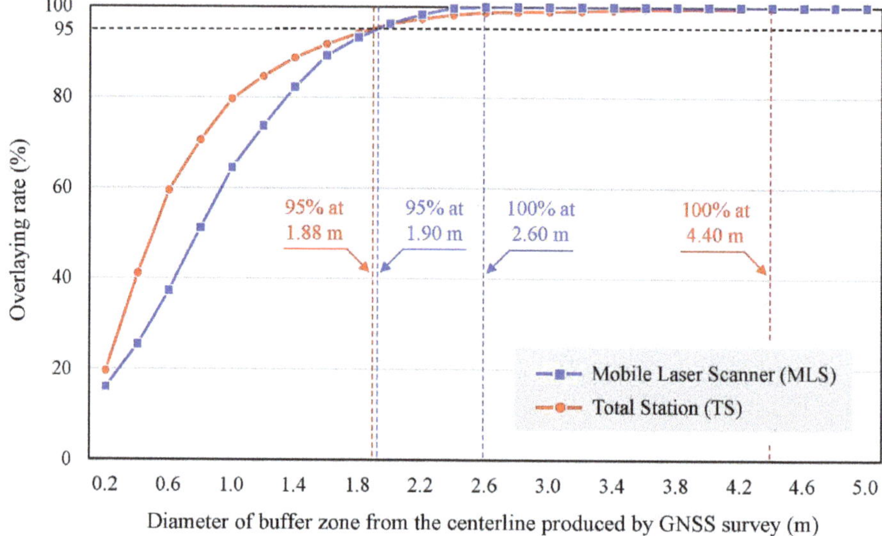

Figure 7. Changes in mean ratios of the comparative maps by Mobile Laser Scanner (MLS) and Total Station (TS) surveys overlaid to the buffer zones of the Global Navigation Satellite System (GNSS) survey reference maps for the four forest road routes.

3.3. Estimated Mapping Time and Cost Using MLS, TS, and GNSS

Table 7 shows that the unit mean time and cost spent for mapping the four forest road routes were different with different surveying methods. The mapping starting with the MLS survey spent the unit mean time of approximately 127 min/km (i.e., 31.7 min/km for field survey, 64 min/km for mapping). In the mapping by the TS survey, the unit mean time spent was approximately 313 min/km (i.e., 195 min/km for field survey, 118 min/km for mapping).

Table 7. Differences in the unit mean time and cost spent depending on the surveying methods in the four forest road routes.

Survey Method	Work	Labor	Working Time (min/km)	Labor Cost Unit (USD/h)	Labor Cost Total (USD)	Equipment Cost * (USD)	Total Cost (USD)
MLS	Field survey	1 professional 1 assistant	32	20.5 10.1	32.0	143.2	175.2
	Mapping	1 technician	64	15.2			
TS	Field survey	1 professional 2 assistants	195	20.5 10.1	155.1	46.8	201.9
	Mapping	1 technician	118	15.2			
GNSS	Field survey	1 professional	128	20.5	56.2	26.3	82.5
	Mapping	1 technician	60	15.2			

* Equipment rental price during the operation time.

The mapping by GNSS survey required a unit mean time spent of approximately 188 min/km (i.e., 128 min/km for field survey, 60 min/km for mapping). Upon considering that three, four, and two persons participated in the mapping by MLS, TS, and GNSS surveys, respectively, the unit mean cost for mapping using the TS survey (201.9 USD/km) was the largest, followed by the MLS (175.2 USD/km) and GNSS (82.5 USD/km) surveys. Here, two assistants for prisms were employed in the TS survey to shorten the expected long working time even if the relevant labor cost was to increase.

4. Discussion

4.1. Precision Density of Mapping by the MLS Survey

Numerous studies (e.g., [1–3,7]) indicate the need to establish a road inventory for maintenance of forest roads. To accomplish this, surveys using TS and GNSS have been conducted worldwide [6]. In recent years, the interest and need for 3D data using high density point clouds using remote exploration, such as LiDAR, has increased [6,10,11,32].

Mapping by MLS survey was shown to have a relatively high density of precision compared to other survey methods (Table 4), and this is not based solely on the findings of this study, which showed that the mean of unit point numbers on the MLS maps (removed vegetation information) is arithmetically 230,565 times those of both GNSS and TS (with 50 point/km). Of the previous studies regarding the establishment of a forest road inventory, Kim et al. [31] and Talebi et al. [2] conducted TS and GNSS surveys, respectively, to acquire spatial information including locations and/or boundaries of road centerlines, cut-and-fill slopes, and drainage and auxiliary facilities. Here, they acquired spatial information from points of 251 point/km and 131 point/km, respectively, and these precision densities are only approximately 0.002% and 0.001% compared to the mean of unit point numbers of the MLS maps produced in this study. In addition, White et al. [3] and Kiss et al. [7] acquired spatial information of the forest roads using an ALS survey, with precision densities of 0.8 point/m^2 and 1.2 point/m^2, respectively, which were only approximately 0.09% and 0.14% of the MLS maps in this study. Based on the results above, it is judged that the quality of point cloud data acquired through MLS survey in this study has a very high precision. In particular, mapping by MLS survey is sufficient to be considered very useful for creating a forest road inventory, in that it can visualize not only the centerline of the road but also all the spatial elements making up the forest roads, such as the road surface, cut-and-fill slope surfaces, and the ditches and drainage facilities exposed to the ground.

4.2. Relative Positional Accuracy of the MLS Map

The accuracy of the MLS survey was also seen in the RMSE values relative to the corresponding points on the reference centerline maps created by the GNSS survey (Table 6). The RMSEs of the MLS survey, which are not significantly different from those of the TS survey, averaged about 0.8 m. These results are likely to be very accurate compared to those of previous studies, which examined positional accuracy of topographical features using ALS (with RMSEs ranging between 1 and 3 m) (e.g., [3,7]) and GNSS (with RMSEs ranging between 1 and 88 m) (e.g., [1–3,6]) in forest roads and/or mountain terrains. Conversely, Kim [22] reported that the RMSEs on the location accuracy of topographical features in urbanized areas ranged between 0.09 and 0.293 m, which is more accurate than the results of this study. This difference may be due to the lower reception rate of the GNSS signals on the forest roads compared to that in urban areas.

For determining positional accuracy of linear objects, such as roads and railways, consistency of the shape itself can also be used as one of the most important criteria [33]. In this study, whereas 95% of both centerlines produced from MLS and TS surveys were nested at similar diameters (i.e., 1.90 m and 1.88 m, respectively) from that produced by the GNSS survey, the buffer diameter of the reference map at which 100% were nested was found to be smaller in the MLS survey (i.e., 2.60 m) than the TS survey (i.e., 4.40 m) (Figure 7). Although it may be difficult to attach a relative meaning to the quantitative differences between these two methods (may be due to user error at turning points in TS survey and/or difference in machine accuracy of two survey instruments), this also reflects the advantages of mapping by MLS survey, as shown in the results of the RMSE analysis, in excluding potential user error during field surveying and in ensuring the positional accuracy of forest roads with a long linear form.

4.3. Efficiency of MLS Mapping for Forest Road Inventory

Numerous studies (e.g., [3,6,7,10,11,32,34]) have suggested that it is necessary to survey using LiDAR (such as ALS, TLS, and MLS) on the basis of the efficiency in terms of working time and cost. Although the high equipment rental price of MLS mapping with high precision and accuracy can be pointed out as a drawback in the deployment of a forest road inventory, the advantage is that labor costs incurred are only 57% and 21% of the GNSS and TS mappings, respectively, due to relatively short working time (Table 7). In particular, as described previously, it may be assessed that the quality and quantity of the point cloud data collected during a MLS survey are large enough to allow the production of a 3D map, including all elements of the forest roads (such as roadbed, cut-and-fill slopes, drainages, auxiliary facilities), and are superior in comparison with mapping by other survey methods.

5. Conclusions

By paying attention to both surveying methodologies and implementation of research results to facilitate sustainable forest road management, this study produced a 3D map for the establishment of a forest road inventory using MLS, and reviewed its precision, accuracy, and efficiency. As a result, the point cloud data acquired by the MLS survey on the study forest roads has very high precision and therefore is sufficient to produce a high-resolution 3D forest road map. Although the equipment rental cost is somewhat high, the fact that information on all spatial elements of forest roads, which are linear objects, can be obtained at a low cost of labor is expected to act as a positive contributor for evaluating the efficiency of MLS survey and mapping. Our findings are likely to be meaningful in that it can provide a quantitative assessment of both maintaining sustainable effectiveness and preventing potential environmental damage of forest roads.

Nevertheless, several technical and academic challenges still remain in the establishment of a forest road inventory through MLS survey and mapping. First, if it is inevitable for forest roads to be constructed on forested area with steep slopes like Korea, spatial information on the fill-slope surfaces, which is obtained at a sharp angle, may not be fully acquired, even if vehicles equipped with MLS drive close to the right edge of the road based on the direction of driving. Thus, technical supplementation is required to overcome this limitation. Second, the heavy canopy covering forest areas acts as a restricting factor for MLS surveys [15,35,36], and thus further studies are needed to develop ways to reduce the loss of point data due to this restriction and to improve the survey accuracy. Third, it is believed that studies to automate or simplify digitization of spatial information during the MLS mapping process will be needed, as forest roads are linear objects that have a slightly regular form (although not as regular as general roads or railways). These studies may refer to recent studies conducted on general roads and railways [21,35]. Fourth, in order to compensate for the disadvantages of a MLS survey with high equipment rental costs, it is necessary to explore ways to diversify the use of spatial information in the form of point clouds that can be obtained from limited spaces (i.e., forest roads). Here, we should take note of the fact that we can effectively create 3D maps including longitudinally and cross-sectionally detailed components of forest roads and their surroundings in a relatively short time using MLS. Solving these four challenges is a necessary condition for enabling rational decision-making for systematic forest road management, which will ultimately lead to the maintenance of the permanent utility of and the prevention of potential environmental damage from forest roads.

Author Contributions: Conceptualization, H.K. and J.-W.L.; Data curation, H.K. and J.I.S.; Funding acquisition, H.K.; Project administration, J.-W.L.; Supervision, J.I.S.; Writing—original draft, H.K. and J.-W.L.; Writing—review & editing, J.I.S. All authors have read and agreed to the published version of the manuscript.

Funding: This research was supported by Basic Science Research Program through the National Research Foundation of Korea (NRF) funded by the Ministry of Education (grant number: 2018R1A6A3A03013244).

Acknowledgments: The authors would like to thank Myeongjun Kim and Seongmin Choi for offering valuable comments on the early drafts of the manuscript. Sincere appreciation goes to the anonymous reviewers that improved the manuscript concision and perspective.

Conflicts of Interest: The authors declare no conflict of interest.

References

1. Abdi, E.; Sisakht, L.; Goushbor, L.; Soufi, H. Accuracy assessment of GPS and surveying technique in forest road mapping. *Ann. For. Res.* **2012**, *55*, 309–317.
2. Talebi, M.; Majnounian, B.; Abdi, E.; Tehrani, F.B. Developing a GIS database for forest road management in Arasbaran forest, Iran. *For. Sci. Technol.* **2015**, *11*, 27–35. [CrossRef]
3. White, R.A.; Dietterick, B.C.; Mastin, T.; Strohman, R. Forest roads mapped using LiDAR in steep forested terrain. *Remote Sens.* **2010**, *2*, 1120–1141. [CrossRef]
4. Laschi, A.; Foderi, C.; Fabiano, F.; Neri, F.; Cambi, M.; Mariotti, B.; Marchi, E. Forest road planning, construction and maintenance to improve forest fire fighting: A review. *Croat. J. For. Eng.* **2019**, *40*, 207–219.
5. Kweon, H. Comparisons of estimated circuity factor of forest roads with different vertical heights in mountainous areas, Republic of Korea. *Forests* **2019**, *10*, 1147. [CrossRef]
6. Kweon, H.; Kim, M.; Lee, J.-W.; Seo, J.I.; Rhee, H. Comparison of horizontal accuracy, shape similarity and cost of three different road mapping technique. *Forests* **2019**, *10*, 452. [CrossRef]
7. Kiss, K.; Malinen, J.; Tokola, T. Forest road quality control using ALS data. *Can. J. For. Res.* **2015**, *45*, 1636–1642. [CrossRef]
8. Höhle, J.; Höhle, M. Accuracy assessment of digital elevation models by means of robust statistical methods. *ISPRS J. Photogramm.* **2009**, *64*, 398–406. [CrossRef]
9. Stereńczak, K.; Kozak, J. Evaluation of digital terrain models generated from airborne laser scanning data under forest conditions. *Scand. J. For. Res.* **2011**, *26*, 374–384. [CrossRef]
10. Akgul, M.; Yurtseven, H.; Akburak, S.; Demir, M.; Cigizoglu, H.K.; Ozturk, T.; Eksi, M.; Akay, A.O. Short term moniterning of forest road pavement degradation using terrestrial laser scanning. *Measurement* **2017**, *103*, 283–293. [CrossRef]
11. Akay, A.O.; Akgul, M.; Demir, M. Determination of temporal changes in forest road pavement with terrestrial laser scanner. *Fresenius Environ. Bull.* **2018**, *27*, 1437–1448.
12. Liang, X.; Hyyppa, J.; Kukko, A.; Kaartinen, H.; Jaakkola, A.; Yu, X. The use of a mobile scanning system for mapping large forest plots. *IEEE Geosci. Remote Sens. Lett.* **2014**, *11*, 1504–1508. [CrossRef]
13. De Agostino, M.; Lingua, A.; Piras, M. Rock face surveys using a LiDAR MMS. *Ital. J. Remote Sens.* **2012**, *44*, 141–151. [CrossRef]
14. Qin, R. An Object-based hierarchical method for change detection using unmanned aerial vehicle images. *Remote Sens.* **2014**, *6*, 7911–7932. [CrossRef]
15. Čerňava, J.; Mokroš, M.; Tuček, J.; Antal, M.; Slatkovská, Z. Processing chain for estimation of tree diameter from GNSS-IMU-based mobile laser scanning Data. *Remote Sens.* **2019**, *11*, 615. [CrossRef]
16. Schwarz, K.P.; Lapucha, D.; Cannon, M.E.; Martell, H. The use of GPS/INS in a highway inventory system. In Proceedings of the FIG XIX Congress, Helsinki, Finland, 10–19 May 1990; Volume 5, pp. 238–249.
17. Kang, I. Method for Improving the Integrity of the Data from Land-Based Mobile Mapping System to Create Multipurpose Precise Road Map. Ph. D. Thesis, University of Seoul, Seoul-si, Korea, 2013. (In Korean with English abstract)
18. Grejner-Brzezinska, D.; Toth, C.; Yi, Y. On improving navigation accuracy of GPS/INS systems. *Photogramm. Eng. Remote Sens.* **2005**, *4*, 377–389. [CrossRef]
19. Qin, R.; Gruen, A. 3D change detection at street level using mobile laser scanning point clouds and terrestrial images. *ISPRS J. Photogramm. Remote Sens.* **2014**, *90*, 24–35. [CrossRef]
20. Cui, T.; Ji, S.; Shan, J.; Gong, J.; Liu, K. Line based registration of panoramic images and LiDAR point clouds for mobile mapping. *Sensors* **2017**, *17*, 70. [CrossRef]
21. Ma, L.; Li, Y.; Li, J.; Wang, C.; Wang, R.; Chapman, M.A. Mobile laser scanned point-clouds for road object detection and extraction: A review. *Remote Sens.* **2018**, *10*, 1531. [CrossRef]
22. Kim, Y. A Study on Calibration Method of Vehicle-Based Mobile Mapping System. Master's Thesis, Sungkyunkwan University, Suwon-si, Korea, 2011. (In Korean with English abstract).
23. Tang, J.; Chen, Y.; Kukko, A.; Kaartinen, H.; Jaakkola, A.; Khoramshahi, E.; Hakala, T.; Hyyppä, J.; Holopainen, M.; Hyyppä, H. SLAM-aided stem mapping for forest inventory with small-footprint mobile LiDAR. *Forests* **2015**, *6*, 4588–4606. [CrossRef]
24. Forsman, M.; Holmgren, J.; Olofsson, K. tree stem diameter estimation from mobile laser scanning using and hyperspectral data. *Sensors* **2011**, *11*, 5158–5182.

25. Buwens, S.; Bartholomeus, H.; Calders, K.; Lejeune, P. Forest inventory with terrestrial LiDAR: A Comparison of static and hand-held mobile laser scanning. *Forests* **2016**, *7*, 127. [CrossRef]
26. Korea Forest Service (KFS). Available online: http://www.law.go.kr/DRF/MDRFLawService.do?OC=foalaw&ID=10317 (accessed on 5 August 2019).
27. Trimble Korea. Available online: https://geospatial.trimble.com (accessed on 15 September 2019).
28. Spectra. Available online: https://spectrageospatial.com/ (accessed on 15 September 2019).
29. Kagawa, Y.; Sekimoto, Y.; Shibasaki, R. Comparative study of positional accuracy evaluation of line data. In Proceedings of the 20th Asian Conference on Remote Sensing, Hong Kong, China, 22–25 November 1999.
30. Ömer, M.; Ayhan, C. Accuracy and cost comparison of spatial data acquisition methods for the development of geographical information systems. *J. Geogr. Reg. Plan.* **2009**, *2*, 235–242. [CrossRef]
31. Kim, M.; Kweon, H.; Choi, Y.; Yeom, I.; Lee, J. Evaluation of horizontal position accuracy in forest road completion drawing. *Korean J. Agric. Sci.* **2010**, *37*, 471–479, (In Korean with English abstract).
32. Yurtseven, H.; Akgul, M.; Akay, A.O. High accuracy monitoring system to estimate forest road surface degradation on horizontal curves. *Environ. Monit. Assess.* **2019**, *191*, 32. [CrossRef] [PubMed]
33. Velkamp, R.C. Shape matching: Similarity measures and algorithms. In Proceedings of the International Conference on Shape Modeling and Applications, Genova, Italy, 7–11 May 2001.
34. Balenović, I.; Gašparović, M.; Milas, A.S.; Seletković, A.B. Accuracy assessment of digital terrain models of lowland pedunculate oak forests derived from airborne laser scanning and photogrammetry. *Croat. J. For. Eng.* **2018**, *39*, 117–127.
35. Arastounia, M. Automated recognition of railroad infrastructure in rural areas from LIDAR data. *Remote Sens.* **2015**, *7*, 14916–14938. [CrossRef]
36. Zhong, M.; Sui, L.; Wang, Z.; Yang, X.; Zhang, C.; Chen, N. Recovering missing tracjectory data for mobile laser scanning systems. *Remote Sens.* **2020**, *12*, 899. [CrossRef]

MDPI

St. Alban-Anlage 66

4052 Basel

Switzerland

Tel. +41 61 683 77 34

Fax +41 61 302 89 18

www.mdpi.com

Remote Sensing Editorial Office

E-mail: remotesensing@mdpi.com

www.mdpi.com/journal/remotesensing